Management
in the Fire Service

Third Edition

by
Harry R. Carter
and
Erwin Rausch

with contributions from Arthur Kiamie

National Fire Protection Association
Quincy, Massachusetts

Product Manager: Jim Linville
Developmental Editor: Pam Powell
Manufacturing Buyer: Ellen Glisker
Production Services: Pre-Press Company, Inc.
Copyediting: Carol Reitz
Indexing: Carol Noble
Interior Design: Glenna Collett
Cover Design: Whitman Design
Composition: Pre-Press Company, Inc.
Printing: R.R. Donnelley

NFPA No.: MFS-98
ISBN: 0-87765-441-7
Library of Congress Card Catalog No.: 89-60368
Third Edition: 60146

Printed in the United States of America
10 9 8 7 6 5 4

Contents

Preface

A new edition of a book often brings updates and relatively minor changes. This third edition of *Management in the Fire Service* is different. Although this edition updates and covers roughly the same ground as the previous editions, it also brings major changes in perspective and treatment.

The first two chapters address fire department organization and the fire officer's responsibilities. Each of Chapters 3 through 12 concentrates on a major fire service function:

Chapter 3, Commanding the Response to an Emergency Incident
Chapter 4, Fire Prevention and Code Enforcement
Chapter 5, Fire and Life Safety Education
Chapter 6, Prefire Planning and Related Loss Reduction Activities
Chapter 7, Management of Physical Resources
Chapter 8, Management of Financial Resources
Chapter 9, Fire Service Personnel Management
Chapter 10, Labor Relations in the Fire Service
Chapter 11, Training as a Management Function
Chapter 12, Fire Department Services Beyond Fire Fighting

Most chapters begin with a hypothetical scenario that portrays a realistic decision-making situation relative to the fire service function covered in the chapter. Each scenario is followed by an analysis that addresses the decision issues from both the fire service and the management/leadership perspectives. These sections suggest decision guidelines that officers can use when working in that function. Increasingly more detailed guidelines are provided for one or more of the three management/leadership requirements for an effective organizational unit (the 3Cs of control, competence, and climate). A comprehensive listing of all guidelines can be found in Appendix A.

To provide more depth on management and leadership issues in this edition, Additional Insight sections have been developed. These sections, which carry forward the issues and concepts introduced in scenario analyses, appear in the following chapters:

Additional Insight 1.1: Participation in Decision Making and Planning

Additional Insight 1.2: Management Theory and Concepts: Origins and Pathways

Additional Insight 2.1: Communications Techniques and Skills

Additional Insight 3.1: Goal Setting and Implementation

Additional Insight 4.1: Management of Potentially Damaging Conflict

Additional Insight 5.1: Enhancing Work Satisfaction (Providing Recognition) and Performance Evaluation

Additional Insight 6.1: Time Management and Delegation

Additional Insight 7.1: Decision Making and Problem Solving

Additional Insight 9.1: Positive Discipline and Counseling

Additional Insight 9.2: Interviewing

Additional Insight 11.1: Learning Concepts in Support of Management of Learning and Coaching

Changes that make this third edition a more useful guide to fire service personnel include the following:

- Updating the fire service function discussions and adding new sections on nontraditional roles of the fire service that are becoming common
- Integrating the fire service function discussions more effectively with the management/leadership segments
- Making the management/leadership sections more *practical* and *useful* so they will help fire officers become better managers and leaders and help those aspiring to officer rank better prepare for the officer exams
- Providing guidelines for fire officer decisions in all fire service functions
- Coordinating, even more thoroughly than the previous editions, with NFPA 1021, *Standard for Fire Officer Professional Qualifications*

As in the previous editions, Study Questions and Activites are included to help readers integrate new information into their work in the fire service and to explore some of the issues that officers might be confronted with during their career. Additional readings and reference lists are provided as further aids to readers who wish to seek further details on relevant topics.

Although this book addresses many modern fire service and management/leadership issues encountered by today's fire service officer, it is not intended as a substitute for a professional skills development program. It provides an opportunity for new and established officers to gain a dual perspective on their role as fire fighters and as managers/leaders. To acquire the thorough competence outlined in NFPA 1021, *Standard for Fire Officer*

Professional Qualifications, resources and opportunities for professional development programs include the following: departmental officer training programs; county or state fire training academies; two-year community college fire science programs; four-year college and university programs; National Fire Academy residence programs in Emmitsburg, Maryland, or open university programs administered by a number of colleges on behalf of the National Fire Academy; "university-without-walls" programs; and self-study books.

ACKNOWLEDGMENTS

The many persons and organizations who provided information for the first and second editions of this book continue to deserve gratitude for their contributions. This edition, because it is a far more extensive revision than the second edition, depended much more on thoughtful reviews and suggestions. The authors are therefore most grateful to the following for the extensive efforts they devoted, graciously, to these reviews and comments:

Lt. Paul Mucha of the Roselle, New Jersey, Fire Department; Inspectors Helge Nordveit and Daniel J. Thorn of the Cranford, New Jersey, Fire Department; and Fire Fighter Mitchell Meyer, Newark, New Jersey, Fire Department.

Credit is also due to the following fire service professionals who contributed ideas, materials, and time to make this book far better than it would have been without their assistance: Captain George Andersen, Newark, New Jersey, Fire Department; Lt. Kenneth Dannevig, Westfield, New Jersey, Fire Department; Captain Wesley W. Ditzel, Cranford, New Jersey, Fire Department; Chief Leonard Dolin, Cranford, New Jersey, Fire Department; Fire Coordinator Tim Eures, Highlands County, Florida; Batt. Chief Kenneth Folisi, Lisle-Woodridge Fire District, Naperville, Illinois; Chief Robert Hill, Roselle, New Jersey, Fire Department; Captain Patrick Kelleher, Roselle, New Jersey, Fire Department; Batt. Chief Kevin Killeen, Newark, New Jersey, Fire Department; Director/Fire Chief Stanley J. Kossup, Newark New Jersey, Fire Department; Inspector Patrick Lynn, Roselle, New Jersey, Fire Department; Fire Fighter Rich Meier, Cranford, New Jersey, Fire Department; Chief Glenn Miller, Garwood, New Jersey, Volunteer Fire Department; Captain Terry Moore, Roselle, New Jersey, Fire Department; Lt. Ernest Ousley, Roselle, New Jersey, Fire Department; Assistant Chief John Peltier (retired), Marlborough, Massachusetts, Fire Department; Captain Fred Roberts, Cranford, New Jersey, Fire Department; and Sandy Pells, RMC, Cranford, New Jersey, Fire Department.

Introduction to Management in the Fire Service

INTRODUCTION

Management in the Fire Service is a basic management text for present and potential officers of the fire service as well as for students and members of
other fire science–related professions. At first glance, management in the fire
service might appear to be totally different from management in other organizations. Certainly the fire service has unique management challenges,
including the need for personnel to be constantly ready for instant changes
from tranquil, routine duties to intense emergency responses. Nonetheless,
the basic principles for managing any activity—a government agency, a gro-
cery store, a bank, a manufacturing concern, a busy household, or any other
business—also apply to management in the fire service. In fact, most of the
challenges that fire service officers face require decisions that are similar, in
many ways, to those made by line managers in government, manufacturing,
service industries, and retail situations. Emergency situations also require
common management skills, as is evident from the analyses and conclusions
of the scenarios in almost every chapter, especially in this chapter's Ox-Bow
Motel fire scenario and the initial example in Additional Insight 1.1, Participation in Decision Making and Planning.

Let's begin, however, with a few words for those of you who are about
to embark on a major change in your work life—from fire fighter to officer.
The transition may be a difficult time in your life. As a new officer you will
have to take charge of a group of former colleagues; you will have to gain a
broad perspective on the many facets of the fire service while taking into
consideration the stakeholders. To become fully established as an officer
and manager, you will have to retain the confidence of the members of the
company and their regard for your fire-fighting skills, while considering
other constituencies and gaining the stature of a leader. Initially you may
also have to overcome any hard feelings that might exist because you were
the one who was chosen for the promotion.

For the most successful career possible, fire department officers need
perspective, and possibly detailed knowledge, of fire service functions with
which they may have had only limited contact in the past. The approach
suggested by this book—namely, attention to guidelines for fire service
functions and for management/leadership—can help new officers make the
transition smoothly and successfully. Considering these two aspects in all
decisions is important for both new and established officers at all levels and
in all areas of fire service work.

As a fire fighter—whether you were one recently or long ago—you were
primarily concerned with fire department functions. As an officer, you are
required also to pay attention to management/leadership considerations in

making decisions. Viewing decisions from a management/leadership perspective is not necessarily more difficult, but it does require a new set of thought habits or disciplines.

GUIDELINES FOR FIRE SERVICE FUNCTIONS AND FOR MANAGEMENT/LEADERSHIP

This book has been designed to help you gain this additional perspective as easily as possible. It presents guidelines for both aspects of decision making. For fire service functions, these guidelines are "prescriptive." In many cases they tell you what you have to consider, and often they even tell you how you should look at a situation. For the management/leadership considerations of your decisions, guidelines are in the form of questions you should ask yourself or your team before you implement a decision or plan. Although you may have little time in a fireground situation, if you have made guidelines part of your thinking—that is, if you have developed the desirable habits—your mind will run through the guidelines with lightning speed. In emergencies as well as in all nonemergency work, you will be a better leader, not only a competent fire strategist and tactician.

In short, the purpose of guidelines is to strengthen the influence of rational thought in every decision and plan—but not at the expense of sound intuitive reasoning. In fact, the better you understand and apply the guidelines, the sounder your intuitive reasoning will become.

Little room exists for you to modify or adapt the fire service function guidelines because they are determined by fire science considerations, codes, and established sound procedures. In contrast, although they stem from a widely accepted model, management/leadership guidelines are not based on an equally codified foundation. You can and should, therefore, modify them to better fit your personal leadership views and style.

One purpose of this book is to make it as easy as possible for you to apply management/leadership perspectives to fire-officer work. Here, the term *manager/leader* applies to all fire officers who have supervisory responsibilities and also those charged with management functions (such as fire prevention), even if the function is performed by a one-person department or division. Although management and leadership are not identical concepts, discussing the many views of the differences is not appropriate here. What is important from the viewpoint of this book is that competent managers should also be competent leaders—as supervisors, as managers of functions with no staff, and in unstructured groups such as meetings of peers.

Another purpose of this book is to help fire officers gain a dual perspective of their roles as fire officers *and* as managers and leaders. As managers and leaders, fire officers play a major role in developing the goals and objectives that fulfill the fire service mission. From a management/leadership perspective, we define *goals* as broadly stated outcomes that are directly related to the organization's mission and that generally have a long time frame. *Objectives* state specific accomplishment that are to be achieved to satisfy goals. From a fire service function perspective, in an emergency situation, *goals* focus on the preservation of life and property; *objectives* focus on specific accomplishments, such as rescuing the second-floor occupants or containing the fire, depending on the needs of the situation.

As you will see in more detail later, a *strategy* is a plan or decision that identifies the objectives that will lead to a specific goal, but it does not specify *how* the objectives will be carried out. *Tactics* identify the methods for achieving the objectives and lead to the development of *action plans* that identify the required nuts-and-bolts tasks. In the fire service function, officers have significant responsibility for goal achievement—that is, for the strategies and tactics—whereas staff members have responsibility for the completion of the action steps. In emergency situations, officers and other personnel rely on the incident command management system, which is an effective approach for managing all resources, including personnel and equipment, to achieve the multiple goals of fire fighting while ensuring fire fighter safety.

Let's take a look at a fire scenario to see what all this means.

SCENARIO
THE OX-BOW MOTEL FIRE

It was still a few hours before dawn, one warm fall night. Dry leaves were in deep drifts all over the ground; it hadn't rained in more than two weeks. As the pumper reached the high ground, heading south, the captain and the three fire fighters could see the smoke even before the road turned into the straight stretch where the Ox-Bow Motel was.

They knew the building was old enough not to have sprinklers in the rooms as the state had mandated several years ago. Located on the main road, the large, two-story, wood frame structure had a stucco veneer to give it an Old West appearance. The office and about 75 units, with large windows, were situated back-to-back under a common roof, in a U-shape around the parking lot that was right off the road. Outside stairs led up to open walkways along the building in the front and in the rear.

Over the years, a number of renovations had been made, not all with proper permits and inspections. The many shrubs and bushes that surrounded the structure and the trees on the property had not been trimmed in years. After severe storms, tree limbs that littered the ground were often not removed for weeks.

"There'll be lots of dry leaves all over the lot," Joe yelled over the roar of the racing engine to Marty who was in the other jump seat.

"I know; won't stop us!" Marty yelled back.

Meanwhile the captain was thinking about the situation he might face. The Ox-Bow Motel had deteriorated considerably, mainly because a major chain had opened a modern motel only a few blocks away and had taken business and vacation customers. As a result, Ox-Bow's tenants, who occupied only about half of the rooms, were transients on welfare or homeless people being sheltered by the county.

The police had relayed the call to the captain's single-unit station and to the chief on duty, whose office was in a larger station with an EMS engine and a truck. The caller had said that the fire was in the south rear corner unit on the ground floor and that it had apparently been set during a fight between two men. After hanging up, he said he would bang on the doors downstairs to get people out. Because the fire seemed to be spreading quickly, he didn't think he could get to the upper floor. There might be people in one or more of those units. The captain knew that Engine 2, the truck, and the chief would probably take a few more minutes to get to the scene.

"Pull to the hydrant at end of the motel, Sue," he instructed the driver. Then, as the pumper slowed down, he yelled to the fire fighters: "Let's get prepared to take two lines, the 2½ and a 1¾. While I do the sizeup, Marty, connect the 1¾ and lay it out so we can take it upstairs, if necessary. Joe, stretch the 2½; you might have to handle it alone for a while. Sue, you stay with the pumper until you've got the flow adjusted, then help Joe set up the line, and then go back to the pump. Marty, you and I may have to go in. Let's take the Halligan bar to force entry, just in case we need it."

"Wouldn't it be better to use two 1¾-inch lines? It should be adequate if you have the other line upstairs." Joe, who was training to become an officer, thought he ought to express his views.

"I think we'll stick with the 2½. Let's go." The engine had come to a halt as the three jumped off. Several people in the small crowd that had gathered pointed to the back of the motel. A woman screamed, "Hurry, hurry! There are people upstairs."

The captain ran to the back where the police officer had just finished knocking on the doors of the units next to the fire. No one answered. Meanwhile Marty and Joe started to lay out the lines.

The situation was worse than the captain had expected. Three units downstairs were fully involved, and he could see fire in the windows of two of the upstairs units.

Apparently the fire had broken through the ceiling or gone up in the wall. "I hope if there's someone in there, they were smart enough to lie low," he thought, as he called the chief on the radio. "Engine 1 portable to chief 3. We have a working fire and a possible rescue problem." Then he helped with laying out the lines.

As soon as the 1¾ line was ready, the rescue operation started. Racing up the stairs that connected to the outside hallway and making a forcible entry through the door of the unit directly above the one where the fire had started, the captain and Marty found a man and a woman on the floor at the rear of the unit, both unconscious. By the time they had carried the victims out, the chief and the other company had arrived, and the chief took over the command of the incident.

While the EMS1 technicians worked to revive the victims, the captain told Marty to assist Joe. The others checked the involved units and the other units near the fire, upstairs and downstairs, but did not find any other victims.

The fire was quickly extinguished, and mopping up began. Inspection of the fire alarms in the units showed that a large number of them, including four of the five in the involved units, were defective.

Before reading the analysis in this section, give some thought to what you consider to be the strengths and possible weaknesses of the strategies and tactics the captain used.

Scenario Analysis: Fire Service Function Perspective

This scenario involves primarily incident command. For the captain in charge of the first engine and his fire fighters, the incident command issues and their guidelines are most relevant.* Fire prevention and code enforcement, prefire planning, and training issues are also involved. They need to be considered later by the fire department.

Every fire service function has goals.

> The primary goals of incident command are to ensure that all is done that can be done to protect people—the civilians at risk and the fire fighters—and that everything is done to preserve and protect property by confining the fire and extinguishing it as quickly as possible.

These goals provide the foundation for the general guideline on which we will base the scenario analysis. For incident command, this guideline, like all guidelines in this book, is in the form of a question:

---

*In this chapter, we will give only the general guideline. In later chapters, we will add specific guideline segments. A comprehensive listing of all decision guidelines is given in Appendix A.

To protect people—both the civilians at risk and the fire fighters—and to preserve and protect property, what do I have to consider in the sizeup and during the operation to assure that I will use the most appropriate strategy and tactics?

In the scenario, the captain did a creditable job in managing the situation, and he clearly had the goals in mind. However, he did not address the guideline fully. If he had been aware of and seriously thought about the guideline question during his sizeup, he might have asked additional questions:

- Is the 2½-inch line the best size to use?
- Do Joe and I have a common understanding on how to direct the stream?
- How can I work with Marty to minimize any potential risks to the rescuers?
- Should Joe or Sue specifically be charged to perform a lookout safety role to carefully observe the fire and smoke conditions and make me aware of any developments that might require a change in tactics?

Deep-seated awareness and regular use of the incident command guideline, especially in conjunction with management/leadership guidelines, might have helped the captain to ask and answer these questions. If he had, he probably would have improved the handling of this fire from an evaluation of "good" to "excellent."

Sometime after the fire, the department needs to address the way the goals of fire prevention and code enforcement, prefire planning, and training are satisfied, based on the experience from the Ox-Bow Motel fire. Let's take these one at a time.

The primary goals of fire prevention and code enforcement are to create a community safe from fire through adherence to codes, construction plan reviews, and field inspections.

The general guideline question is

What else needs to be done to ensure thorough adherence to codes—specifically, what should be changed with respect to relations with architects, enforcement, competence development of inspectors, and communications with stakeholders (architects, engineers, property owners, and contractors)?

If the department were to address this guideline, what questions might it raise as a result of the Ox-Bow Motel fire?

Based on the description of the scenario, the possible enforcement lapses did not contribute to injuries (or possible fatalities). However, an analysis of the fire prevention and code enforcement procedures might have led to questions about these issues:

- The frequency of inspections, especially with multiple occupancies
- The procedures used to follow up on violations, especially those pertaining to fire alarms
- The need for an ordinance to require all multiple occupancies to connect fire alarms directly to the fire department

Now, let's look at prefire planning and related functions.

> The primary goals of prefire planning and related functions are to ensure that department members have thorough plans for attacking fires most effectively and that preparations are made so that members are knowledgeable, skilled, and equipped to implement the plans competently.

The general guideline is more detailed:

> What else needs to be done to ensure that adequate information is available to responding companies, appropriately analyzed and formulated into plans, that these plans are used in staff development, fire ignition sequence investigations, and water supply review and testing, and that the information management systems are as effective as possible?

Based on the description of the scenario, any inadequacies in prefire planning and related functions did not adversely affect the attack on the fire or the rescue operation. However, a later review of prefire plans might have led to questions about each of the components of the guideline:

- Adequacy of the available information
- Adequacy of the analyses
- Appropriateness of the plans
- Use of the plans in staff development
- Adequacy of fire ignition sequence investigations, water supply reviews, and testing
- Updating of the information management systems

Finally, let's look at the goals of the training function.

> The primary goals of training are to ensure that all members of the department have high-level competence for all their functions and that officers are competent in management of learning and training and in coaching fire fighters.

These goals lead directly to the general guideline:

> What else needs to be done to ensure that all members of the department have high-level competence for all their functions and that officers are competent in management of learning and training and in coaching fire fighters?

Here, too, nothing in the scenario description indicated any inadequacies in training. In fact, Joe's suggestion to use the 1¾ line indicated that training was quite thorough. Nevertheless, reviewing the guideline from time to time, especially in conjunction with the management/leadership guideline on competence (as discussed later), would lead to questions about each component:

- How fire officer and fire fighter competence for all their functions could and should be reviewed
- What could or should be done to take advantage of competence strengths of fire fighters and officers, and how any weaknesses could best be reduced or eliminated
- How officer competence in management of learning and training and in coaching fire fighters could be enhanced

Scenario Analysis: Management/Leadership Perspective, Part 1

In his role as officer, the captain had to address management/leadership considerations. They are based on the following reasoning:

1. It is the function of managers to see to it that things get done—that an organizational unit works effectively to reach its goals and objectives.
2. Getting things done requires involving people or affects people.
3. When people are involved or affected, it is necessary to consider their reactions.

In effect, managers must somehow align the characteristics and needs of the task or organization with the needs and characteristics of the people involved or affected.

On the basis of these three points, we can conclude that it is the responsibility of a fire officer, as manager and leader, to see to it that the organizational unit's goals and objectives are achieved by ensuring effective *control*, by ensuring *competence*, and by creating and maintaining a positive *climate* conducive to effective teamwork.

THE 3CS MODEL

The 3Cs or Linking Elements Model, which is built around control, competence, and climate, is a sound comprehensive management/leadership model that is solidly based on the research and writings of management scholars.* The simple diagram in Figure 1.1 identifies these 3Cs (control, competence, and climate). They spell out the characteristics and needs of an effective organization. In effect, they define the arenas where an officer uses management skills (the linking elements).

- *Control* concerns the definition, communication, and coordination of direction to ensure that they are clear and understood and that warnings will be raised quickly when progress is unsatisfactory. Control, as used in the 3Cs model, is not control by the officer but rather control by the organizational unit as a result of appropriate participation in all relevant decisions, as suggested by the model.
- *Competence* concerns the knowledge, skills, and abilities (KSAs) required for the functions, including the management/leadership competence of officers and sometimes even the competence of other stakeholders such as architects, engineers, and teachers.[†]
- *Climate* concerns the environment in which staff members and other stakeholders can find the greatest possible satisfaction from participating or being affected.

The bottom half of the diagram in Figure 1.1 identifies what staff members bring to the work scene—that is, their attitudes toward the control policies and procedures, their competencies, and their needs. It is the job of officers, as managers and leaders, to reach the best possible alignment between

---

*The model has been used widely in management development programs in such diverse organizations as the American Management Association, U.S. Air, Cabrini Hospitals, U.S. Federal Prison System, U.S. Federal Office of Personnel Management, U.S. Army, U.S. Navy, Government of Alberta, General Electric, Girl Scouts, and JCPenney. The model has been published in various books and publications. (See Rausch and Washbush 1998; Office of Military Leadership 1976; Didactic Systems 1977; Heyel 1982; Jones and Lieverman 1978; Rausch 1978; Rausch 1980.)

[†]It might be useful to point out here that NFPA professional qualifications standards refer to Job Performance Requirements (JPRs), which list job tasks, requirements, and evaluation parameters. For example, for the task of ventilating a pitched roof, the material requirements are an ax, a pike pole, an extension ladder, and a roof. The evaluation parameters are such that a 4-feet-by- 4-feet hole is created; all ventilation barriers are removed; ladders are properly positioned for ventilation; ventilation holes are correctly placed; and smoke, heat, and combustion by-products are released from the structure. KSAs identify the knowledge, skills, and abilities that are needed to perform JPRs.

An effective leader links:

control ~ attitudes
competence ~ K.S.A's
climate ~ needs

```
                    ┌──────────────────────┐
                    │   The organization   │
                    └──────────────────────┘
                                │
                          Performance
                                │
        ┌───────────────────────┼───────────────────────┐
      Control                Competence                Climate
        ▼ * ▲                 ▼ * ▲                     ▼ * ▲
      Attitudes         Knowledge and skills            Needs
        └───────────────────────┼───────────────────────┘
                    ┌──────────────────────┐
                    │    The individual    │
                    └──────────────────────┘
```

Figure 1.1 Basic Model of the 3Cs of Management, or the Linking Elements Model. Asterisks (*) indicate the skills (linking elements) a manager must apply to facilitate alignment between the needs and characteristics of the organization and the individual.

the characteristics and needs of the organization and those of the people in the organization and the other stakeholders. Officers must provide the linking elements, or the leadership and management skills that will reduce or eliminate the gaps between the opposing arrows in the diagram.

This book can help you learn how to reach that alignment as manager and leader. Later on it can serve as a convenient reference on related skills as you endeavor to gain greater competence.

You may think that management requires different skills and guidelines from those required by leadership or that management and leadership skills apply only to positions with supervisory responsibilities. Neither is true. Decision skills are almost the same for both sound management and effective leadership. The guidelines that will make you a good manager are almost certain to make you a better leader at the same time. They apply at all organizational levels and even to officers in functions without direct reports because, almost without exception, management of functions affects people—the stakeholders.

Aside from decision-making skills, the main difference between a good manager who is also a good leader and one who is not is certain personal characteristics. Some of these traits can be acquired, but most are part of one's nature. If you accept at least some of the guidelines suggested here as a starting point and then gradually shape them to fit your personal style, you will have won most of the battle. You can then concentrate on acquiring the traits of more effective leaders that you don't already possess.

The guidelines for competent managers/leaders pertain mostly to matters that do not show up regularly on the "things to do" lists of fire officers. These matters are easily overlooked because they do not ring bells or otherwise make themselves visible on a regular day. If ignored, however, they can lead to nasty crises, usually after damage has already been done. For the competent officer who tries to anticipate future problems, the 3Cs guidelines can be of great help.

Control, Competence, and Climate (3Cs) Guidelines

Decisions in the Ox-Bow Motel scenario, like other officer decisions, involve all three management/leadership considerations—control, competence, and climate (the 3Cs)—in addition to the fire service function guidelines, which we presented earlier. Decisions and plans that take the 3Cs guidelines into consideration are likely to result in a better alignment between the department and company and the fire fighters. The more you know about these guidelines and what they imply, the better your decisions are likely to be. More detailed questions to consider when using the guidelines will be explored in later chapters.

1. *The general control guideline: Are things going right?* What else needs to be done to ensure effective control and coordination so that the decision we are considering will lead to the outcome we seek, and so we will know when we have to modify our implementation or plan because we are not getting the results we want? In other words, how can we gain better control or coordination over the process of "getting there"?

2. *The general competence guideline: Does everyone know what to do, and can they do it?* What else needs to be done so that all those who will be involved in implementing the decision and who will otherwise be affected (all the stakeholders) have the necessary competencies to ensure effective progress toward excellence in fire department operations and service to the community?

3. *The general climate guideline: How will the stakeholders react?* What else needs to be done so that the reaction of the various groups and individuals who have to implement the decision or plan and those who will be affected by it (all the stakeholders) will be in favor of it or at least have as positive a view as possible, so there will be a favorable climate?

Note the strong interrelationship among these three guidelines: Actions taken to strengthen control affect competence and climate, for better or for worse. The same comprehensive interaction is true of actions taken to improve competence and climate.

Before you read this scenario analysis, give some thought to how the captain handled the management/leadership aspects of decisions.

Scenario Analysis: Management/Leadership Perspective, Part 2

How well did the captain satisfy the requirements of these three guidelines? With respect to control, he communicated direction (through his instructions) effectively. He assured coordination by the specific assignment of tasks. His personal involvement in all critical activities ensured that appropriate action would be taken if it became likely that the tactics would not achieve the objectives of the operation.

The competence of the company members for the tasks in the attack on the Ox-Bow Motel fire is assumed to be adequate, based on the fire fighter training they had undoubtedly received. The captain showed some gaps in his leadership skills, however, especially in the way he responded to Joe. He should give thought on a continuing basis to how he responds to reasonable suggestions from fire fighters.

The overall climate may or may not have been good. The scenario did not provide many clues on which to base judgments. It is likely, however, that the captain gave little thought to one of the primary issues in relation to climate—what he could do to increase the satisfaction that fire fighters obtain from their work.

One issue that is very important to control and to climate (where there is a hint in the scenario description) is participation in decision making. Appropriate participation in decision making and planning (not too much, not too little, not too soon, and not too late) leads to better decisions that make use of the special knowledge and views of several people. At the same time it enhances communications and relations between company members and the officer by making members aware that their opinions are respected and that they will have as great a voice as possible in matters that affect them. (See Additional Insight 1.1, Participation in Decision Making and Planning).

When the captain ignored Joe's suggestion to use the 1¾ line, he missed a chance for appropriate participation. True, there was little time, but many possible responses would have required little or no time. Before responding to Joe, the captain should have asked himself either the control or the climate guideline question, or both. If he understood their implications, he would have realized that he had several ways to respond to Joe's desire to participate in the line size decision. To choose among these options, he might have asked himself these questions:

1. Since there is not enough time for a discussion, should I tell him that we'll use the 2½ line and I'll discuss it with him as soon as we get back to the station?
2. Should I ask him to tell me quickly why he thinks that 1¾ is better and then make my decision accordingly?
3. Should I tell him that we'll stick with the 2½, but add, "If it looks like we can make it with the 1¾, I'll let you know, but we may need all the flow we can get."
4. Should I accept his suggestion?

In the scenario, the captain had to choose a response to Joe. All four of the choices listed are good responses. Choice 4 might be better if it came after choice 2. If there were enough time, choice 2 would be the best because it elicits information that might be useful; we are all human, and the captain might have overlooked something that he should have considered. If there were not enough time, choice 1 would probably be better than choice 3 because it shows more respect for Joe's opinion than choice 3 does.

If the captain had thought about it, he would have used the 1¾ line; according to most experts, it is more appropriate. In so doing, he would have enhanced control because appropriate participation encourages the expression of other views and may result in better decisions. He also would have improved the climate slightly, and he would have strengthened his leadership image.

CONCLUDING REMARKS

All competent officers work to develop potentially successful directions and then see to it that progress is made toward the desired results. If they review the fire service function guideline questions and apply the 3Cs guideline questions to every decision, officers can use strategies that can help them reach their goals faster, often with less effort, and using fewer resources. The organization works smarter.

The fire service function guidelines are based on the latest fire science knowledge. The management/leadership guidelines have proven their worth in hundreds of cases. Managers in various private and public development programs have been asked to use them for reviewing decisions or plans. Without exception each manager has made some change—sometimes a small one, sometimes a significant one—after considering just the basic form of the guidelines. The important thing to keep in mind about decision guidelines is **use**; use them with every important decision and plan. They provide reminders of the issues to consider and serve as a shortcut to better decisions. It is not necessary to adhere to the management/leadership guidelines suggested in this book. If you prefer, you may change them or use others that serve the same purpose of providing a check on the quality of your decisions. Guidelines should, of course, be sufficiently specific for your needs, and you should be able to articulate them.

In many ways the fire service is unique; no other profession has the same characteristics. The fire service has dramatic variations in workload. On the one hand are the emergency incidents with life-threatening danger, enormous time pressures, and the need for rapid and precise performance. On the other hand is the nonemergency work in the station and in the field, where the pace is leisurely and the requirements for accomplishment are often ill defined. The work is based on the initiative of the company commander, who follows general guidelines issued by the department. Still, much of the nonemergency work, such as fire prevention, fire safety education, prefire planning, and water supply monitoring, is very important to the well-being of the community and the safety of the fire fighters. The training on and maintenance of equipment are also crucial. Productivity, though it does not have the same tangible character as units rolling off the assembly line, has the same importance in that work.

In the nonemergency environment, the competence of the officer as a professional manager and leader makes the critical difference between an energetic, motivated team that is productive and proud and one that gets by with minimum effort and has equally low morale and self-respect. Using the 3Cs guidelines for every decision or plan, as suggested in this book, or as modified to fit the specific needs of the department and individual officer, can be of great help to a professional in building and maintaining high-level management/leadership competence. In a similar vein, most chapters suggest guidelines for the specific fire service function, for use as suggested or as modified. These guidelines, when used with every relevant decision and when reviewed from time to time, will gradually lead to improved operations and greater competence for the officer who heeds them. A comprehensive listing of all guidelines is given in Appendix A.

CHAPTER 1

STUDY QUESTIONS AND ACTIVITIES

If you are working alone, prepare your written responses to these questions. If you are studying in a team or working as a member of a class, discuss the questions with others and write a consensus answer.

1. Using examples of your own, describe how a department chief could encourage appropriate participation when setting goals for the department and for a company.
2. Without referring to the text, state the incident command guideline. Then describe how well the fire department in the Ox-Bow Motel scenario satisfied the guideline.
3. In the Ox-Bow Motel scenario, what specific instructions should the captain have given Joe?
4. In the Ox-Bow Motel scenario, how should the 1¾ line have been used to minimize risk to the rescuers?
5. Without referring to the text, state the 3Cs control guideline. Then describe how it applied in the Ox-Bow Motel scenario, if at all.
6. Without referring to the text, state the 3Cs competence guideline. Then describe how it applied in the Ox-Bow Motel scenario, if at all.
7. Without referring to the text, state the 3Cs climate guideline. Then describe how it applied in the Ox-Bow Motel scenario, if at all.

CHAPTER 1

REFERENCES

Office of Military Leadership. 1976. *A Study of Organizational Leadership*. Harrisburg, PA: Stackpole.

Bennis, Warren, and Nanus, Burt. 1985. *Leadership: Strategies for Taking Charge*. New York: Harper & Row.

Blanchard, Kenneth H., and Johnson, Spencer. 1982. *The One-Minute Manager*. New York: Morrow.

Boulding, Kenneth. 1958. *Principles of Economic Policy*. Englewood Cliffs, NJ: Prentice-Hall.

Covey, Stephen R. 1989. *The Seven Habits of Highly Effective People: Restoring the Character Ethics*. New York: Simon & Schuster.

Didactic Systems (Erwin Rausch, ed.). 1977. *Management in the Fire Service*. Quincy, MA: National Fire Protection Association.

Hersey, Paul, and Blanchard, Kenneth H. 1969. "Life Cycle Theory of Leadership." *Training and Development Journal* 23(2) (May).

Heyel, Carl, ed. 1982. *The Encyclopedia of Management*, 3d ed. New York: Van Nostrand.

Jones, Betty, and Lieverman, Harvey. 1978. "Linking Elements: A Comprehensive Approach to Management Training." *Canadian Training Methods*, February.

Koontz, Harold. 1980. "The Management Theory Jungle Revisited." *Academy of Management Review*.

Lakein, Alan. 1973. *How to Get Control of Your Time and Your Life*. New York: Wyden.

Maier, Norman R. F. 1967. "Assets and Liabilities in Group Problem Solving: The Need for an Integrative Function." *Psychological Review* 74(4):240–241.

Mayo, Elton. 1933 and 1946. *The Human Problems of an Industrial Civilization*. Boston: Division of Research, Harvard Business School.

Moorhead, Gregory, and Ricky W. Griffin, 1992. *Organizational Behavior: Managing People and Organizations*, 3d ed. Boston: Houghton Mifflin.

Rausch, Erwin. 1980. *Management in Institutions of Higher Learning*. Lexington, MA: Lexington Books.

Rausch, Erwin. 1985. *Balancing Needs of People and Organizations*. Washington, DC: Bureau of National Affairs.

Rausch, Erwin, and Carter, Harry. 1989. *Management in the Fire Service*, 2d ed. Quincy, MA: National Fire Protection Association.

Rausch, Erwin, and Washbush, John B. 1998. *High Quality Management: Practical Guidelines to Becoming a More Effective Manager*. Milwaukee: ASQ Quality Press.

Stogdill, Ralph M. 1974. *Handbook of Leadership: A Survey of Theory and Research*. New York: Free Press.

Tannenbaum, Robert, and Schmidt, Warren H. 1958. "How to Choose a Leadership Pattern." *Harvard Business Review*, March/April 1958; revisited May/June 1973.

PARTICIPATION IN DECISION MAKING AND PLANNING

It's not my place to run the train, the whistle I can't blow,
It's not my place to say how far the train's allowed to go,
It's not my place to shoot off steam or even clang the bell,
But let the bloody thing jump the track, and see who catches hell.
—Anonymous

Appropriate participation—that is, allowing the right amount of authority: not too much, not too little, with the right people, at the right time—has a major impact on all 3Cs (control, competence, and climate). It is so critical to effective management/leadership that it deserves exceptionally careful attention.

Sharing the decision making has a far greater impact on organizational performance than is commonly assumed. Obviously, appropriate participation leads to better decisions because it focuses the expertise of several minds on the issues. It ensures a better informed staff. It creates a more motivating climate, more open communications, and usually a higher level of mutual trust. At the same time it satisfies the expectations of people that their views are at least considered.

AN EXAMPLE

It is often difficult to see whether or not an officer encourages appropriate participation. Take the example of a fire. The battalion chief arrives first, moments before the first two fire engines get to the scene. The chief already knows who the captains are from his two-way radio contacts. The building is heavily involved, with fire pouring out of windows on the first and second floors. Two people are framed in third-floor windows, screaming for help. On the radio the chief gives these orders: "Captain, Engine 1, there are two people on the third floor. Do not hook up (the water hoses). Get these people out." Then he continues: "Captain, Engine 2, hook up and protect Engine 1." Was this participative decision making, or was it a set of autocratic orders from a "boss"?

Even though the example deals with a crisis situation in which most people consider "orders" to be appropriate, the communication could be either participative decision making or orders, depending on the relationship of the people involved. On the one hand, if the captain of Engine 1 heard that he had to get these people out or there would be unpleasant consequences, or if either captain felt that the assignments should have been reversed and was reluctant to point that out, then there was little or no participation.

On the other hand, it is possible that the captain interpreted the message as: "Do your best to rescue these people. I know the task couldn't be in better hands."

Furthermore, if both the captains knew that, even in a crisis situation, they could make a quick plea for changing or even reversing the assignments and their point of view would be given full consideration, then the decision making was highly participative, with strong mutual trust.

Participation, then, is like a book—it can't be judged by the cover—or like beauty—can be skin deep but usually isn't. The mutual trust that is built up from many instances of appropriate participation is the key to this most important management/leadership skill.

Participation in planning and decision making is not a "yes or no" issue. Instead it is a series of continuums:

- Who should be invited to participate (from no one to everyone)?
- How much weight should the views of staff member(s) be given (from the slightest consideration to letting the staff's views control the decision)?
- When should the staff members be brought into the decision-making process (at the very beginning or later, and only after limits for their participation have been set)?
- Who should be informed of progress toward the decision or plan, and how and when (from no one to everyone, verbally, individually, at meetings, in bulletins, with individual memorandums, etc.)?

A number of factors should be taken into consideration to make these participation decisions. In the next section, we discuss these factors in the form of guidelines.

PARTICIPATION GUIDELINES

Before making any significant decision, ask yourself whom to involve, how much authority to grant that person or group, and when they should participate. The objective is to make these participation decisions so they will result in the highest level of success. You can think in terms of five subordinate guidelines:

1. Selection of decision participants
2. Level/extent of participation
3. Technical and acceptance quality
4. Work maturity of participants
5. Other elements of the situation

Selection of Decision Participants

Who should participate actively, and when? Who should be kept informed of what is happening, and who should merely be informed after the decision has been made? Managers should make the selection on the basis of the relevant considerations in

ADDITIONAL INSIGHT 1.1

the following subsections. They should also consider to what extent individuals feel they should be involved in the decision-making process.

Level/Extent of Participation

How much of a voice should participants have in the decision? This question also involves when each person should participate, which covers the whole range of possibilities, from requesting opinions only after the decision has been made tentatively to full participation in almost every step of the process, starting right from the beginning.

An excellent diagram on this subject was developed by Tannenbaum and Schmidt (1973). Figure 1.2 depicts the many possible combinations of authority by the manager and freedom for subordinates. The range extends from complete control by the manager over decisions to the other extreme where the members of the staff have wide freedom in determining what should be done. According to this model, the most effective leaders are those who choose the most appropriate point between "boss-centered" leadership and "subordinate-centered" leadership, the terms used originally by Tannenbaum and Schmidt in 1958 (Rausch and Carter 1989, 58; Tannenbaum and Schmidt 1973). More elaborate models have been built on this foundation, but for practical purposes, this picture is the easiest to remember and apply.

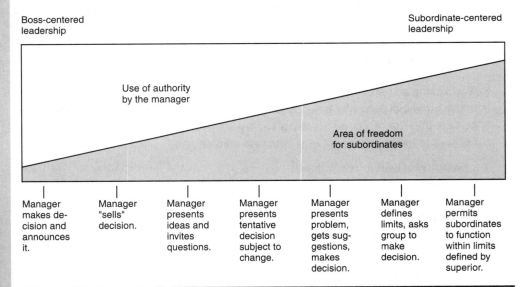

Figure 1.2 The Tannenbaum and Schmidt Continuum
(Source: Tannenbaum and Schmidt 1973)

ADDITIONAL INSIGHT 1.1

Technical and Acceptance Quality

How much technical expertise is needed for a sound decision? To what extent do people (all stakeholders) feel they should have a voice in the decision? To answer these questions, a manager must consider the expertise/ knowledge that is needed to make a decision (technical requirement) and the extent to which acceptance of the decision will bring about its successful implementation (acceptance requirement). These two considerations are illustrated by Norman Maier (1967) of the University of Michigan in the grid in Figure 1.3. Like the Tannenbaum and Schmidt diagram, the Maier diagram is not the latest word on the aspects of participation, but it is a clear and helpful way to express these ideas.

The grid shows four spaces (quadrants): C, low acceptance requirement/low technical requirement; A, high acceptance requirement/low technical requirement; D, low acceptance requirement/high technical requirement; and finally B, high acceptance requirement/high technical requirement.

Low acceptance requirement/low technical requirement decisions generally involve matters that nobody really cares about. Everyone wishes that somebody else would make the decision and go on with the business at hand. There are not too many decisions of this type in the work environment. A manager who does not see to

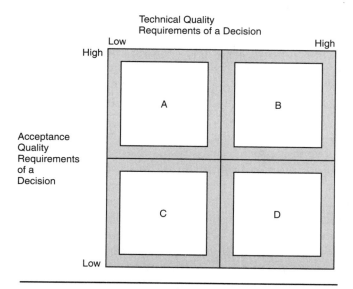

Figure 1.3 Technical and Acceptance Requirements for Participation in Decision Making
(Source: Maier 1967, pp. 240–241)

ADDITIONAL INSIGHT 1.1

it that a decision in this quadrant is made quickly, either by delegating it or by making it, is seen as a procrastinating decision maker.

Examples of matters with *high acceptance requirement/low technical requirement* are plans to reduce absenteeism, increase the attention to certain quality details, reduce waste, and encourage better customer communications. Most matters that involve a significant amount of effort or attention by staff members are in this area, with the exception of those that require high technical knowledge as well. For instance, the decision to reduce absenteeism by 20% does not require technical knowledge. Everybody knows what the implications of the decision are and to what extent it will affect the people who have to implement it. Some trivial decisions, such as allocating parking spaces or changing the coffee break time, also fall into this quadrant.

In decisions with high acceptance requirement/low technical requirement, staff members or their representatives should have the primary voice. In any decision in which people believe they can make a useful contribution and they will be affected seriously by the outcome, high levels of participation will lead to better results. If group members have differences of opinion, the manager must lead the group to a joint decision. Such a decision should minimize resentment so that all individuals will exert the greatest effort to make the decision or plan successful.

Decisions and plans with *low acceptance requirement/high technical requirement* clearly need the involvement of people (from inside or outside the organization) who have the necessary technical expertise. Often these people can recommend a plan and the manager can approve it (after confirming that the acceptance requirement is really very low), possibly without consulting staff members. All that may be necessary is to inform the staff in sufficient detail. Whether a decision is on the low end of the acceptance requirement or near the high end makes a big difference. If a manager is in doubt, it is better to assume that a higher level of acceptance is required.

Finally, decisions and plans in the *high acceptance requirement/high technical requirement* area require the greatest skill on the part of the manager. Apparatus selection decisions, recommendations for special regulations pertaining to a unique industrial facility, and the review of construction plans for such a facility call for considerable technical knowledge. For effective use of the apparatus and for sound prefire plans and inspections for such facilities, though, a high level of acceptance is also necessary. If the decisions made by people with technical expertise have high acceptance among the people who will implement the decisions, then the probability of achieving success is much higher than if such acceptance is lukewarm or lacking. After obtaining the advice of the experts, the manager therefore should convince staff members that the recommended plan or decision alternative is indeed the best course of action.

ADDITIONAL INSIGHT 1.1

One important expansion of the Maier concept is that the acceptance requirement does not involve only staff members. If a decision will have an impact on other stakeholders, managers should consider the issues that may affect their acceptance as well as the implications of rejection by any one group of stakeholders. Managers may find it helpful to consult with representative members of these other stakeholders.

Work Maturity of Participants

To what extent can participants be counted on to accept responsibility for their input into the decision and for their respective roles in implementation? According to the Hersey–Blanchard Situational Leadership Model, which is based on the authors' original work on the "life cycle" needs of staff members, the least mature staff members or teams need the maximum help and guidance. Their experience, achievement motivation, and willingness and ability to accept responsibility are the lowest. The greatest freedom is appropriate for workers at the highest maturity level (Rausch and Carter 1989, 116; Hersey and Blanchard 1969, chap. 1).*

Other Elements of the Situation

To what extent should other aspects of the situation influence participation? Other aspects include the time and cost of participation (to both the organization and the participants), the extent to which potential participants expect or want to be involved, the likelihood of conflict, the information that is available or can be made available, the extent to which the decision is predetermined by procedures and policies, the impact of the decision on the participants, and the urgency and importance of the decision. (See Additional Insight 6.1, Time Management and Delegation.)

Managers should appreciate the importance of appropriate participation in achieving better control. Staff members have greater freedom to make independent decisions, but it clearly identifies the limits. Experienced and competent staff members are given wider limits than those who are new or less competent. By establishing and revising these limits, managers are involved in the control process. Better control, when understood as a joint activity and appropriately implemented, also leads to higher levels of knowledge, skills, and abilities (KSAs), and performance, as well as greater work satisfaction for staff members, thus further strengthening effective control.

---

*Although much has been written on situational leadership, the original Hersey and Blanchard collaboration still represents the basic idea in clear form. To give credit to others or to only Blanchard, who continued to popularize the concept as part of his business strategy, would not be fair to Paul Hersey and would not be the best way to present this idea.

ADDITIONAL INSIGHT 1.1

MANAGEMENT THEORY AND CONCEPTS: ORIGINS AND PATHWAYS

> *The problem is to organize*
> *This monumental enterprise*
> *So that, to see that all are boarded,*
> *Both need and reality are rewarded.*
> —*Adapted from Kenneth Boulding*
>   *(1958)*

In this section we trace the evolution of management theories and thus provide a foundation for the concepts discussed in this book.

IN THE BEGINNING

It all started toward the end of the Industrial Revolution, as many organizations and businesses grew to a considerable size. Frederick Taylor, an American engineer who is often seen as the father of what is called either "management science" or "scientific management," began to study worker productivity. The objective was to find optimal ways to design the jobs of production workers and to determine how those workers might best be selected, trained, paid, and supervised. The effort led to standard costing, method study, time-and-motion study, worker performance standards, and other measurement techniques. In an indirect way, Taylor is also the father of Total Quality Management (TQM) concepts, which evolved from the Value Engineering approach that was popular during the 1950s.

LOOKING AT THE MANAGER

Soon questions arose about how managers and supervisors contributed to productivity. Henry Fayol, a French mining engineer, led the way. He depicted the manager's function as a cycle of planning (defining ends and means); organizing (providing for the necessary equipment, resources, and people); commanding (supervising subordinates); coordinating (ensuring that equipment, resources, and people are effectively interacting); controlling (ensuring that outcomes are consistent with plans); and planning again. The inclusion of "commanding" sounds strange to us today, but in the early part of the twentieth century, managers were expected to issue orders, to command. In a similar vein, today we do not speak of "subordinates," as was the practice well into the 1950s; we refer to people who report to a manager as "staff members" or "associates."

Most modern management books continued to be organized around versions of Fayol's cycle. Although the terms *planning* and *organizing* have usually remained, some of the other terms have been replaced with words like *executing, implementing, staffing, leading,* and *follow-up.* In a way these words reflect more current views of the functions of managers. Fayol also defined a number of "principles of management," including division of labor and unity of command (one person should report to only one manager or supervisor). These concepts, too, have been modified or partially replaced.

World War II and the emergence of the Cold War era stimulated development of the managerial uses of quantitative decision analysis methods, statistical inference, and computer technology. Techniques such as linear programming, probability-based decision analysis, game theory, dynamic programming, and simulation are direct descendants from this period. The modern digital computer continues to make powerful techniques such as these readily available to managers at all levels of organizations. The development of modern computer information systems and software tools is also a result of management science.

A SHIFT IN DIRECTION

On a different track, one branch of management theory arose by accident from a famous attempt to determine how working conditions affected productivity. At Western Electric Company's Hawthorne plant in Chicago, during the late 1920s, among other experiments, six young women were detached from a department with hundreds of assemblers. In their separated room they were observed closely and given friendly attention. Their work environment was altered, first favorably with longer breaks and other changes such as better lighting, and then unfavorably, becoming less and less desirable. At one point, rest periods were eliminated entirely. Through it all, the women's work output increased.

The extensive Hawthorne experiments studied the various aspects of human behavior in the work environment in depth for the first time. Topics included group formation and development (formally and informally), behavioral influences on productivity, communications, and the sources of morale. As interpreted by Elton Mayo (1933 and 1946), these studies heralded the beginning of the human relations theory of management, often referred to as the behavioral sciences in management. Today the two primary branches of this body of theory are leadership and motivation.

The impact of motivation and leadership theories on management education and development reached a peak during the 1950s and 1960s when the most significant, broad-gauged research was performed. Since then, scholars and other researchers have concentrated on correcting or adjusting for inadequacies in the

ADDITIONAL INSIGHT 1.2

original theories and on introducing some highly subjective and even controversial new ideas. In effect, what has happened since then is a fractionalization of behavioral theories, with overlaps and internal contradictions. Only sporadically have there been significant attempts to create bridges between these theories and practical application and use. Total Quality Management is one of the most successful of these bridges.

DISAPPOINTMENTS

Unfortunately, behavioral studies continue to offer more questions than answers. They are full of competing and conflicting models and theories. As Moorhead and Griffin (1992) wrote so aptly,

> The field of organizational behavior, still in its infancy as a science, remains full of competing and conflicting models and theories. There are few laws or absolute principles that dictate proper conduct for organizational members or predict with certainty their behaviors. The role of human resources in the long-term viability of any business or not-for-profit enterprise is nevertheless recognized as enormously significant. Other resources—financial, informational, and material—are also essential, but only human resources are virtually boundless in their potential impact (positive or negative) on the organization. [p. xix]

Leadership theories, in particular, are having a hard time. In *Handbook of Leadership: A Survey of Theory and Research*, Stogdill (1974) noted that leadership has as many definitions as there are persons who have attempted to define the concept. In *Leadership: Strategies for Taking Charge*, Bennis and Nanus (1985) referred to leadership as the most studied and least understood topic of any in the social sciences. Colleges and universities offer a plethora of courses and workshops on the topic of leadership, undaunted by the challenge of teaching an idea that cannot be defined and is not yet well understood. Unfortunately, managers who take such courses often emerge with a frustrating sense that leadership is akin to the Abominable Snowman, whose footprints are everywhere but who is nowhere to be seen.

The question of whether or not a discipline of management has developed continues to occupy the concern of both academics and practitioners. Surely, if such a discipline exists, we should have a common definition of management and a body of consistent literature. This, unfortunately, is not the case. Many have argued for understanding management as a general concept rooted in organizational theory, as human behavior in organizations, as problem solving, as decision making, and as a social process. None of these approaches can stand alone. After a lifetime of study, the

ADDITIONAL INSIGHT 1.2

distinguished management scholar, Harold Koontz (1980, 5:175–187), lamented the continuing "management theory jungle" composed of these schools of thought: empirical (case study), interpersonal behavior, group behavior, cooperative social system, sociotechnical systems, decision theory systems, mathematical (management science), contingency (situational), managerial roles, and operational (management function). Koontz even noted that the situation had actually deteriorated in the 20 years since his earlier review of the state of the art.

Part of the reason for the lack of progress in management studies may be that there is not much of a comprehensive nature that new research is likely to uncover. There are more rewarding ideas to concentrate on. During the 1980s and 1990s, in addition to the continuing growth of computer-related concepts, attention shifted to helping businesses overcome increasingly fierce domestic and foreign competitive pressures and assisting government agencies to cope with severe budget cuts. Management focused on organizational culture and strategies; quality improvement, which brought Total Quality Management and quality circles; productivity with reengineering; restructuring and mergers, with their almost inevitable downsizing; global marketing; and the most recent interest in organizational learning.

THE SITUATION TODAY

The many and frequently bewildering theories often confuse more than clarify and do not provide practical guidelines for decisions and action, especially in the leadership aspects of management. The search for theories continues to lead us to seek solace in those who promise easy answers to complex realities. Management development programs have focused on topics such as time management (Lakein 1973, 11), one-minute management (Blanchard and Spencer 1982), habits of effective people (Covey 1989), transformation, and total quality. None of these is irrelevant, but they all seem to have their day, deliver less than they promise, and lead us to start the search for insight all over again.

All these insights can help in some situations; however, they fail to concentrate on the practical application of what we learn from the theories. We need answers to two questions: What does management/leadership mean to me? How can I best manage and lead in a given situation? To answer these questions you need something that is important at every managerial level: a comprehensive approach that fits all decisions and enhances managerial competence as well as leadership competence. A thorough understanding of the management/leadership guidelines discussed in this book and of the 3Cs model can satisfy some of that need (Rausch and Washbush 1998)

ADDITIONAL INSIGHT 1.2

Management Functions in the Fire Service and Fire Officer Responsibilities

Introduction

Management Functions in the Fire Service:
  Areas of Fire Department Responsibility
  Fiscal Management
  Personnel Management
  Productivity
  Public Information and Community Relations
  Equipment, Apparatus, and Buildings
  Public Fire and Life Safety Education
  Important Non-Fire-Fighting Functions

Organizational Structure of a Fire Department
  Division of Work
  Coordination
  Lines of Authority
  Unity of Command
  Management Levels

Fire Officer Responsibilities
  Responsibilities of the Chief
  Responsibilities of Intermediate-Level Officers
  Responsibilities of Chiefs' Aides
  Responsibilities of Company Officers

Concluding Remarks
Study Questions and Activities
Chapter 2 References
Chapter 2 Additional Readings
Additional Insight 2.1: Communications Techniques and Skills

INTRODUCTION

The traditional goals for effective management in the fire service are related to fire suppression (Cote 1986, p. 15-3):

- Preventing fires from starting
- Preventing loss of life and property when fires start
- Confining fires to their place of origin
- Extinguishing fires

However, "in recent years, the role of the fire service has expanded far beyond fire suppression. The name 'fire department' doesn't begin to cover the services that progressive organizations are providing to their communities" (p. 10-5). New and changing roles are placing demands on the time and talent of the fire service. These new services, as discussed in Chapter 12, are

- Emergency medical/paramedic services
- Confined space and specialized rescue operations, including vehicle and railroad crashes and structure collapses
- Hazardous materials response
- Airport rescue/fire-fighting services
- Community emergency consultation
- Community disaster planning, preparedness, and response, including such diverse emergencies as floods, earthquakes, tidal waves, riots, tornadoes and hurricanes, terrorist attacks, and other natural or human-caused incidents (the customer service concept)
- The use of facilities and human resources to assist the community in other ways, such as shelters for abused children or neighborhood medical centers

These wide-ranging functions are responses to the fire service's drive to meet the needs of the public. In fact, public service itself has emerged as an operational imperative and now includes the concept of customer service, with the public as the ultimate consumer of fire service "products/services."

Clearly, today's fire chiefs and officers are more than fire-scene leaders. They must also be knowledgeable in all those newer functions that apply to their departments and skilled in managing human, physical, and economic resources. Possibly most important, they must be flexible to adapt to rapidly emerging technological developments and the equally rapid changes of the political and social environment.

To be effective in this multifaceted environment, fire officers need to understand the organizational characteristics of their own department and how they differ from those of other fire departments. They also should be aware of productivity issues, research and planning, the administrative responsibilities of higher-level officers, and public and community relations activities. Such awareness can help them perform their duties more effectively and participate in decisions as informed members of the department.

In this chapter we present the topic of management functions in the fire service—areas of fire department responsibility.

MANAGEMENT FUNCTIONS IN THE FIRE SERVICE: AREAS OF FIRE DEPARTMENT RESPONSIBILITY

The primary tasks of fire departments are fire suppression, fire prevention, and fire loss reduction. NFPA 1201, *Standard for Developing Fire Protection Services for the Public*, sets forth the purpose of a fire department:

> The fire department shall have programs, procedures, and organizations for preventing the outbreak of fires in the community and to minimize the danger to persons and damage to property caused by fires that do occur. The fire department also shall carry out other compatible emergency services as mandated. [1994, Section 2-1]

To carry out these purposes effectively, a fire department performs many specific functions that are not directly related to fire-fighting activities. Local, state, or provincial legislation specifies the functions and the reporting relationship of the fire department to other government bodies. Legislation generally places the fire department under the jurisdiction of a county government, a municipal government, or a special district that is empowered by statute to perform the overseeing function.

Operating a fire department and providing a given level of service are usually the responsibility of local government; in the case of a fire district, fire department operations may be its only function. The government body is not concerned with the technical aspects of fire department management, but it is involved in three areas:

> (1) fiscal management, (2) personnel management, and (3) productivity. In general, fiscal management practices follow those used by the government agency supporting the department and include budgeting, cost accounting, personnel costs (including payroll), and purchasing or procurement costs. [Cote 1986, p. 15-11]

Fiscal management refers to the economic activities involved in the operation of a fire department. In addition to accounting and budgeting, planning and research is required to assess current and future developments and needs so that budget proposals can be formulated. After a budget is submitted, fiscal management addresses decisions about purchasing equipment and supplies and maintaining the facilities for storing these supplies.

Fire department *personnel management* is involved to some degree in the recruitment, selection, and promotion of personnel needed to fill positions in the organization. These matters are largely governed by local or state law or both; by personnel agencies, including civil service authorities; and by direct decisions of the government agency that operates the fire department.

Productivity in the fire service differs from fiscal and personnel management because it applies to all functions and is difficult to measure. With respect to the basic goals of the fire service, the protection of life and property, management cannot easily assess the number of fires and the suffering that fire department activities prevent. The same is true of the value of lives that have been saved or injuries that were prevented. The effective measurement of productivity, which is calculated by dividing the value of the service by its cost, would require placing a value on the benefits of fire protection and prevention.

Fiscal Management

All departments maintain an accounting system for financial administration. An efficient system maintains a record of funds received by the department and funds expended and provides data for a continual analysis of how the department's funds are spent. A thorough and relevant analysis helps the fire department manage its available financial resources effectively.

Budgets are an important element of a dynamic system. They outline the financial plans of the fire department. The operations budget, sometimes called the expense budget, lists the expected income and expenses for department activities. A separate budget, the capital budget, lists the planned expenditures for acquisition or renovation of facilities and equipment. (We will discuss financial controls and budgeting in greater detail in Chapter 8, Management of Financial Resources.)

Purchasing and Storing (or Logistical Considerations). In most municipalities, purchasing departments procure fire department equipment and supplies. Records of procurement must be maintained and storage of acquisitions must be authorized when necessary. Common supplies are usually requisitioned from the purchasing agency and charged to the appropriate fire department account. Items of a specialized nature, however, require purchasing specifications to be made by the fire department, followed by approval by the purchasing department and advertisement for bids. The fire chief, with the advice of the appropriate local government official, usually determines whether or not proposals submitted by bidders adequately meet specifications.

Even for emergency purchases, most departments still require estimates from several suppliers. If the expenditure is small and the budget has funds available, then the chief can authorize the expenditure. If the budget does not have funds available, authorization and additional funds must be obtained from the municipal manager or finance officer.

the chief in at the mercy of the county

B. **Planning and Research.**   Planning for the future needs of a fire department is one of the most important jobs of top-level fire department managers. Inadequate planning increases the potential for crises. All departments, therefore, need to develop long-range plans that are flexible and reflect changes in the local community (Cote 1986, p. 15-11). To plan effectively, fire departments should maintain a close working relationship with local and regional planning groups.

The responsibility for planning varies among fire departments. For example, in some jurisdictions, all department heads are required to submit estimates of their capital equipment needs for five years. Some rapidly developing communities, and some of the larger fire departments have planning staffs to assist in the planning for new fire stations, the replacement and possible relocation of old stations, and the purchase of apparatus. In the vast majority of communities, however, planning is much less sophisticated. City administrators might approve citizens' demands for better protection of areas far from existing fire stations and then merely consult with the fire department about a suitable location. Outside consultants might be employed to recommend the addition, relocation, or consolidation of fire stations. They may use fire department annual reports, population projections, census tract data, planning board approvals, and pending applications as the basis for recommending improvements.

Sound planning should be based on research. Research falls into two major areas. The first area concerns activities of all functions of a fire department and uses reviews and analyses of records of responses, inspections, and other information on department services. The second area of research concerns the investigation of fires and the use of the research findings in fire service activities. The fire investigation is basic to good management because it identifies actions the fire department can take to lessen the number and severity of fires in the future. Data from fire investigations contribute to the improvement of inspection procedures, public education programs, and fire suppression activities. (We will discuss this type of research in Chapter 6, Prefire Planning and Related Loss Reduction Activities.)

Few departments are adequately staffed or financed to support significant research activity. Most fire departments are relatively small organizations that lack sufficient personnel to plan the ongoing activities involved in furnishing fire protection. True research into such areas as efficient equipment design and improved turnout equipment is generally beyond the capability of most departments.

(c) **Records and Reports.**   An effective and efficient record-keeping system is the basis for all planning, budgeting, and risk-analysis programs. With the use of computers, stored data and records can easily be presented as tables, flowcharts, or graphs to illustrate department functions. Exhibit 2.1 shows a breakdown by percentage of fire department calls and a classification of fires that may be used for planning and procuring apparatus or for allocating inspection resources to properties that are fire risks.

The modern fire department needs a computerized record-keeping system that meets both current and anticipated needs. This system must provide the data for evaluating the department's effectiveness in all phases of its operation, including data for developing reports on accomplishments and other performance measures as well as recommendations for organizational changes and capital investments. An effective record-keeping system includes hardware, software, and reporting procedures for personnel files; training records; maintenance records; hose, pump, and ladder test data; and the standard response and inspection information. Officers can find

Exhibit 2.1 Sample of Statistical Data for Fire Department Management

Fire department calls (%)		Classification of fires (%)	
Structural fires	25.0%	Dwellings	35.2%
Emergencies other than fires	12.0	Other buildings	10.2
EMS response	56.0	Rubbish outdoors	10.8
Hazardous materials incidents	2.0	Trees, brush, grass	21.0
False alarms	5.0	Miscellaneous fires outdoors	5.0
	100.0	Vehicles	17.5
		Aircraft	0.2
		Ships and boats	0.1
			100.0

guidance in NFPA 901, *Standard Classifications for Incident Reporting and Fire Protection Data.*

No matter what type of record-keeping system a department has, it must satisfy the legal requirements in the particular state or province. Chapter 4, Fire Prevention and Code Enforcement, gives examples of typical management records and reports with respect to that function. Officers may find a list of available research sources (in addition to the department's records) in the *Fire Protection Handbook*, which gives the officials or agencies responsible for gathering fire data in each state (Cote 1997, Section 10-169). The data usually conform to NFPA 901. The NFPA's Fire Research and Analysis Division can also benefit fire departments by providing comparison data.

Personnel Management

To operate effectively and efficiently, a fire department must fill vacancies with individuals who meet the appropriate standards. Normally, recruitment of personnel is not a paid fire department responsibility because it is handled by the local government's personnel agency. NFPA 1201 recommends that the fire chief prepare the personnel policies and standards for the department and issue the orders necessary for administering personnel procedures. In some departments, the chief designates an assistant chief as department personnel officer. Specifically, the 1994 edition of NFPA 1201 states that "the fire department shall have a human resource policy . . . that ensures fair and equitable . . . orders necessary for administering personnel procedures" (pp. 1201–1208).

The actual personnel activities of a fire department depend on which personnel services are provided by municipal or state personnel or civil service agencies. State, provincial, and civil service commission legislation might set specific standards of pay, hours, working conditions, working schedules, and other features of personnel policy that limit the authority of the municipality and the fire department. (See Chapter 9 for a more detailed discussion of fire service personnel management.)

Personnel Standards. The personnel standards of a fire department should be designed to establish and maintain a competent and well-trained force by recruiting highly qualified individuals and by providing a satisfying career from recruitment to retirement. An excellent way to meet this task is to use the professional qualification standards published by the National Fire Protection Association.

Ⓑ **Training.**   Training is an integral part of any fire department's operations. It is an invaluable tool for developing the human resources of the department to meet the challenges of providing effective fire protection. We will discuss features of comprehensive fire department training programs in Chapter 11, Training As a Management Function. You may find additional guidance for training in various NFPA professional qualifications standards as discussed in Chapter 11.

③ **Productivity**

Let's first clarify the meaning of productivity and its relationship to efficiency and effectiveness. The word *productivity* is generally synonymous with its more common meaning, efficiency, which is measured as the ratio of outputs to inputs. In a fire department, productivity usually relates to the extent of services delivered within budget. Higher productivity means more services delivered with the same budget or the same services delivered with a smaller budget.

A broader definition considers effectiveness, which includes long-term and qualitative issues and is even more difficult to evaluate. An organization's effectiveness is also determined by its ability to improve efficiency over time as well as factors that cannot be assigned numerical values, such as suffering. Sometimes managers must sacrifice immediate productivity improvements to gain greater future benefits.

Productivity is a management responsibility that is of great interest to the governing body and to the top levels of fire department management because it is directly related to quality of service, community satisfaction with the department, and the cost of providing the services. Businesses, whether they manufacture a product or provide a service, usually measure productivity as the value of the output divided by the resources devoted to production. This calculation is often not as precise as it may sound, however. The quality of the product or service must be considered along with other issues such as resources (training and development of better methods, for example) devoted to improving the quality or quantity of production in the future.

Measuring Productivity.   In the fire service, the problem of measuring productivity is vastly compounded by the difficulty of determining the department's output. To take an extreme example, if the number of fires extinguished were the primary measure, a department could easily increase its productivity by setting some fires and then putting them out. More seriously, if productivity were measured by the number of inspections, inspectors could conduct more inspections by being less thorough.

In addition to efficiency, which is usually the primary measure of productivity, the fire service must consider lives, injuries, and human suffering from the loss of possessions. The public image of the department is a component of productivity as well as an aspect of the quality of service, at least as perceived by the public and funding sources. The quality of service also includes promptness of response, concern for citizens' needs, and thoroughness of efforts to return property to a normal condition as possible after an emergency. These important elements of productivity further complicate its measurement, especially when little visible output can be evaluated.

Productivity can be measured to some extent by comparing the key achievements of departments of similar size and quality of service, in similar situations, with data obtained from research in a process called *benchmarking* (Cote 1997, p. 10–29). For instance, a department could evaluate the productivity of its fire hydrant inspections by comparing the average fire fighter hours it spends on ten hydrants to the time spent by a similar size, high-productivity department, or by comparing the adherence and time requirement to satisfy NFPA 1410, *Standard on Training for Initial Fire Attack.*

Since public perception is critical for public organizations, fire departments must report their productivity objectively. Rather than reporting a loss of $9,000 from a structural fire in a $150,000 home, a more realistic way to present the event is to show the loss but also report that $141,000, or 94 percent, was protected and not damaged by fire. This type of reporting reflects the effectiveness of a particular fire department more factually and fairly.

Detailed records of specific activities of a department's various functions, analysis of these records, and plans for improvement, function by function, can provide evidence of increasing productivity. For instance, response time is a component of productivity. Records of response times may involve records of communications, not only of emergency calls from the public but also calls en route, fireground communications among fire department personnel, alarm and signaling systems from individual properties, and street box alarm signals. Such records establish levels of relative productivity to compare response times over time or with other departments.

Equipment effectiveness and sound procedures also improve response time and thus one element of productivity. Effective procedures and training, including drills, lead to finely honed communications practices. Regular testing and inspection of all communications systems can help to ensure that equipment is in peak working order and result in the best possible response time for effective fireground activity. (See Chapter 7, Management of Physical Resources, for a more detailed discussion of this hardware component of productivity.)

Managing Water Resources to Assure Productivity.   Adequate amounts and effective use of water resources have a major effect on the productivity of a fire department, especially for fighting fires but also for the maintenance of water facilities and equipment. The 1984 edition of NFPA 1201 listed recommendations for water resources (pp. 45–46), which were reviewed and revised in the 1994 edition. These are the 1994 recommendations for fire departments (p. 19):

- Establish minimum fire flow requirements for representative locations in the municipality or fire protection district.
- Carry out and maintain a program for evaluating all sources of water supplies and delivery systems for fire fighting within the community.
- Facilitate the delivery of adequate water supply consistent with the fire risk and the fire department capabilities.
- Carry out a continuing program of evaluation for all water supplies for fire fighting, maintaining a liaison with the water authorities on fire protection with water supply matters.
- Assess the adequacy or weakness of water supplies in relation to the fire risk thoughout the community in conjunction with a prefire inspection or planning program.

In addition, to maintain high productivity, fire departments should determine water flows for new construction proposed for the district and specify a mechanism for their delivery.

NFPA 1201 has more recommendations with regard to fire department water supply operations:

- The fire chief should assign a full- or part-time water officer to maintain regular contact with the managers of public and private water supply systems. This officer must keep the fire department apprised of all water supply sources available for use and recommend new facilities as advisable, based on data available in the fire department and responsible local government agency.

• Training should be provided to all personnel in the use of water supply system equipment and facilities, and regularly scheduled drills should be conducted to develop a thorough working knowledge of the system's capabilities, including information and instructions on available water sources, and field exercises to ensure that fire fighters are aware of hydrant locations for fighting potential fires.

NFPA 1201 (p. 19) also specifies that each company in the fire department should maintain a water resources map (see Figure 2.1) for the respective first-due area. The map should show these features:

• Location and size of mains
• Areas of insufficient flow and/or pressure
• Areas that require special operations to supply water
• Hydrant locations and capacities
• Location, capacity, and accessibility of auxiliary water supply systems

Following these procedures can bring about the highest possible level of productivity.

Public Information and Community Relations

A positive public perception is vital to a fire department's ability to protect the public. As a public agency supported by public funds, the fire department needs public support. A department that is esteemed by the public will

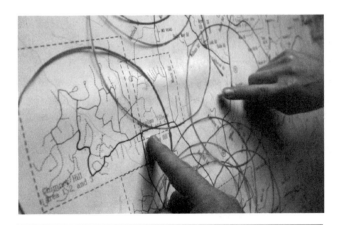

Figure 2.1 Water Resources Map (Source: Chief Brian Johnson, Wilmington Fire Department, Wilmington, VT)

find it easier to obtain a large enough budget and also to enlist volunteers and gain public cooperation with code enforcement, for example. To get a high level of support, a fire department needs to devote both time and talent to making the public aware of the benefits it recieves from expenditures on fire protection in the community. The public's understanding and cooperation can enhance fire department programs and enable the department to function effectively within the political environment of a particular community. Public information and community relations programs are therefore an important management activity.

Two main goals of a public information program are: (1) to promote public awareness of the fire department and (2) to promote public understanding of fire. An effective public relations program sets up procedures to keep the public informed of department operations. Important developments and newsworthy events should be transmitted regularly to the public through newspaper releases and other media. To gain maximum readership, the department should deliver information in a lively and informative manner, geared to the target audience in the community.

Community relations is a year-round activity, and the department's efforts will bring the best results if they are targeted to the public's actual needs. For example, the Newark, New Jersey, Fire Department has a community relations division with a community fire safety center to meet the fire safety needs of the neighborhoods. Public fire safety education programs are available for all age groups. Interested business people receive fire training sessions designed for their specific organization. A staff member provides information about codes. The financial support for this center comes from both budgeted money and donations from the business community.

Equipment, Apparatus, and Buildings

Fire department managers are responsible for evaluating and upgrading department facilities, apparatus, and equipment to ensure good performance and safety on the job. (See Chapter 7, Management of Physical Resources, for an explanation of management's responsibility in assessing and providing adequate facilities and equipment for effective department operation.)

Public Fire and Life Safety Education

An emerging service of great importance is truly effective public fire and life safety education programs. NFPA 1035, *Standard for Professional Qualifications for Public Fire and Life Safety Educator*, defines public fire and life safety education as "comprehensive community fire and injury prevention programs designed to eliminate or mitigate situations that endanger lives, health, property, or the environment" (1993, pp. 1035–6). As that definition suggests, public fire and life safety education differs from community relations because it goes beyond making people aware of fire department activities; it arms them with ways to prevent fires, save lives, and minimize damage. Citizens do not know enough about how to live in a fire-safe manner or how to act appropriately in case of fire. By providing continuing educational messages in various ways, a fire department can raise the knowledge level. Where these programs have been used, the result has been fewer fires, less injury to people, and less damage to property. (We will discuss fire safety education more thoroughly in Chapter 5, Fire and Life Safety Education.)

Important Non-Fire-Fighting Functions

Chapter 12 presents nontraditional roles of the fire service. A fine line exists between nontraditional and non–fire-fighting functions. Fire fighting is traditional; but non–fire-fighting functions are also part of the traditional services, such as water removal during flooding conditions. The line is not sharp, so we need to mention both terms.

The role of a modern fire department is not limited to fire suppression and associated functions. As we mentioned earlier, considerable resources of many fire departments are devoted to various services. (We will discuss these topics in Chapter 12, Fire Department Services Beyond Fire Fighting.)

ORGANIZATIONAL STRUCTURE OF A FIRE DEPARTMENT

The organization of fire departments varies depending on the size and population of the communities they serve. NFPA 1201 provides some guidance: "This standard contains requirements and recommendations on the structure and operation of organizations providing public fire protection services" (1994, pp. 1201–5).

Figures 2.2, 2.3, and 2.4 show typical organizational structures of small, medium, and large fire departments, respectively. These structures adhere to the organizational principles discussed later, and they define the roles of the individuals and units within the organization.

Four organizational principles that underlie effective organizational structures are division of work, coordination, clearly established lines of authority, and unity of command.

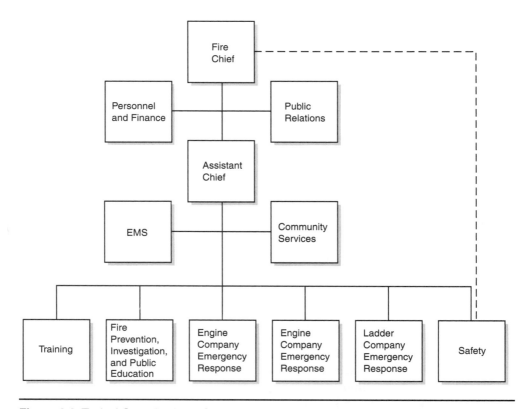

Figure 2.2 Typical Organizational Structure of a Small Fire Department

Division of Work

The most basic organizational principle for a fire department is the division of work according to a plan that defines the relationships among the operating units and their individual members. The division should be based on the functions that must be performed, such as fire prevention, training, and communications.

Coordination

The need for internal coordination becomes greater as a department grows in size and complexity. A small department usually has a simple organizational structure that allows frequent personal contact among individuals, thus ensuring efficient coordination. Because the structure of a larger department does not allow such frequent personal contacts, however, more extensive coordination of the operating units is necessary.

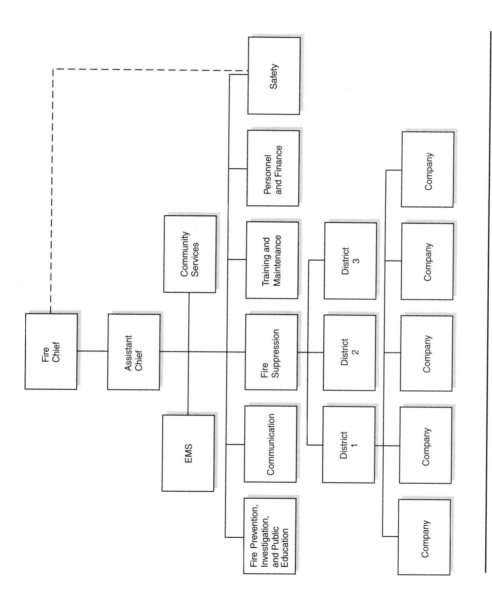

Figure 2.3 Typical Organizational Structure of a Medium-sized Fire Department

43

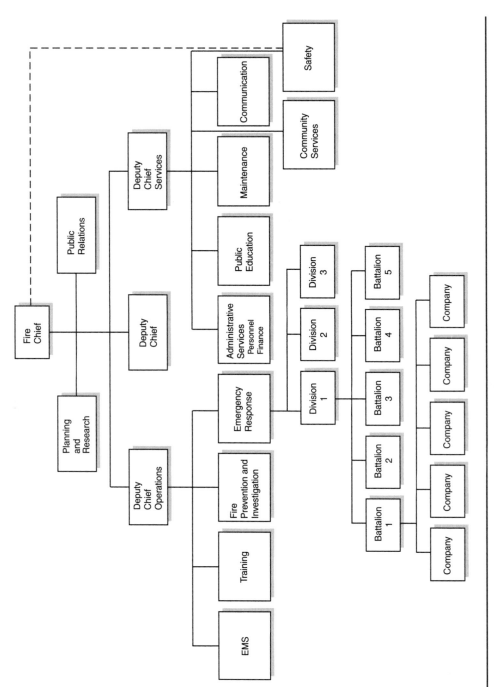

Figure 2.4 Typical Organizational Structure of a Large Fire Department

Lines of Authority

Lines of authority define both the extent of input into the decision-making process by individuals and their responsibilities and authority when working on assigned tasks. Sometimes individuals are given responsibility for performing certain tasks, but they are not given the authority to make decisions necessary to complete the tasks. This lack of decision-making authority tends to restrict the performance of the department because these individuals must frequently consult their immediate supervisors about decisions.

Unity of Command

Conflicting orders from several higher level officers can result in confusion and inefficiency. With unity of command, a fire fighter receives orders from only one officer; thus, he or she can usually perform more efficiently. Unity of command also presents an inherent problem, though: Too many people reporting to an officer can compromise effective supervision and the officer's ability to perform other duties.

Management Levels

Fire department officers bear the primary burden of managing the fire department. NFPA 1021, *Standard for Fire Officer Professional Qualifications,* recognizes the various levels of management in its definitions and Job Performance Requirements (JPRs) for Fire Officers I, II, III, and IV:

- A Fire Officer I functions at the supervisory level.
- A Fire Officer II functions at the supervisory/managerial level.
- A Fire Officer III functions at the managerial/administrative level.
- A Fire Officer IV functions at the administrative level.

Titles differ from fire department to department. Fire Officer I may also be called a "company officer," Fire Officers II and III may be called "intermediate-level officers," and Fire Officer IV personnel are usually "chief officers."

The Job Performance Requirements in NFPA 1021 describe specific tasks, list the items necessary to complete the tasks, and define measurable or observable outcomes and evaluation areas for each specific task. As we noted in Chapter 1, JPRs focus on tasks and thus differ from KSAs, which focus on the knowledge, skills, and abilities required to accomplish tasks.

NFPA 1021 sets forth Job Performance Requirements for each officer level in the areas of human resource management, community and government relations, administration, inspection and investigation, emergency service delivery, and safety. Exhibit 2.2 provides the JPRs in the area of human resource management for each of the four managements levels covered by NFPA 1021.

FIRE OFFICER RESPONSIBILITIES

To be most effective, an organization must delegate various roles and individual responsibilities to ensure a proper division of labor. The fire chief and other officers must understand each member's relationship with the department and with its management. Just as managers in the private sector have assistant managers or department heads to aid them in their decision-making and leadership roles, fire chiefs are assisted by officers and sometimes aides in making decisions and performing other fire department functions.

Responsibilities of the Chief

Every department is headed by a chief, who may hold some other title, such as director (see Figure 2.5). The chief is usually assisted by two types of officers: intermediate officers and company officers. The chief of a fire department has two related major roles. First, the chief is the primary representative of the department in its relations with the governing body and the community. Second, as the highest-level manager, the chief is the manager/leader of the entire department and is responsible for establishing the direction of the operation of each unit and the members of the department. This direction should guide the department to achieve high levels of control and competence and a productive, satisfying climate (the 3Cs and their guidelines). The job requirements described for Fire Officer IV in NFPA 1021 should be considered.

To meet the difficult leadership challenges, the policies of the department should be geared to bring all officers to a high level of competence with respect to all aspects of the 3Cs model (see Chapter 1) and its guidelines, or equivalent guidelines. Unfortunately, training programs directed at officer KSAs (knowledge, skills, and abilities) are not sufficient. The chief has to use all possible approaches, including personal example, perseverance, and regular sessions on exploring issues from the 3Cs perspective. Complicating this task is the fact that officers serve on several rotating shifts and are rarely together at the same time.

Furthermore, the chief should involve all constituencies and work in conjunction with community administrators to determine the long-range

Exhibit 2.2 Sample Job Performance Requirements in Human Resource Management

Fire Officer, Level I
- Assign tasks or responsibilities to unit members, given an assignment under nonemergency conditions at a station or other work location, so that the instructions are complete, clear, and concise; safety considerations are addressed; and the desired outcomes are conveyed.
- Direct unit members during a training evolution, given a company training evolution and training policies and procedures, so that the evolution is performed safely, efficiently, and as directed.
- Recommend action for member-related problems, given a member with a situation requiring assistance and the member assistance policies and procedures, so that the situation is identified and the actions taken are within the established policies and procedures.
- Apply human resource policies and procedures, given an administrative situation requiring action, so that policies and procedures are followed.
- Coordinate the completion of assigned tasks and projects by members, given a list of projects and tasks and the job requirements of subordinates, so that the assignments are prioritized, a plan for the completion of each assignment is developed, and members are assigned to specific tasks and supervised during the completion of the assignments.

Fire Officer, Level II
- Initiate actions to maximize member performance and/or to correct unacceptable performance, given human resource policies and procedures, so that member and/or unit performance improves or the issue is referred to the next level of supervision.
- Evaluate the job performance of assigned members, given personnel records and evaluation forms, so each member's performance is evaluated accurately and reported according to human resource policies and procedures.

Fire Officer, Level III
- Establish personnel assignments to maximize efficiency, given knowledge, training, and experience of the members available in accordance with policies and procedures.
- Develop procedures for hiring members, given applicable policies and legal requirements, so that the process is valid and reliable.
- Develop procedures for promoting members, given applicable policies and legal requirements, so that the process is valid and reliable.
- Describe methods to facilitate and encourage members to participate in professional development to achieve their full potential.

Fire Officer, Level IV
- Appraise a grievance program, given appropriate data, to determine if the program is effective, consistent, and produces resolution at the appropriate level.
- Establish and evaluate a list of education and in-service training goals, given a summary of the job requirements for all positions within the department, so that all members can achieve and maintain required proficiencies.
- Appraise a member-assistance program, given appropriate data, to determine if the program, when used, produces the desired results and benefits.
- Evaluate an incentive program, given appropriate data, so that a determination is made regarding achievement of the desired results.

Source: NFPA 1021, *Standard for Fire Officer Professional Qualifications*, 1997.

Figure 2.5 Chief Ted Lowden of Evesham Township, NJ
(Courtesy of Harry Carter)

goals for the department. The chief is the primary person concerned with long-range goals. The goals and objectives of other officers concern primarily medium- to short-range accomplishments.

Long-range goals and plans help a department prepare for challenges in the years to come. They determine what steps the department should take for the changes that will occur. For example, a new fire station takes years to build after the department recognizes a need for it, or a department might have to wait three to six years before large pieces of apparatus are delivered. In the interim, the department needs plans to cope with deficiencies in the existing facilities or equipment and to lay the foundation for the transition to an additional station or more complex equipment.

Obviously, long-range goals need short-range goals (or objectives) to support them. (See Additional Insight 3.1, Goal Setting and Implementation.) Short-range goals and objectives are immediate and deal with fairly concrete and easily forecasted matters. They often involve officers and fire fighters who might be several levels removed from the chief. The chief should guide and lead the department members to establish and achieve lower-level goals and objectives.

In addition to goals and objectives, a department needs operational procedures such as duty schedules, standard operating procedures (SOPs), or standard operating guidelines (SOGs), and response procedures. You may find a list of requirements in NFPA 1201 (Chapter 2). To meet the changing needs of the department, these procedures should be regularly reviewed, revised, and thoroughly communicated to all department members. The procedures should, of course, be set in consultation with the appropriate members of the officer staff.

Responsibilities of Intermediate-Level Officers

For guidance on job requirements at this level, you may want to consult the job descriptions for Fire Officers II and III in NFPA 1021.

All fire departments need at least one officer who is responsible for department operations in the absence of the fire chief. This officer will also assume administrative and operational command responsibilities at fires. The rank of the individual charged with this duty varies; it can be a deputy chief, a battalion chief, or a district chief. In some small, paid fire departments, shift officers perform dual roles as company officers and duty-shift commanders. In other departments, shift commanders respond in a command vehicle to all alarms. The shift commander is in charge of operations and has the authority to call off-shift help or to request mutual aid assistance—that is, the shift commander is in charge of a department in the absence of the fire chief.

Deputy chiefs, battalion chiefs, and officers in staff positions, such as those in bureaus concerned with buildings or apparatus, provide direction and recommendations to the department. They should apply guidelines and the 3Cs concepts to their decisions and leadership techniques.

Intermediate-level officers who are specialists in certain fields can be valuable to a department. Specialists are likely to be more effective in performing their respective functions than officers assigned on a part-time basis. They provide expert input for decisions. In addition they can conduct training sessions to acquaint officers and fire fighters with matters related to their special fields. Their expertise adds value to the instruction.

Responsibilities of Chiefs' Aides

The chiefs' aide, sometimes called a fire technician, is an important position for fire department efficiency. The position does not exist in every department. When there is no aide, the functions may be performed by a full-time officer or fire fighter or by a company officer or fire fighter who has been assigned on a part-time basis. Full-time, experienced aides are assigned to work directly as assistants to various chief officers rather than as members of a company team.

Aides perform valuable services in the operation of the incident command system. The functions of aides differ from department to department. They may operate the command car, handle and channel radio communications with the alarm center, and assist the chief in other ways, including helping to assess a fire and placing fire companies as determined by the chief.

Responsibilities of Company Officers

For guidance on job requirements at this level, you may wish to consult the job responibilities for Fire Officer I in NFPA 1021. Every fire company or firefighting unit should be supervised by a qualified company officer when in quarters, when responding to alarms, and when making in-service inspections. In many fire departments, company officers hold the rank of fire captain, and a fire captain is assigned to each duty shift of each fire company. The captains also are used as relief chiefs when one or more of the chiefs on the shift is absent and as station commander and coordinator of operations in the absence of a chief officer. In some jurisdictions, the rank of lieutenant designates the commander of an individual unit or station.

Some fire departments also have a rank of sergeant for fire officers. Sergeants are officers in charge of work shifts in their assigned companies. Although sergeants often function as lieutenants, they rank below lieutenants in the chain of command, both in quarters and at the incident.

Company officers have to keep the 3Cs and their guidelines in mind. With respect to control and competence, they translate the goals of the department into goals and objectives for their units. This job requires considerable coordination among shifts. Officers work not only with operational objectives for the activities of the company (such as maintenance of facilities and equipment) but also training and development objectives. Company officers may bring in staff specialists in various fire department functions (regardless of rank), including the training officer, as well as knowledgeable representatives of equipment manufacturers and other suppliers to help increase the competence of a company's members so that they can gain maximum benefit from new equipment, supplies, and techniques.

At the same time, the company officer is responsible for the third of the 3Cs, the climate. Climate issues include helping fire fighters gain the greatest possible satisfaction both from their work on maintenance, housekeeping, and so on, and in the use and purchase of equipment and supplies. This may seem to be a difficult task, but many opportunities exist for providing social satisfaction and higher esteem through allocating work assignments that are seen as desirable, through encouraging appropriate participation in decisions, and through providing opportunities for recognition. Fire fighters gain recognition when the general public and special groups are given the chance to inspect the station and equipment during Fire Prevention Week or at other times. [For a more detailed discussion of ways to enhance work satisfaction, see Additional Insight 5.1, Enhancing Work Satisfaction (Providing Recognition) and Performance Evaluation.]

For specific examples of goals and objectives of officers at all levels, see Additional Insight 3.1, Goal Setting and Implementation.

CONCLUDING REMARKS

In summary, the main topics discussed in this chapter and in Chapter 1—the functions of a fire department and the management/leadership principles that enable officers to succeed—serve as the basis for the remaining chapters of this book. To acquire the thorough competence outlined in NFPA 1021, officers need much more in-depth knowledge, which they can gain only through a systematic professional development program. Some of the many resources and opportunities available for learning are listed here:

- Departmental officer training programs
- County or state fire training academies
- Two-year community college fire science programs
- Four-year college and university programs
- National Fire Academy residence programs in Emmitsburg, MD, or Open University programs that are administered by a number of colleges on behalf of the National Fire Academy
- A wide range of "University Without Walls" programs
- Self-study in books or with correspondence courses aimed at meeting the standards of NFPA 1021

CHAPTER 2

STUDY QUESTIONS AND ACTIVITIES

If you are alone, prepare your own written responses to these questions. If you are studying as part of a team or working on these activities as part of a class, discuss the questions with other members of your team or class and write a consensus answer.

1. In what ways have the traditional goals of a fire department been revised and expanded?
2. Briefly explain the structure of a fire department.
3. What is productivity? Why is it difficult to define? In what ways can a fire department ensure high productivity?
4. Identify which functions of a fire department are involved in each of the following tasks:
   a. Procuring a new pumper
   b. Retiring a fire officer
   c. Preparing a budget
   d. Ensuring adequate water supplies
   e. Training a new member of the department
5. Select at least two job performance requirements (JPRs) from NFPA 1021, *Standard for Fire Officer Professional Qualifications* for your work and at least two for the work of a company officer. Prepare a list of knowledge, skills, and abilities (KSAs) for each of these JPRs and then discuss them with the officer to whom you report.

CHAPTER 2
REFERENCES

Berne, Eric. 1967. *Games People Play—The Psychology of Human Relationships.* New York: Grove Press.

Bramson, Robert M. 1981. *Coping with Difficult People.* Garden City, NY: Anchor Press.

Clary, Thomas C.; Lieberman, Harvey; and Rausch, Erwin. 1974. *Transactional Analysis—Improving Communications: A Didactic Exercise.* Cranford, NJ: Didactic Systems.

Cote, A. E., ed. 1986. *Fire Protection Handbook,* 16th ed. Quincy, MA: National Fire Protection Association.

Cote, A. E., ed. 1997. *Fire Protection Handbook,* 18th ed. Quincy, MA: National Fire Protection Association.

Fast, Julius. 1970. *Body Language.* New York: Evans and Company.

Gray, John. 1992. *Men Are from Mars, Women Are from Venus: A Practical Guide for Improving Communications and Getting What You Want in Your Relationship.* New York: Harper and Collins.

Harris, Thomas A. 1976. *I'm OK, You're OK—A Practical Guide to Transactional Analysis.* New York: Avon.

Luft, Joseph. 1970. *Group Processes: An Introduction to Group Dynamics.* Mountain View, CA: Mayfield.

NFPA 901, *Standard Classifications for Incident Reporting and Fire Protection Data,* National Fire Protection Association, Quincy, MA.

NFPA 1021, *Standard for Fire Officer Professional Qualifications,* National Fire Protection Association, Quincy, MA.

NFPA 1035, *Standard for Professional Qualifications for Public Fire and Life Safety Educator,* National Fire Protection Association, Quincy, MA.

NFPA 1201, *Standard for Developing Fire Protection Services for the Public,* National Fire Protection Association, Quincy, MA.

NFPA 1410, *Standard on Training for Initial Fire Attack,* National Fire Protection Association, Quincy, MA.

Nichols, Ralph. 1962. "Listening is Good Business." *Management of Personnel Quarterly* 1(2):2–9.

Rausch, Erwin. 1971. The Effective Organization: Morale vs. Discipline. *Management Review.* New York: American Management Association.

Rausch, Erwin. 1978. *Balancing Needs of People and Organizations—The Linking Elements Concept.* Washington, DC: Bureau of National Affairs (Cranford, NJ: Didactic Systems, 1985).

Rausch, Erwin, and Rausch, George. 1971. *Leading Groups to Better Decisions: A Business Game.* Cranford, NJ: Didactic Systems.

Tannen, Deborah. 1990. *You Just Don't Understand.* New York: Ballantine Books.

CHAPTER 2
ADDITIONAL READINGS

Coleman, Ron. 1978. *Management of Fire Service Operations.* Duxbury, MA: Duxbury Press.

Cote, Arthur. 1997. *Fire Protection Handbook,* 18th ed. Quincy, MA: National Fire Protection Association.

Drucker, Peter F. 1954. *The Practice of Management.* New York: Harper Brothers.

Drucker, Peter F. 1974. *Management: Tasks, Responsibilities, Practices.* New York: Harper & Row.

Granito, John, ed. 1988. *Managing Fire Services,* 2nd ed. International City Management Association.

Hughes, Charles L. 1965. *Goal Setting—Key to Individual and Organizational Effectiveness.* New York: American Management Association.

Odiorne, George S. 1961. *How Managers Make Things Happen.* Englewood Cliffs, NJ: Prentice-Hall.

Odiorne, George S. 1968. *Management Decisions by Objectives.* Englewood Cliffs, NJ: Prentice-Hall.

Odiorne, George S. 1987. "Measuring the Unmeasurable: Setting Standards for Management Performance." *Business Horizon* 30(4): 69–75.

Peters, Tom. 1987. *Thriving on Chaos*. New York: Alfred A. Knopf.

Peters, Tom, and Austin, Nancy. 1985. *A Passion for Excellence: The Leadership Difference*. New York: Random House.

Peters, Tom, and Waterman, Robert H. Jr. 1982. *In Search of Excellence: Lessons from America's Best-Run Companies*. New York: Harper & Row.

Rausch, Erwin. 1980. "How to Make a Goals Program Successful." *Training and Development Journal*. 34:3.

Senge, Peter M. 1990. *The Fifth Discipline—The Art and Practice of the Learning Organization*. New York: Currency and Doubleday.

Torbert, William. 1992. "The True Challenge of Generating Continual Quality Improvement." *Journal of Management Inquiry* 1(4): 331–336.

COMMUNICATIONS TECHNIQUES AND SKILLS

The most important obstacle to effective communication is the illusion that it has been achieved.

—Anonymous

This saying clearly articulates why we have devoted a section of this book to communications concepts, techniques, and skills.* Communications affect every aspect of control, competence, and climate. To fully understand and appreciate thorough, open, two-way communications in all managerial and leadership activities takes many years or exceptional concentration on the subject.

Even the best communications do not necessarily bring agreement. Achieving agreement requires more, much more, than just good communications. Mutual understanding may be a necessary condition for preventing serious conflict, but even with mutual understanding, agreement may still be impossible if one or both of the parties to the conflict stand fast on unreasonable or irrational positions. If that were the case, it wouldn't matter that both parties were skilled in the techniques—that is, good at asking questions, experienced in observing and interpreting nonverbal signals, and skillful listeners. Nevertheless, mutual understanding is likely to reduce the harshness of the conflict. Also sound communications can lead to more satisfying relationships and probably greater mutual trust.

After a brief discussion of two theoretical concepts useful for thorough communications, we will outline the knowledge, skills, and abilities (KSAs) that officers should sharpen so they can better establish and maintain comprehensive, open, two-way communications in their units and with other stakeholders. These include conducting effective meetings, writing, speaking, listening, probing, seeking and providing feedback, and being aware of nonverbal communications.

TWO THEORETICAL CONCEPTS

Transactional Analysis

Transactional analysis is a psychological theory that has practical applications for communications. It became a management development fad during the 1970s with two best-selling books (Berne 1967; Harris 1976), but it faded quickly, in part because of the excesses of the programs. Its proponents explored issues far beyond

---

* An interesting aspect of communications, not covered in this Additional Insights Section is discussed under the heading "Communications at the Emergency Incident," in Chapter 3.

communications and emphasized the psychological foundations, the games people play (on, not with, one another), and life scripts. With that, the theory lost its useful focus. Still, many people have retained an understanding of the three ego states of the individual and apply them effectively when communicating.

Briefly, transactional analysis (TA) gets its name from the idea that every set of messages between two people is a "transaction." In transactional theory, everyone can speak from three different levels: as a "parent," as a "child," and as an "adult." These levels (*ego states*) exist in everyone, together, no matter how old the person is.

- As "parent" we preach, moralize, and act as though we are superior. The "parent" in us is the stuffy know-it-all, the self-righteous part that has an answer for everything.

- As "child" we show emotions easily; we like to play, have fun, explore, rush into things. Our "child" is insecure, relatively weak, unsure, and sometimes rebellious.

- As "adult" we are rational. We recognize our emotions, but we channel them so they do not interfere with what we are trying to say or do. In a communication we think of the other person's needs. We ask appropriate questions, listen carefully, obtain clarification, observe the nonverbal signs, and seek and provide feedback; in short, we are consummate communicators.

There is an appropriate time to let each of these three levels dominate our behavior. In serious communications, whether at home, in social interactions, or at work, the "adult" should have the upper hand most of the time. When we are playing, we have the most fun when we give the "child" a long leash. Sometimes the "parent" is entitled to have the upper hand: when we are expected to teach or preach, or when we are commiserating with another "parent" about how terrible it is that there is so much crime or so little free time—not like in the "old days"!

Communications are a series of transactions. Recognizing the other person's ego state allows us to communicate more effectively. The book by Berne (1967) and especially the one by Harris (1976) can provide ideas on how to use TA for better communications. You may also refer to a didactic simulation game, which is even more specific on management/staff member communications (Clary, Lieberman, and Rausch 1974).

The Johari Window

The Johari Window diagram is a useful tool, even for people who are not too keen on diagrams. It shows how communications techniques interact to achieve fully open communications. The title comes from the names of the two men who developed it, Joseph Luft and Harry Ingham (Luft 1970).

The Johari Window was also popular during the 1970s. It was often presented as an intensely personal device for achieving self-awareness through self-disclosure and seeking feedback. It first gained great popularity in management development programs when self-analysis was a major theme, and then it quickly lost favor. The Johari Window has much broader application, however, when used to create a climate of open communications between two people or two groups, as discussed in a book on linking elements (Rausch 1978) and in the didactic simulation game referred to earlier (Clary, Lieberman, and Rausch 1974).

The diagram resembles a window separated into four squares, like the old-fashioned, double-hung windows (see Figure 2.6). Every discussion has two Johari windows, one for each person, and at the start of every conversation or discussion, the four squares are all the same size.

1. The first square is the "open communications" ("arena") area, where both sides to the discussion—whether it be an interview, a coaching or counseling session, a conflict, or just an exploratory conversation—begin with the same amount of information about the topics.
2. The second square is the "blind spot" area, which represents the information that the person who seeks to open communications does not know about the matter under discussion.
3. The third square is the "hidden" ("closed mouth") area, or what the other person or persons do not know. The person who seeks to open communications has not yet disclosed or does not want to disclose this information.
4. The fourth area is the "unknown" area, the relevant information that neither party has but that would be useful.

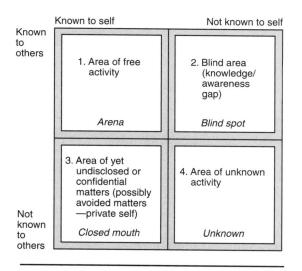

Figure 2.6 The Johari Window (Source: Luft 1970)

The objective for each person is to enlarge the "open communications" square as much as possible. Strictly speaking, during any one discussion, the "unknown" area cannot be shrunk (but the desired information may be available at the next discussion), and the "open communications" square is enlarged at the expense of other squares, the "blind spot" and "hidden" areas (see Figure 2.7).

The point of the diagram is that open communications can best be achieved in two ways:

- Seeking information, including information about what the other person does not yet know on the matter under discussion, requesting feedback, asking questions, and listening effectively will help to enlarge the "open communications" square at the expense of the "blind spot" area (the relevant things that are not yet known).
- Providing information, feedback, and self-disclosure about feelings will help to enlarge the "open communications" square at the expense of the "hidden" area (what the other person or persons do not yet know).

The Johari Window applies to both parties; each side has its own window. Every discussion starts with a new diagram because after some time has passed, the parties may be more knowledgeable than they were the first time around, and each area therefore has different content.

KNOWLEDGE, SKILLS, AND ABILITIES (KSAS) FOR EFFECTIVE COMMUNICATIONS

Effective communications are based on seven KSAs: conducting effective meetings, writing and speaking, listening, probing, practicing empathy, communicating with nonverbal signs, and seeking and providing feedback. We will discuss each topic.

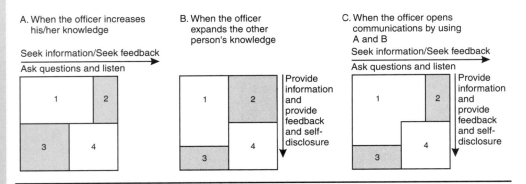

Figure 2.7 The Johari Window Under Various Conditions (Source: Luft 1970)

Conducting Effective Meetings

Let's begin by taking a look at Rausch's tongue-in-cheek "laws of meetings":

- More meetings are held than are needed.
- Most meetings take longer than necessary to bring the desired result.
- More people are asked to attend most meetings than are needed.
- Few meetings are adequately prepared.

Are you lonely? Work on your own? Hate making decisions?

*Hold a meeting; you can: * See people * Draw flowcharts*

** Feel important * Impress your colleagues * Eat donuts*

All on department time

Meetings: The practical alternative to work

—Anonymous

All managers want to avoid problems with meetings, including making participants dissatisfied and frustrated by wasting their time. To conduct meetings that are both productive and satisfying, you should keep these rules in mind:

1. *Prepare carefully for your meeting.* If complex decisions have to be made, participants at meetings should not have to start by defining the alternatives. It will usually be preferable for you to present one or a few alternatives and ask the participants to revise the alternatives and choose the best one. Sometimes an entirely new alternative emerges.

2. *Don't postpone an urgent meeting because a critical person is unavailable, especially if you cannot get everyone together soon.* Often you may consult with that person after the meeting, make whatever adjustments come from the discussion, and inform everyone affected. Occasionally you may want to call a follow-up meeting, but that meeting is likely to be short. In the meantime, you may be able to make some progress on the issue at hand.

3. *Ensure that those who are invited or expected to attend the meeting receive timely reminders.*

4. *State the objective of the meeting at the beginning, and then keep the meeting focused on that outcome.* Don't allow yourself to get sidetracked; you must deal effectively with difficult people (Bramson 1981; Rausch 1971; Rausch and Rausch 1971).

5. *Make sure the meeting ends with a clear plan for action.* At that time, all participants should understand the plan, and every person who walks away with an assignment should know it and accept it willingly.

ADDITIONAL INSIGHT 2.1

6. *Ask yourself whether the matter could be addressed without a meeting.* If your answer is yes, don't call a meeting, but handle the matter in a different way, by phone, via the network, with memos, and so on. (Participation in decisions does not require face-to-face discussion.)

7. *Call only the minimum number of people to the meeting—only those who are necessary for the technical and acceptance requirements.* (See Additional Insight 1.1, Participation in Decision Making and Planning, and Additional Insight 7.1, Decision Making and Problem Solving.)

Writing and Speaking

The arts of writing and persuasive public speaking are not the focus of this book. These skills are important, but they are functional competencies and only tangentially relevant to effective management/leadership. Furthermore, given that local colleges and even adult education programs provide courses in writing and possibly also in public speaking, an officer who is interested in improving these competencies has many options available. Many programs also offer instruction in the preparation of promotional materials, technical procedure manuals, communications with clients, and formal oral presentations.

What is most important is that officers, and especially chief officers, are competent in the verbal and written communication that pertains to the 3Cs. We will consider two situations: communicating information to groups and communicating information to individuals.

Communicating Information to Groups.   When deciding what and how to communicate to groups, an officer must first consider the needs of the audience. An officer who regularly thinks of the 3Cs guidelines as a reminder of issues to consider will have little difficulty realizing when and to whom information about new procedures and policies should be communicated. Equally important is the need to keep staff members informed about events that they consider useful or interesting. Sharing such information can be very helpful in reinforcing mutual trust.

Two opposing forces shape each individual decision about communication with groups. On the one hand, the more the leader can communicate to the staff and to other groups, the more trust and confidence that group will have in the leader. On the other hand, it is almost impossible to share confidential information with a group and expect it to remain confidential. Case-by-case review is therefore necessary, with the bias in favor of sharing information as early as possible.

Whether to communicate in writing or orally is primarily a question of logistics— which is more convenient or more useful to the audience. For instance, for procedures and policies, either just writing or both writing and verbal announcements and explanations may be appropriate.

Communicating Information to Individuals. Individuals' needs go beyond the needs of the group, so officers need more skills to communicate with them. In addition to what individuals want as members of the group, they want information on how their performance is seen, and they want signs that their efforts are noticed and appreciated. Other needs may arise as a consequence of conflicts, negotiations (a form of conflict), counseling, and coaching.

Although most officers are aware that written signs of appreciation are more powerful and satisfying than verbal thanks or commendations, recognition in writing is not widely used. In fact, few officers and other managers are highly competent in providing psychological rewards as frequently and as effectively as they should. [See Additional Insight 5.1, Enhancing Work Satisfaction (Providing Recognition) and Performance Evaluation.]

Listening

We can paraphrase the anonymous quote at the beginning of this section, considering the widespread complaints by staff members that "management does not listen to us":

The most important obstacle to effective listening is the belief that it has been achieved.

Officers need to be aware of their listening competence, so that they can practice, reflect, assess their strengths and weaknesses, and improve their habits. They also need to be keenly aware that staff members have a view of listening that is broader than just the act of hearing and even understanding. Their view includes the action that results from hearing and understanding—some form of action, even if it is no more than a follow-up that explains what is being done or why more is not being done.

Listening involves not only the hearing of words but also an active search for their meaning and their relationship to what is already known. It also provides the basis for questions that help to clarify what has been said. It is the key to uncovering needs or getting cues to hidden objections or conflicts. You need to listen when the other person is speaking, of course. You also need to listen while you are speaking, to pick up the other person's reaction (primarily from nonverbal signals).

You should be aware that effective listening occurs at two levels: passive and active listening. Passive listening is merely paying attention to what the speaker is saying. Active listening includes empathy and the responsibility for understanding the speaker's full thought. Active listening requires two-way communications and the ideas expressed by the Johari Window.

ADDITIONAL INSIGHT 2.1

Nichols (1962) identified ten bad listening habits:

- Calling the subject uninteresting
- Criticizing the delivery
- Getting overstimulated
- Listening for only facts
- Outlining everything
- Faking attention
- Tolerating or creating distractions
- Evading the difficult
- Submitting to emotional words
- Wasting thought power

To break these bad habits, Nichols suggested using the power of the mind through three mental manipulations to promote active, effective listening: (1) anticipating the speaker's next point, (2) identifying elements, and (3) making mental summaries.

How effectively do you listen? You may want to take the self-assessment questionnaire in Exhibit 2.3 or a commercially available program on listening followed by coaching to sharpen your listening competencies.

Probing

Probing is another technique that can clear the way to better communications. Though not widely understood, probing can help clarify meaning and can help determine whether the speaker's message was understood. Probing may also help a listener ensure that he or she understood a given message correctly.

Most people are not aware that asking appropriate questions is a skill. In fact, asking good questions at the right moment is not easy at all; it requires conscious awareness and practice. A wide range of professionals to whom communications is of paramount importance, spend considerable effort to learn and sharpen their probing skills. The professionals include psychologists, trainers, mediators, case workers, detectives, and technical sales representatives, as well as all others whose work involves exchanging information.

Programs that teach probing are not readily available, so an officer may have to set aside some time for instruction in staff meetings. Four techniques to discuss and practice in such sessions are closed questions, open questions, information-seeking statements, and moments of silence. Expectations for probing also need to be clarified: Even the most competent probing cannot get people to talk about matters that they simply do not want to share. Nevertheless, good questions can produce far more information than inappropriate questions; sometimes they can even bring surprising, unexpected disclosures.

ADDITIONAL INSIGHT 2.1

Exhibit 2.3 Listening—A Self-Analysis Questionnaire

How effectively do you believe you listen?
Please answer this question with a percentage between 0% and 100%: _____%

After you have entered your estimated percentage, please place check marks in the appropriate columns in answer to the following questions.

	Always	Usually	Sometimes	Rarely
1. Do you look the speaker in the eye?				
2. Do you look for nonverbal signs with which the speaker communicates?				
3. Do you concentrate on ideas when you listen?				
4. Do you determine, while you are listening, whether you agree or disagree?				
5. Do you begin to phrase your responses while you are listening?				
6. Do you accept the responsibility for making certain that you have received a message correctly and completely?				
7. Do you listen while you are speaking?				
8. Do you try to summarize, in your mind, what the speaker has said before you shape your response?				
9. Do you concentrate on feelings when you listen?				
10. Are you affected in the way you listen by your relationship with the speaker?				

When you finish, check whether your marks are consistent with the percentage you wrote down.

Source: Rausch and Washbush 1998.

ADDITIONAL INSIGHT 2.1

Closed Questions.   Closed questions can be answered with a single word or brief phrase like "yes," "no," "8:30," "Monday," "I think so," or "That's true." Closed questions are not very useful for obtaining information, but an occasional question like "When was that?" or "Where was it?" is much better than a grunt to show that you are listening and that you are trying to understand fully. Closed questions can also help you set the stage for further probing or for the continuation of your explanation:

- Do you mind if I ask another question?
- Am I correct in saying that . . . ?
- All right if I continue?
- Do you agree that . . . ?

Closed questions such as "Do you like . . . ?" and "Do you agree that . . . ?" can be effective bridges between different thoughts. Other types of closed questions are good for obtaining confirmation: "It was three o'clock when we got there, wasn't it?" or "There were about twenty of us, right?"

In some unique situations, a closed question will elicit significant information. The question may be used to "gain a foothold" by searching for areas to explore: "Where do you think the problem could be—with your inventory records or with your schedules?" or "Did you go to the library?" This type of closed question begs for a very specific answer that is at the core of the information being sought. If the person answering the question is honest and forthright, he or she will answer with a few words at most. If trying to hide something, the person answering is on the spot.

Closed questions are more effective than direct statements for conveying some messages. When you suspect that the other person may have tried to make you believe something that is not true, you may want to let that person know, politely, that you are aware of the truth. "Wasn't the price you quoted $249?" That's much less threatening or challenging than the direct statement, "The price you quoted me was $249." Closed questions can also be a polite way to ask someone to do something. "Should we add a few words of thanks to the letter?" is an example.

Closed questions have a negative side, however. They can shut a person out and make it difficult to continue. The trick is to avoid questions like "Do you want to tell me . . . ?," "Would you like to . . . ?," "Shall we talk about . . . ?," or "Do you know why?" The other person can end the discussion by replying with a cryptic no. Even two of the questions mentioned, "Do you mind if I ask another question?" and "All right if I continue?," should be used with care.

Questions that often lead to trouble are "Do you understand?," "Was that clear?," and even sometimes "Do you have any questions?" The other person may be reluctant to admit a lack of understanding, being unable to think of a smart question, or, heaven forbid, failing to pay attention. The worst question is "Do I make myself clear?" because too often it sounds almost like a threat.

ADDITIONAL INSIGHT 2.1

Closed questions, then, can be useful to test whether a message has been received and for agreement or disagreement; they can elicit a specific item of factual information, or convey it; and they can sometimes uncover areas to explore. If your purpose is to obtain information, you should use closed questions with care and sparingly.

Open Questions.   Open questions are much more effective than closed questions for obtaining information that the other person is willing to share. They draw people out, stimulating them to tell what they know and what is on their minds. In a group setting, they promote participation in the discussion. For instance, instead of asking "Do you understand?," you can ask "How do you feel about . . . ?"

Most open questions ask for a specific explanation. They usually start with "What," "How," or "Why." Other examples are "What happened?," "What is that all about?," "What did you do then?," "What are your reasons?," "What would you do . . . ?," "How would you approach this situation?," "How do you feel about . . . ?," "Why do you think that happened?," "Why were you there?" All these questions must be used tactfully, especially the "why" questions. "Why did you . . . ?" and "Why didn't you . . . ?" may sound more like accusations than questions. When phrased and asked properly, however, "why" questions are perfectly good open questions.

The choice of questions depends on the relationship between the persons involved in the conversation. In an environment of mutual trust and confidence, questions are not likely to be misinterpreted. If there is only casual contact or a doubt about motives, the phrasing of questions can be crucial.

"Where" questions may be either open or closed. On the one hand, "Where did you go?" is mainly closed because it can be answered with "Out" or "To the doctor." On the other hand, "Where could we go?," "Where might we hold the meeting?," and "Where could he have gone?" are questions that beg for information.

Another way to obtain more information is by rephrasing something that another person has said into a question. "You don't agree, then, that it would be a good statement to add?" is likely to elicit an extensive response. If the other person is still reluctant to speak, you may want to use information-seeking statements.

Information-Seeking Statements.   Some kinds of statements effectively ask questions. Instead of "You don't agree, then, that it would be a good statement to add?," you could say, "Please tell me why you feel that way." "Tell me more about . . . " is not very different from "What is this all about?" Other information-seeking sentences are "Let's talk about that a little more," "I believe you said that . . . ," "I was wondering what . . . ," "If you don't mind, I'd like to . . . ," and "Then you agree that. . . . "

Moments of Silence.   After someone has responded to an open question, you may feel that more could or should have been said. You may be tempted to ask another question; however, you may find it more effective to hold such a follow-up question for a while. Sometimes a brief silence will bring amplification or another thought that a premature follow-up question might have aborted.

Moments of silence should not be overdone; they should be neither too long nor too frequent. Silences should seem natural, or they will raise doubts about your motives.

A Word of Caution About Probing.   Several questions, information-seeking statements, and moments of silence in a row can easily give the other person the feeling of being "pumped" for information or of being in an interrogation. Sharing information is therefore an essential companion to effective probing, and you should provide input between questions or short series of questions.

Practicing Empathy

Communications without empathy are shallow. Especially when you are discussing a delicate topic, nothing is more important than being empathetic, tuning in, putting yourself in the shoes of the other person. If you are empathetic, then your questions become more relevant, and they are less likely to be threatening. Empathy can also help you to sense when you are in danger of stepping across that vague line that separates prying from probing.

Empathy is not sympathy. Sympathy adds feelings to understanding. Sympathy with the joys and sorrows of others cannot be learned like a skill but empathy can. Sympathy is a feeling. You either have it or you don't. The ability to be sympathetic does, however, grow with age and life experience.

It's different with empathy. Thorough communications do not need sympathy, but empathy is essential. When you are empathetic, you understand the other person's situation and position. You need not agree with it or even share the feelings. All you have to do is ask yourself, What is the other person thinking? If you listen carefully, you will either confirm what you thought or learn more about the needs, concerns, and feelings of the other person. Doing this often will help you acquire deeper levels of empathy.

Communicating with Nonverbal Signs

Much has been written about the interpretation of nonverbal (nonvocal) signs—that is, about the voluntary and involuntary messages carried by them and about what they reveal. Listeners send messages via these nonverbals, and senders reinforce or contradict their words with them. Nonverbals include these signals:

- Eye contact and expression
- Other eye movements
- Head movements
- Facial expressions like smile, smirk, seriousness, sternness, pensiveness
- Gestures and other body movements, including those of fingers, hands, limbs, and shoulder
- Poise and posture
- Voice strength and emphasis
- Other overtones like questioning, relaxed, sarcastic, sharp, harsh, happy, cynical, laughing, smooth
- Hesitant speech, careful choice of words, long (possibly pregnant) pauses

If you could only read these signals as well as you can "read" words, you would gain a wealth of additional information. A person may not express feelings and lack of interest in words, but you can detect them by observing nonverbals. Some feelings are easy to recognize in most people; anger, joy, apprehension, lack of interest, and sadness, especially when strong, reveal themselves readily.

Experts like Julius Fast (1970) write as though they can interpret nonverbal messages accurately. They may read eye movements or other facial expressions, leg positions, or specific body movements like looking away, tapping fingers, crossing arms, and leaning forward. Because there probably is a fair amount of similarity in the way people reveal their emotions, experts can offer some useful insights.

A healthy dose of skepticism about the interpretation of most nonverbals is always advisable. Although the similarities among people may be greater than the differences, nonverbals are not reliable indicators unless or until you can confirm them through adroit probing.

Seeking and Providing Feedback

With feedback you give your reaction to a message or elicit a reaction to what you have said. Verbal feedback is usually in the form of questions or statements in response to a message. Nonverbal feedback can be especially valuable because it is usually spontaneous, sometimes involuntarily. Often the person sending the nonverbal message is not even aware of it. Feedback can confuse, be misunderstood by, or intimidate or hurt the person to whom it is directed. All that can happen even when feedback is solicited.

Although most nonverbal feedback comes naturally, seeking and giving verbal feedback require skill. When seeking feedback, you have to be willing to accept it without becoming defensive. You also have to keep in mind that the other person may not be fully candid out of concern that the feedback may have negative impact on you. When giving feedback, you should not interrupt at the wrong time or be too abrupt, blunt, unclear, or untimely.

Here are some rules for providing feedback:

1. Feedback to statements should be factual, based on what the sender has said and not on how the receiver interpreted it. Interpretations and other assumptions should be clarified with probing questions before being used in feedback.
2. Feedback should be given calmly. When given in an excited manner, it can be perceived as disagreement or criticism.
3. Feedback should be as timely as possible. It should refer to something the speaker has just said, not to something that was said several minutes ago.
4. Feedback should concern only things that are currently under the control of the speaker. Telling someone that he or she does not know enough or can't comprehend will not lead to better two-way communication.

You may find useful ideas for feedback and empathy in Tannen (1990) and Gray (1992); both are enjoyable reading.

ADDITIONAL INSIGHT 2.1

Commanding the Response to an Emergency Incident

INTRODUCTION

The efficient deployment of resources to any emergency scene is usually dictated by a standardized response plan or system that specifies the orderly assignment of the resources (people and equipment). The first arriving company officer, or the person in the right front seat, should have an initial game plan. That first commander must assume command of the incident, determine the initial strategy for the incident, and assign tactical objectives. The initial strategy has a great impact on the outcome of the incident. As someone once said, "The first five minutes on the scene will dictate the next five hours."

The incident commander is responsible for determining the specific goals of the incident. As we noted in Chapter 1, these goals are supported by objectives, or specific accomplishments, that will achieve the goals. The strategy the commander develops on the scene dictates the objectives needed for the situation at hand. Then the commander spells out the tactics (the methods for achieving the objectives) and the action plans (the nuts-and-bolts tasks).

In this chapter we focus on management/leadership in the context of incident command and fireground operations. The responsibilities of a company officer extend beyond assuming command at emergency incidents. In today's fire service, 80% to 90% of an officer's time is spent working with people. The officer's ability to communicate with staff and other personnel in both emergency and nonemergency environments is critical. Officers must also encourage and develop leadership skills among staff. As you will see in this chapter, setting and achieving goals and objectives are complex responsibilities, requiring that both officers and staff members participate and realistically accept their responsibilities.

SCENARIO

THE FACTORY FIRE

At about 1:30 A.M. on a cold February night, the watchman at a furniture manufacturing company was preparing to leave the security office to make his hourly rounds of the administration building, the factory, and the warehouse. Suddenly the fire alarm sounded, indicating a fire in the factory. The watchman immediately placed a 911 call to report the fire and rushed to the factory to see what he could do. From the factory manager's office, which was separated from the rest of the factory by a masonry wall with a large window looking out over the production floor, he could see that the fire, which seemed to come from the painting area, was spreading. It seemed to him that the sprinklers were working, although he could not be certain because of the dense smoke.

The watchman again called 911 to report the location of the fire, its rapid spread, and that he had opened the gate to the plant. When asked about risk to people, the watchman said he was the only person within the fenced area.

The initial alarm assignment, which included two engine companies, an aerial ladder, and the on-duty chief officer, was dispatched. The engine companies were staffed with an officer and two fire fighters, the aerial responded with a driver/operator, and the duty chief responded in his command vehicle.

The chief reviewed his memory of the location and the prefire plan. The entire compound was surrounded by a fence that had an electrically operated gate near the center, on the side of the road. Three buildings faced the road behind the parking area:

- The small administration building, near the west side of the property
- The manufacturing plant, in the middle, with a front of about 150 feet and a depth of about 200 feet plus a storage addition and loading dock of another 50 feet or so
- The warehouse, the largest building, about 200 x 300 feet, at the east end

The buildings had approximately a 40-foot separation, mostly asphalt driveways. In the rear of the property was some additional parking, but mostly access for trucks to the loading docks of the factory and warehouse.

The telecommunicator provided the en route units with the following information:

- The manufacturing plant, like the administration building, was an old masonry structure.
- The storage and loading dock addition was wood planking on wood frame.
- The warehouse, which had been built last, was a steel structure with aluminum siding.
- All buildings were equipped with automatic sprinkler systems.
- The nearest hydrant was across the street from the gate, about 20 feet to the west.
- The Siamese connectors were at the front center of each building.
- The painting booths were on the east side near the center of the building.

Upon arrival, the captain of Engine 2 assumed command and gave an initial report. He gave orders to the second-due engine: "Engine 3, hook up to the hydrant and then, during the sizeup, supply the sprinklers and support Engine 2." The message was repeated and confirmed. Then he radioed the aerial apparatus driver to stage the apparatus in the parking area, pending further instructions.

When the chief arrived, the captain and the fire fighters from Engine 2 were about to enter the door of the factory on the east side, near the front. The chief conferred with the captain of Engine 2, who transferred command of the incident to the chief. Engine 3, which had stopped briefly at the hydrant, was at the end of the factory, near the sprinkler connection. Meanwhile, the aerial ladder had pulled into the parking area and had staged.

The chief quickly sized up the fire, which was burning through the roof. Thick smoke was pouring out of the paint booth exhaust stacks. The sprinklers were working, so the chief decided that, with adequate ventilation and with the fire department supplying the sprinkler system to control further spread of the fire, he did not require additional assistance.

The chief stepped back from the building to gain a broader perspective on the scene. That's when he noticed that the driver/operator of Engine 3 was helping to deploy the attack line that the lieutenant and a fire fighter were taking through the door. No other line was coming from Engine 3. A crew member of Engine 3 was not following orders to supply the sprinklers. The chief quickly corrected the problem by instructing the driver/operator to hook up the other line to supply the sprinkler connections. The combined efforts of the two companies and the fire sprinklers soon extinguished the fire without any further incidents.

When the companies returned to the fire station, the chief called them together for a tailboard postincident analysis. He reviewed what had taken place, without mentioning the delayed connection to the sprinklers, and thanked everyone for the hard work and successful completion of the job. Then he asked the lieutenant of Engine 3 to join him in his office.

Once they were alone, the chief asked the lieutenant what had gone wrong. Why had the driver/operator helped with the attack line before hooking up the line to feed the sprinklers?

The lieutenant was surprised. He said that he had given the driver/operator the instructions, and when he came out of the building, the connection was there. The chief told the lieutenant what had happened and asked him to discuss the problem with the driver/operator and to report his findings.

Near the end of the next shift, the chief had not heard from the lieutenant, so he asked him into the office again. The lieutenant explained that he had not yet decided what to do because of extenuating circumstances. The driver/operator apparently had been preoccupied when she got the instructions. She claimed she did not hear them, and the lieutenant had failed to give them emphasis in light of the standard operating guideline. (Note that this department uses standard operating guidelines rather than standard operating procedures to allow more judgment at the scene. The SOG calls for hooking up a sprinkler or standpipe feed line. The driver/operator had a full load at the fire scene. Her primary focus was on connecting the supply line. In the rush of things, she had forgotten the SOG. The lieutenant felt that he should not blame her because she was a reliable person, though fairly new.

"Don't you think I ought to be involved and you should discuss this with me, so we can decide jointly what should be done?" the chief asked.

"I think I can handle it myself. It's not such a big deal—just a miscommunication."

The lieutenant clearly did not want the chief to make this an issue with his people or with anyone else.

"There's more to this, apparently, than you think. If there is even the slightest chance that vital instructions don't get through and SOGs are overlooked, we have to determine what can be done to prevent similar incidents in the future. Everyone has to know about that."

"OK, if you insist." The lieutenant was not happy, but he knew he couldn't stop the chief.

"All right, then. What did the driver/operator have to say?"

"I told you, chief. She said she had not heard me give her the instruction to hook up to the feeder, and she is sorry that she forgot about the SOG."

The chief continued his inquiries. "Did you accept that explanation?"

"Not quite. I told her to review all SOGs and to see me when she is ready for us to review them together. What else can I do? That should take care of it. No sense getting into an argument with her over who said what. That can only bring problems in the company."

"So you're going to drop it?"

"No, of course not. I'll just be more careful in the future and make sure I repeat instructions so they are definitely heard."

"Is there anything else you think should be done?" asked the chief.

"Possibly," replied the lieutenant. "I guess I can make sure I always get confirmation that I was heard."

"Those seems like pretty good solutions. Should we hold a meeting on the incident to make sure the SOGs are high in everyone's mind and to strengthen the resolve to get some confirmation of every message, in both directions, especially when we are in action?"

"I would prefer that you don't do that. It'll be embarrassing for the driver/operator and for me."

"All right, the chief said. "Let me think what else we can do to get these messages across without causing any undue problems."

Alone again, the chief considered what he should do. He decided to let the shift commanders come up with recommendations. He sent them a memo that explained the need to strengthen awareness of SOGs and to develop a program that would emphasize the importance of confirming messages, especially during responses to emergencies. He asked that recommendations include some specific goals.

In response to the memo, the chief received recommendations from the shift commanders. The commanders of shifts 1, 2, and 3 recommended that the department add an SOG on the confirmation of oral messages during emergencies and that each shift should conduct three training sessions on SOGs—one a week during the following three weeks. The commander of shift 4 made the same suggestion and added these points:

- A test showing that each fire fighter has full knowledge of all SOGs should be given in the fourth week to all fire fighters.
- Each shift commander should set a goal to review fire fighter knowledge of SOGs regularly and ensure full knowledge.

Scenario Analysis: Fire Service Function Perspective

Before you read the scenario analysis, you may want to give some thought to the strengths and weaknesses of the strategies and tactics the chief used. Because this scenario involves primarily incident command, incident command issues and their guidelines are relevant.

This situation requires more detailed guidelines than we used in Chapter 1. They are based on the same general guideline for achieving the primary goals:

> To protect people—both the civilians at risk and the fire fighters—and to preserve and protect property, what do I have to consider in the sizeup to ensure that I will use the most appropriate strategy and tactics?

In incident command situations, this goal leads to three additional questions:

1. What's there?
2. What does the situation need?
3. What have I got?

The answer to the first question provides an assessment of the incident scene, which serves as the basis for the second question on what the situation needs. The final question identifies the personnel and equipment available. We will look at these questions in more detail later in this chapter. For now, we assume that the chief considered these questions because his tactics were appropriate in light of the situation.

Scenario Analysis: Management/Leadership Perspective

Before you read this scenario analysis, you may want to give some thought to how the characters in the scenario used the three management/leadership considerations: control, command, and climate (the 3Cs). The factory fire scenario analysis in this chapter will concentrate on the control guideline. The next two chapters will emphasize competence and climate, respectively.

The Control Guideline.   The control guideline in Chapter 1 asks this question:

What else needs to be done to ensure effective control and coordination so that the decision we are considering will lead to the outcome we seek, and so that we will know when we have to modify our implementation or plan because we are not getting the results we want? In other words, how can we gain control or coordination over the process of "getting there"?

A more detailed version makes use of these additional questions:

1. Are the goals appropriate and effectively communicated to all stakeholders?
2. Is there appropriate participation in decision making?
3. Are coordination, cooperation, and inter- and intra-unit communications being stimulated?
4. Is full advantage being taken of positive discipline and performance counseling?

We will review the scenario from the perspective of these questions related to the control guideline.

Are the goals appropriate and effectively communicated to all stakeholders? In preparation for and during the actual battle with the fire, members of the fire department use strategies and tactics designed to achieve the goals of each fire service function. As we stated in Chapter 1:

The primary goals of incident command are to ensure that all is done that can be done to protect people—the civilians at risk and the fire fighters—and to preserve and protect property by confining the fire and extinguishing it as quickly as possible.*

The scenario did not describe the communication of the incident command goals to stakeholders, including officers and fire fighters, property owners, the public at large, and governing bodies. We assume the goals were communicated.

Let's look at what other issues this guideline question might bring to the surface. At the fireground in the scenario, there seemed to be no need to set or communicate specific fire department goals. Everyone clearly understood the appropriate strategies and tactics. Nevertheless, the chief could have asked himself whether he should set any incident goals. Incident goals are often set at the transfer of command, but the factory fire scenario contains no indication that this happened.

---

*Note that these goals conform to the purpose of the incident command system: The success of a fire-fighting operation depends on the ability of a fire department to manage the available resources effectively and efficiently to protect lives and property

However, department goals on full, continuing knowledge of SOGs and on confirmation of messages are different. In support of these goals with much longer time frames, short-term objectives (with strategies and action steps to achieve them) can, and probably should be set. (See Additional Insight 3.1, Goal Setting and Implementation.)

Goals and objectives should not be confused with action steps (tactics). The distinction has significant implications for an officer's actions. For instance, a recommendation that all fire fighters should be aware of all SOGs and adhere to them is a goal. Officers cannot really be held responsible for the full accomplishment of that goal because external influences may interface (fire fighter memory during stress, time allowed for drills of all SOGs, even attitudes of fire fighters toward some SOGs). The officers can be held accountable for using sensible strategies, expressed in the best possible action steps, to achieve the goal. Examples of action steps are developing a realistic and challenging drill plan based on SOGs; coaching fire fighters who are not keeping up with the others in learning, remembering, and applying SOGs; counseling fire fighters who show signs of dislike for some SOGs; and requesting help or support if the other action steps do not seem to accomplish the goal. These steps are all fully under the control of the officer. (See Additional Insight 3.1, Goal Setting and Implementation.)

Useful goals and objectives should be set through participation, and they should be realistic and challenging. Staff members should be expected to work on only a limited number of them at any one time. Therefore, the goals and objectives for an organizational unit should focus on matters that are very important to a major project (such as very short-term objectives for an emergency) or that improve the department. At the same time, staff members must continue to do the normal work that is not covered by a goal as well as it has always been performed. Quality performance demands full attention to all work that is part of one's normal job and extra attention to work on goals and objectives.

Action steps are the nuts-and-bolts tasks in working toward a goal or objective. They are not seriously affected by external conditions. An officer or fire fighter who commits to taking an action step can almost always achieve it and should be held accountable.

Please note that, besides the incident command objectives, only two goals were contained in the recommendations sent to the chief:

- That the tests show that each fire fighter has full knowledge of all SOGs
- That each shift commander review fire fighter knowledge of SOGs regularly and ensure full knowledge

These two objectives support the long–term goal of full, continuing knowledge of SOGs, which is based on both the primary goal of incident command and the management/leadership goal of competence.

Nobody can ensure positively that fire fighters will have "full knowledge." All that a shift commander can do is take appropriate steps to get the company close to that objective. The chief has the responsibility to check, from time to time, on the shift commanders' actions and to recommend changes if desirable or necessary.

The shift commanders' recommendations in the scenario to add an SOG and to conduct three training sessions are action steps. They are fully under the commanders' control, and each commander can therefore be held accountable for their achievement. (See Additional Insight 3.1, Goal Setting and Implementation.)

Goals should be realistic, challenging, and measurable, and they should be communicated. Seeing to it that fire fighters have full continuing knowledge of SOGs is realistic and yet challenging. It can be communicated to all members of the department. The requirement that the goal be measurable, however, is not fully satisfied in the sense that its achievement cannot be determined. When do fire fighters have "full" knowledge? Only when every one of them gets 100% on the test? Is 98% adequate? To be fully measurable, the goal should specify what percent of the fire fighters should achieve a particular grade: 90% of the fire fighters should earn a grade of 90% or higher on the test. Such measurable targets are given in the Job Performance Requirements (JPRs) used in the professional qualifications standards. (See Exhibit 2.2 in Chapter 2.)

Is there appropriate participation in decision making? (See Additional Insight 1.1, Participation in Decision Making and Planning, for clarification of the meaning of *appropriate*.) In the factory fire scenario, the chief had a good sense for appropriate participation. He did not interfere with the captain of Engine 2, who was the first to arrive at the scene. In his discussions with the lieutenant about the delayed sprinkler connection, the chief gave him a strong voice in the decision making. The chief approached the two solutions (confirming all instructions through feedback from the recipient, and stronger awareness of SOGs) in a way that avoided assigning blame. Without specifically saying so, he reserved the right to discuss the issues during a meeting with the officers and then jointly to decide on a strategy.

The chief's failure to discuss the hookup problem during the postincident analysis may strike many fire service professionals as being too soft, possibly showing weak leadership. This approach is, however, a useful alternative that officers may consider when reviewing an incident, and it is

guideline becomes second nature, and recalling the guideline automatically brings up specific issues. As we pointed out earlier, you need not accept the guidelines as written here; adapt them to your own needs, situation, and style. Using the guidelines is important, not their specific ideas. A comprehensive listing of all guidelines is given in Appendix A, Decision Guidelines.

Also keep in mind that officers use the fire service function guidelines and the 3Cs guidelines in different ways. Whereas the 3Cs guidelines (or other management/leadership guidelines you may have developed to take their place) should be used with *every* decision, each fire service function guideline needs to be considered only with relevant decisions and should be reviewed from time to time.

INCIDENT COMMAND AND FIREGROUND OPERATIONS

Of all the responsibilities of a fire department, emergency situations involving fire are traditionally the most important. When a fire department suppresses fires its effectiveness is put to the ultimate test. All of the activities used by a fire department must come together at the fireground. The results of planning and preparing for emergency operations provide the foundation for developing strategy and tactics at the fire scene.

The success of a fire-fighting operation depends on the ability of a fire department to use the available resources effectively and efficiently to protect lives and property. The *incident command management system* for emergencies has evolved as the most effective approach for managing the available personnel and equipment to achieve the multiple goals of fire fighting and other emergencies at the incident scene.

Establishing Incident Command

Incident command is an effective system for meeting emergencies. For additional information and more detail, you may refer to NFPA 1561, *Standard on Fire Department Incident Management System*.

From the arrival of the first unit at any incident, one person is in command, with the responsibility and authority to direct all phases of the operation. The officer of that first unit is the incident commander (IC) until command can be transferred one or more times as higher-ranking chief officers arrive and assume command. As the incident grows, in either size or complexity, the incident management system provides for modular expansion to effectively manage the incident.

If substantial mutual aid comes from a much larger jurisdiction than the one in which the fire occurs, the highest-ranking officer of that mutual aid department does not take charge. The IC from the fire department that is responsible for the location remains in charge, although extensive consultation with the top officers of mutual aid responses may take place.

The IC is responsible for the direction and control of operations and, in more complex situations, for all elements of the incident, including logistics (resupply of materials and equipment), communications (between sectors and staff at the scene), and even public relations (communications with the media and with other departments). In large-scale, complex incidents, the IC may delegate certain functions such as logistics, planning, finance, safety, public information, and liaison with other agencies and departments. The IC should establish a command post and announce its location over the radio. The command post is the place where the IC makes decisions and sets up communications and plans. It is a quiet place out of the elements where the IC can think and work.

For simple emergencies, such as a residential fire that poses little threat to adjoining structures, the incident command system is no different from the *fireground command system*, the system that is still used in many parts of the country. For large-scale incidents, however, managing the emergency requires a more complex organization structure at the scene. Many companies and possibly hazardous materials and medical emergencies may be involved, and the fire may have spread over such a large area that it can no longer be observed from a single point. Then management must think in terms of "incident command." (Later in this chapter we will present a more detailed discussion of the incident command management system and a comparison with the fireground command management system.)

The IC deploys resources to meet the changing conditions of the incident. For the line functions of coping with the emergency, the IC may assign one or more officers to serve as sector/division/group officers, to take charge of the tactical operations in geographic segments, and to carry out specific functional assignments at the scene. In large-scale emergencies, other officers may be called on to take charge of one or more supervisory levels or specific tactical sectors, such as rescue and medical emergency sectors (see Figure 3.1). The incident commander may assign still other officers or fire fighters to provide liaison with other agency representatives, assuring that the logistical needs for additional apparatus, equipment, and supplies are met and providing safety and public information functions. With this system, fewer people report directly to the IC, which reduces the span of control to manageable proportions.

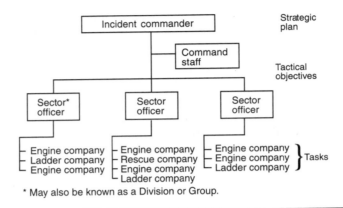

Figure 3.1 In a common fireground organization, the incident commander develops a strategic plan and assigns tactical objectives to the sector officers. The sector officers assign companies to perform specific tasks to accomplish the tactical objectives. (Source: Cote, *Fire Protection Handbook*, 1997, p. 10–11)

Prefire Planning and Other Preparations for Fighting Fires

Adequate preparation is necessary for fighting fires effectively, for ensuring that water supplies are sufficient to extinguish fires, and for reducing the extent of fire spread to minimize loss. Similar preparation is required for non-fire emergencies, such as disaster planning and responses to diverse emergencies like floods, forest fires, earthquakes, tornadoes and hurricanes, tidal waves, other natural disasters, structural collapses, civil disturbances, spills of hazardous materials, train or multiple car wrecks, and terrorist incidents.

In fire emergencies, preparation is

- Prefire planning for efficient distribution of apparatus and personnel to make the best use of available equipment, while providing sufficient coverage for contingencies in each situation.
- The ability of available personnel and apparatus to extinguish any potential fire. This ability results from training with drills, prefire planning, and the availability of adequate supplies of extinguishing agents (mainly water).
- The provision for applying the greatest possible tactical power with the first-alarm units. If these companies do not succeed in confining the fire, the response of additional personnel and equipment might not be rapid enough to prevent loss of life or serious damage.

Effective prefire planning together with adequate preparation and training gives a first-alarm company a head start in its attack and helps to ensure that each unit makes efficient use of resources during actual fire-fighting operations. The preparation is especially important in fire fighting in hazardous occupancies, such as refineries or chemical plants.

Developing the Strategy for Incident Attack

Attacking an emergency involves decisions at three levels: strategic, tactical, and task initiation and assignment. Exhibit 3.1 gives examples of decisions at each level.

Exhibit 3.1 Example of Incident Command Decisions for Fighting Fires

1. **Strategy Decisions**
   - Strategic goal: Protect occupants and contain the fire to the area involved on arrival.
   - Strategic goal: Take necessary precautions to ensure the safety of personnel.
2. **Tactical Decisions**
   - Tactical objective: Search and rescue inhabitants at risk.
   - Tactical objective: Attack with interior hose lines.

3. **Task Initiation and Assignment Decisions**
   - Company 1 search and rescue in immediate fire area.
   - Company 1 provide ventilation.
   - Company 2 extend interior hose lines to attack fire.
   - Company 2 back up attack hose lines.
   - Company 3 search and rescue on floor above fire.
   - Company 3 protect adjacent exposure.

Source: Modified from Cote, *Fire Protection Handbook*, 1997, Table 10–1A, p. 10–11.

The incident commander makes the strategic decisions and possibly some of the tactical ones. Sector and company commanders make the remaining tactical decisions and determine specific tasks.

Strategic decisions address the basic plan for dealing effectively with the situation as it presents itself and as it changes. They identify the overall goals and prioritize the goals for the units at the scene, based on the situation, risk, and available resources. Strategic decisions identify the priorities for committing resources to various tactical positions and activities based on standard approaches and on prevailing conditions. The generally accepted priorities for strategic decisions are life safety, incident stabilization, and property conservation.

Tactical decisions address the methods used to implement the strategic plan. The tactics define specific functions, approaches to be used by companies or groups of companies operating either directly under the incident commander or indirectly under sector/division/group officers (see Figure 3.2). These functions are critical to achieving the strategic plan.

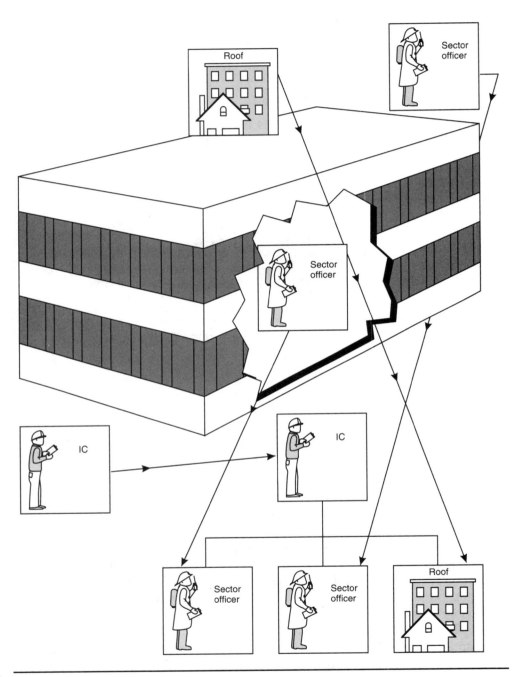

Figure 3.2 Geographic sector officers are responsible for all general fire-fighting activities in an assigned area. (Source: Modified from Brunacini 1985, fig. 3.5.5, p. 91. Copyright © 1985 by Don Sellers, AMI)

Task determination and assignment decisions address the specific steps that companies and fire fighters take. *Standard operating procedures* (SOPs) specify how certain tasks should be performed. SOPs apply to both tasks that have to be performed frequently and those that have to be performed from time to time. Some fire departments limit SOPs to outline the steps for tasks, whereas others have developed them for goals and objectives that are set repeatedly, such as tactical objectives in emergencies. SOPs, of course, are not for emergency tasks only. They also apply to work such as inspections and prefire planning and tasks at the station that pertain to apparatus, maintenance, water supply records, and so on.

To be most useful, emergency procedures should be both flexible enough to allow fire fighters to react to different situations and incremental to permit adjustment to the scale of the incident. Nonemergency procedures, except those for the most routine tasks, need to be modified to fit the situation. To emphasize this flexibility, some departments have begun to use the term SOG (*standard operating guidelines*) instead of SOP.

Developing the strategy for an emergency requires the same steps used in reaching any other decision:

- Define the problem and obtain the data.
- Identify useful alternative courses of action.
- Select the alternative that is best for a given situation.
- Consider the probabilities of unforeseeable events.*

Developing a strategy appears to be a complex and time-consuming process. The experienced fire officer thinks through it all during sizeup, however, often in seconds. In complex situations, the officer repeats this thought process frequently to check on the initial assessment and decisions and to adapt the strategy to the changing situation. He or she might consider the same issues several times. For example, an officer must evaluate the vertical and horizontal ventilating approaches early in the attack and again later in the attack.

It is sometimes even desirable to revisit earlier decisions to see whether revisions might not bring better results. If the initial decision is to delay ventilation procedures, the officer may have to repeat the decision-making process later in the fire-fighting operations. A strategy that has been implemented can lead to results that differ from those that were anticipated. Then the approach must be reevaluated, which again has repercussions on operational tactical decisions. Thus, the process is repeated until the fire is under control.

---

*This is an ongoing form of risk management.

Defining the Problem and Obtaining the Information

In the remainder of this chapter, we will concentrate on fire incidents. Very similar approaches are used for other emergencies such as floods, structural collapses, and earthquakes.

Fire officers must define the problem and obtain information for all the goals and objectives of the incident. The primary goal in fighting a fire is to ensure the safety and protection of both civilians at risk and emergency response personnel and preservation of property. To protect civilians, officers must know whether and how many people are at risk and where they are located. To protect fire fighters, officers must evaluate the dimensions of the fire, obtain structural information, and carefully observe developments at the scene. The use of an incident command system, a personnel accountability system, and adequate personnel to rotate and rehabilitate fire fighters is increasingly important.

Officers should gather specific information for defining the problem at the fire scene either in advance or on the way to a fire. In addition to considering this information, they must also apply applicable fire science elements to the situation. Fire science segments include such items as the behavior of fire and the methods of heat transfer that have an impact on the magnitude of the fire and how it moves through a structure.

For instance, officers can get information about specific features (the terrain and the structures) from maps, prefire surveys, and prefire plans. This information may be carried on the apparatus or relayed on the radio or by telephone. Information that cannot be obtained in advance but might influence fireground decisions includes traffic conditions and the degree to which open hydrants or dry weather might reduce water pressure in the summer. Officers acquire additional information related to the fire situation at the fireground itself during the initial sizeup process.

Fire officers define the problem and obtain much of the data by answering the three questions listed in Exhibit 3.2. As we explained earlier in this chapter, the core questions in Exhibit 3.2 are the ones that the incident commander asks when sizing up the incident scene. Now we will look more closely at these questions to see how an officer can use them to define a problem. (See also Appendix B for more discussion on using these question to define a problem.)

The answers to question 1 help to define the problem by identifying the critical elements of the situation. In some situations, these answers point out the need for further information, such as the percentage of the structure involved in the fire, the potential for structural collapse, and the environmental hazards that are involved.

Exhibit 3.2 Questions to Consider in Defining a Problem

1. **What's there?** This question assesses the elements of the situation.
   - Threats to life and safety of both the civilians at risk and the fire fighters
   - The involved structures
   - The fireground
   - The fire itself
   - Special hazards
   - The exposures
   - Weather and time of day
   - Terrain
2. **What does the situation need?** This is the needs assessment given the specific elements in the situation.
   - Rescue
   - EMS support
   - Exposure protection
   - Confinement
   - Extinguishment
3. **What have I got?** This question assesses the resources to meet needs.
   - Apparatus
   - Personnel
   - Equipment of all types (protective, lighting, for access and evacuation for applying extinguishing materials, communications, etc.)
   - Water and other extinguishing agents
   - Materials for toxic and hazardous conditions
   - Hose and nozzles
   - Knowledge by the IC and other personnel assigned to the incident

The answers to question 2 further define the problem by identifying needs related to the critical elements uncovered by question 1. For example, does the situation require only extinquishment capabilities, or are rescue teams and EMS support needed for life-saving operations?

The answers to question 3 provide the rest of the information necessary for thoroughly defining the problem. They identify what resources are currently available on the scene and what resources can be expected to arrive within given time frames.

When as much information as possible is available, the fire officer can develop the strategy for attacking the emergency. Although the process is complex, during sizeup, the competent officer quickly compares the available resources with the needs of the situation, reviews several alternative courses of action, and selects the one that seems to be the best. This ability to make judgments is based on thorough preparation, practice, and experience. Because the situation is constantly changing, the officer may have to revise the answers to the questions regularly as the situation progresses.

Defining and Evaluating Strategy Alternatives

Strategy alternatives consider primarily the allocation and deployment of resources. New alternatives emerge as the situation changes and information is received. For that reason, continual reevaluation and adjustment must take place.

To determine the strategy alternatives, the officer in charge takes these steps:

- Consider whether aggressive and defensive options are more likely to be most effective.*
- Evaluate the probability of success of possible alternatives.
- Look at the possible allocation, deployment, and augmentation of resources.

Figure 3.3 shows the trade-offs in resource allocation during sizeup. When more resources are committed to confinement, fewer are available for attack. At the left side of Figure 3.3, the officer arrives on the scene and sees that no exposures need to be protected. All available resources can therefore

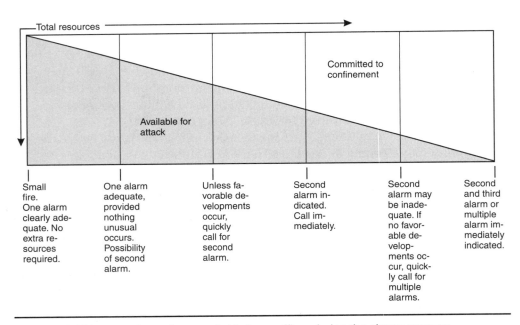

Figure 3.3 These are the options available to an officer during the sizeup process.

---

*The officer should use a thorough, logical risk management process. (See NFPA 1500, Section 2–2; Kipp and Lofflin 1996; for a discussion of defensive strategies. See also Appendix D in this book.)

be used for direct attack on the fire. At the other extreme, at the right side of the diagram, the structures are so involved that all available resources on the first alarm must be devoted to confinement, leaving none for direct attack. In that situation, an immediate call for several additional alarms is indicated. Figure 3.4 shows what happens when additional resources have been called.

The officer in charge must select a strategic combination from those in Figure 3.3. The best combination is the one that is most likely to satisfy the overall fire incident goals of safety for civilians and fire fighters and preservation of property. The officer needs to identy what can be saved without undue risk to personnel (Cote 1997, pp. 10–12).

The officer should focus not only on the attack alternatives that will hold fire damage to a minimum but also on those that will result in the least damage from the fire-fighting operations. For instance, a very poorly trained

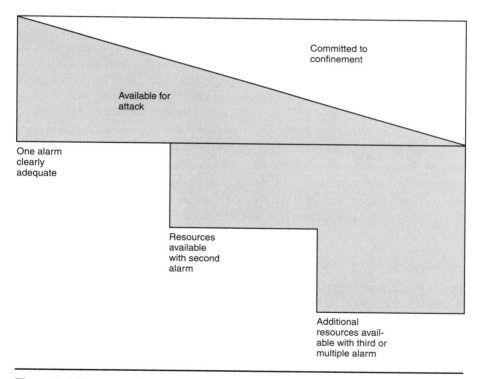

Figure 3.4 These are the options available to an officer during the sizeup process when reinforcements are on the scene.

company might apply such an excessive amount of water to a weak structure that the weight of the water added to the existing load collapses the building. Sometimes the water damage to contents (such as furnishings or carpets) is greater than the damage that would have been caused by smoke and fire if a less aggressive attack plan had been followed. In certain circumstances, the officer may chose to allow a fire to burn a little longer and use less water. This decision has great impact on tactics like the type and volume of streams. Water damage can be held to a minimum when fire fighters are trained in skills such as knowing when to turn off sprinkler valves as the fire is coming under control.

When identifying and evaluating alternatives, the fire officer should forecast the results that can be expected from each of the two or three possibilities that appear most useful. The probabilities of success have to be weighed. These probabilities concern events that might affect the course of the fire and the kinds of resources needed to bring the fire under control. Probabilities change during the course of fireground operations as they are influenced by weather, time, features of the involved structure, and actions of the fire fighters.

The officer in charge of the sizeup must forecast various possibilities about the future course of the fire. For example, what is the likelihood of a fire spreading from the third floor to the fourth and fifth floors and even leapfrogging to the top floor of a seven-story building? In preparing to make strategic decisions (that is, the allocation of resources), an officer must assess probabilities that are part of the environment and the changing fire scene. The officer must evaluate the effects of various fire-fighting steps, including both probable successes and setbacks, such as the possibility that resources might be lost through breakdowns, fatigue, or injury. These assessments—which cannot be discussed in this volume—are a function of the fire-fighting experience and training the officer has accumulated.

Allocating Resources: A Major Consideration in Evaluating Strategy Alternatives

Strategy options depend not only on the extent of the task but also on the allocation of available resources to priority elements of the task, such as the needs for rescue, exposure protection to reduce life hazards, exposure protection (confinement) to reduce property damage, and fire attack (extinguishment). The IC must consider each of these fireground operations and decide, for each alternative, what proportion of the resources should be allocated. Sometimes concentrating on extinguishing the fire takes care of both rescue and exposure protection.

The officer must consider both available and potentially available resources. When regular mutual aid is exhausted, additional support may be obtainable from departments not usually called on. Then there is the possibility of reductions in resources. The latter include apparatus malfunction and reduction in personnel for relief in large-scale fires.

The officer may have difficulty deciding whether, when, and how much help is needed. Many years ago, the late Warren Y. Kimball (1966) offered a critical guideline that remains valid today:

> Fire experience has shown that in a relatively high percentage of cases where only one or two additional companies have been called, second or third alarm has been required subsequently. In general, recommended practice is to promptly sound a second alarm in all cases where the first alarm response is not adequate. Calling companies piecemeal frequently has permitted further extension of a fire. [p. 88]

Standard operating procedures for the local department and mutual aid companies determine which and how many companies respond to each alarm. The IC, however, must answer two questions:

- How many alarms are needed?
- What special equipment, extinguishing agents, or both might be needed?

Both questions are difficult to answer in those borderline cases in which it appears that the resources on the scene might be sufficient to control the situation. When there is doubt, the officer should summon the additional resources.

In general, the wise approach is to be on the safe side and to ensure that extra resources are available, or "staged," and can be called into action quickly. Even though it is costly to bring additional resources to the scene, too many pieces of apparatus are better than too few. Emergencies can arise, such as equipment breakdown, injury to fire fighters, sudden detrimental changes in wind conditions, or unexpected combustibles near the fire.

The decision to have excess resources available is reversible because sending reinforcements back when it is clear that they will not be needed is certainly easier than to be caught short if an emergency does arise. A decision not to call for reinforcements is essentially irreversible, however, because additional fire damage will result if extra help is not at the fire scene when it is needed. Only when the probabilities are extremely small that reinforcements will be needed is it safe not to call on them right after sizeup.

Very often one resource is in limited supply. Then the IC has to balance the resources to make the best possible use of the limited one. For example, in some situations, the IC can conserve personnel by using a heavier flow of water from deluge nozzles. If water or extinguishing agents are the limited resource, then the decisions are more difficult. The IC can consider various options, such as calling for water tender/tanker trucks, running a water shuttle operation to the fire, or adopting a more defensive strategy. When the water supply is limited, decisions about how best to use available supplies can be made in advance and integrated into prefire plans, thus avoiding indecision at the fire scene.

Identifying and evaluating alternatives in emergency situations are not scientific processes, as is selection of a location for a new fire station. In the latter situation, decision makers have time to study the likely impact of various possible sites and to evaluate them against response time criteria. In an emergency situation, the officer's intuition has to play a much larger role. For that reason, officers should develop the habit of thinking along guidelines. They can gain a valuable edge and save time when they use the critical thinking discipline that decision making with guidelines develops. Officers who think along guidelines that they have adapted to fit their needs are more likely to make wise decisions during the planning process, whether that be prefire planning for a specific location or general planning pertaining to limited water resources in certain areas.

Selecting the Best Alternative Strategy

In emergency situations, the evaluation and selection of alternative strategies should be based on specific, timely information. The IC looks at the whole situation and pieces together the various features of the strategy that might do the job best. The selection might start with choosing either an offensive or defensive approach based on an assessment of the risks and the capabilities of the available resources. Next the IC selects one plan from the choices in either approach. Based on a forecast of likely results, the IC decides what additional resources should be called, which should go into action immediately on arrival, and which should be staged, pending need. Finally, the IC makes the specific allocation of resources. The preferred strategy thus emerges piecemeal, though very rapidly.

In formulating a strategy to protect life, the IC needs to consider the degree of danger to which occupants are exposed and the resources available for effecting rescue. The following questions are likely to be helpful:

- What is the risk to the safety of the fire fighters involved in a rescue mission? What is the threat of injury and possible loss of life? Is the likely benefit worth the risk?

- Are resources sufficient for all rescue operations? If not, what rescue efforts can be undertaken? (Insufficient personnel might make it impossible to proceed with all desirable rescue operations at the same time and have all of them succeed.)
- Which rescue operations should receive priority? Which people are in the most imminent danger, either from the fire itself or from other hazards? (The rescue of those most threatened should, of course, receive highest priority, except where danger to rescuers is excessive.)

In addition to direct rescue attempts, the IC has to consider protecting those exposures in which potential risk to life is serious enough to warrant immediate attention. After considering rescue needs and personnel safety in selecting the strategy, the IC should consider exposure protection, confinement, and extinguishment. However, the allocation of resources for one purpose—that is, to achieve one goal or objective—cannot be made in isolation from the others that have to be considered in an emergency situation.

Every operation and each segment of an operation require resources. The IC must therefore coordinate allocation of these resources so that each operation receives an appropriate share. The IC's understanding of this process, combined with the ability to perceive the relationship between a particular decision and the whole plan, determines the overall success of managing the incident.

As a strategy evolves and portions of the total resources are committed to various activities, the IC must estimate whether the resources for exposure protection, confinement, and extinguishment are sufficient for a direct attack against the fire, or whether they must be devoted in part or completely to protecting against fire extension until additional resources arrive. If additional alarms have been sounded, the IC must integrate reinforcements into the overall plan as they arrive. The strategy may be redirected from a defensive to a more offensive one, or to a more comprehensive attack on the fire than had been possible previously.

Strategies for exposure protection, confinement, and extinguishment revolve around the best methods of applying the required water flow to the fire. The line between strategy and tactics is not clear here. Although the choice of elevated streams as opposed to hand streams, for instance, might be a strategic decision because it involves the general approach to the fire attack, the ladder work required is a tactical consideration.

The IC must decide on strategy first, before any tactical decisions are made, because strategy concerns the allocation of resources. The officer who begins to make tactical decisions before the resources have been allocated could end up with inadequate resources available to execute tactical decisions, thus necessitating inefficient changes in the field.

Developing Fireground Tactics

> Tactics are the methods the incident commander uses to implement the strategic plan. The tactical objectives define specific functions that are assigned to groups of companies operating under sector officers. The achievement of these objectives contributes to the strategic goals and must be compatible with the overall strategic plan. [Cote 1997, p. 10–18]

Tactics are the steps used to implement the strategy. They identify the who, what, and how for achieving an objective. Tactics prescribe the exact apparatus, its location, and the specific way that lines are run. They dictate the specific assignments of personnel to the various tasks.

Unless the fire emergency is so small and routine that it can be handled by a single company, the tactical decisions and their implementation generally fall to the officers in charge of individual units. Several tactical operations can be carried out simultaneously during multicompany operations. Every company must therefore be trained for all basic operations and thus be prepared to contribute to tactical objectives when possible.

ORGANIZATION FOR MANAGEMENT OF FIREGROUND OPERATIONS

The U.S. fire service currently uses two incident management systems: the incident command system (ICS) and the fireground command system (FCS). Work has been under way for some time to combine them. The ICS is mandated by the federal government under the Superfund Amendments and Reauthorization Act of 1986 (SARA) for use at hazardous materials incidents. Many state regulations also mandate its use for specific types of emergencies. Progressive fire departments have adopted it at all incidents.

Incident Command System (ICS)

The incident command system was developed as a result of experience gained during wildland fires that consumed large acreages in southern California in the 1970s. The ICS uses a flexible organizational structure, with fixed position titles and descriptions and prescribed operational functions that must be addressed during an emergency situation. The ICS is designed to ensure smooth operations and efficient interaction at the scene of an emergency, regardless of its magnitude, location, or cross-jurisdictional nature. When using the ICS, operational forces work together effectively and efficiently, regardless of how many people, agencies, or communities are involved. Common operational methods, terminology, functions, and position responsibilities throughout the course of an incident ensure as much cooperation and coordination as possible.

Under the ICS structure, someone is responsible for each of the functions considered necessary for control of the emergency at hand. Operational control is thus distributed among a number of people, with responsibilities clearly delineated, in a structure that tracks directly back to the one person who is responsible for overall command (see Figure 3.5). Thanks to the methodical nature of ICS, the incident commander is not likely to overlook a particular task or function or suffer from information overload because of working with too many facts, figures, or people.

Fireground Command System (FCS)

The fireground command system was developed by Chief Alan V. Brunacini of the Phoenix, Arizona, Fire Department (Brunacini 1985). The FCS uses a structured approach to emergency operations that is similar to the ICS. By

Figure 3.5 Command personnel staff are operating inside the command post. The command post provides quiet communications and planning facilities. (Source: Courtesy of Phoenix Fire Department)

defining operational guidelines and establishing standard operating procedures, the FCS assigns responsibility for all anticipated emergency operating tasks.

This system, like the ICS, is designed to apply basic management principles to the hectic nature of fire-fighting operations. Chief Brunacini states,
"The entire command system is an attempt to somehow intellectualize a
fast-moving and violent event" (p. 259). To achieve this end, Chief Brunacini
has standardized fire department responses to emergency operations so that
each member knows what role to play, what procedures to use, and when
the procedures should be applied (see Figure 3.6). His purpose is to "rou-
tinize" as many decisions as possible.

COMMUNICATIONS AT THE EMERGENCY INCIDENT

Good communications are vital at emergency incidents. To hold misinter-
pretation to a minimum, officers must give clearly understood directions.
Because time is of the essence on the fireground, concise and clear communications are essential at all levels of management.

The Meaning of Words and Their Transmission

Plain words are the clearest communication. People from other municipalities
may be unfamiliar with local codes, and even codes like 10-4, which are widely
used, should be avoided. Plain language minimizes communication problems.

The clarity of a verbal or nonverbal order depends on the symbols,
words, sounds, codes, and mental pictures used to relay the order as well as
the way they are used. Words can have widely different meanings in different situations and for different people. For example, the word *truck* produces images that range from a toy truck to an ice cream truck, from a tow
truck to a delivery truck, from an emergency repair truck to a pumper.

Successful verbal communication depends on the ability of the speaker
to use words that transmit the thought to the listener. The speaker must be
able to choose the correct words in the time available for explaining the
thought and consider how the thought should be presented so the receiver
will get the message in the clearest possible way. Sometimes the words used
can mean something slightly different from what the speaker intended. At
other times words can convey a substantially different message because the
speaker is distracted either by an external element (such as loud noises) or
by other thoughts.

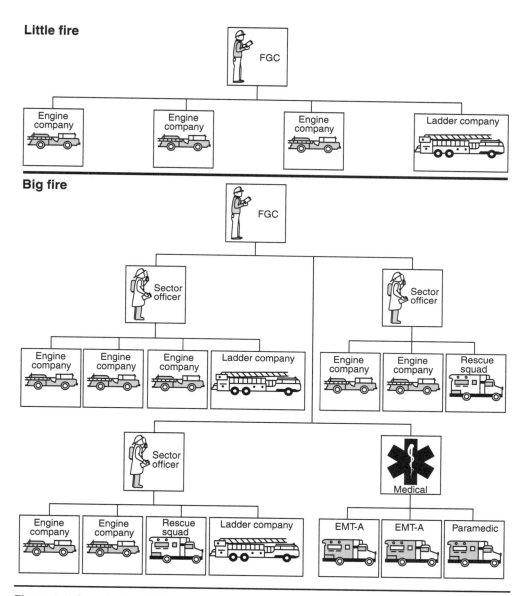

Figure 3.6 Organizational differences exist between little and big fires under the fireground command system. (Source: Modified from Brunacini 1985, figure 3.5.3, p. 89. Copyright © 1985 by Don Sellers, AMI)

Intervening factors affect the message that reaches the receiver. The speaker's voice may be so soft that the receiver cannot hear the full message. Loud noises may interfere. In most instances, words are accompanied by a facial expression or a gesture. Nonverbal cues become part of the message; they can change the meaning of the message by reinforcing or weakening it. For example, words that are spoken with a smile carry a different meaning from words accompanied by a frown.

When the words reach the listener, they arrive as a complete package with gestures and other influences such as the relationship between the speaker and the listener and the situation that surrounds them at the moment. Visual or audible distractions, including other people who might be speaking at the same time, may jeopardize effective communication. What the listener hears and sees of the message might be only part of what the speaker transmitted.

The following safeguards help to avoid misinterpreted and misunderstood communications:

- Good understanding of the communications process by senders and receivers, fire fighters as well as officers
- Joint drills with attention to standardized orders or instructions to ensure that for both the senders and the receivers understand the same meaning
- Sender–receiver feedback on the way the message is understood by the receiver

Improving Communications with the Ladder of Abstractions

Training, especially drills, helps fire fighters move immediately to perform tasks as soon as instructions are given. As much as possible, officers should give orders in the form of general guidelines for the course of action to be taken, so that the fire fighters themselves can make the specific decisions. Clearly, training can simplify communications at the emergency incident. The communications needed in a command for a poorly trained or new fire fighter are much different from those needed for a well-trained, veteran fire fighter.

Communications that transmit orders or messages are more effective when their meanings are as specific as is advisable for the receiver's level of understanding. The ladder of abstractions concept illustrates that the meaning of an order or message can be made clearer by increasing its degree of specificity (see Figure 3.7).

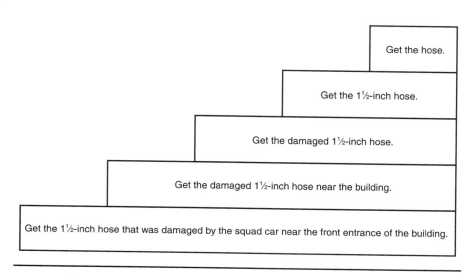

Get the hose.

Get the 1½-inch hose.

Get the damaged 1½-inch hose.

Get the damaged 1½-inch hose near the building.

Get the 1½-inch hose that was damaged by the squad car near the front entrance of the building.

Figure 3.7 Ladder of abstractions show an increasing degree of specificity to help prevent misunderstandings.

When the people involved are experienced and have a high level of mutual understanding, officers can give instructions in fairly broad terms high up on the ladder of abstractions. With less experienced people or with those who do not usually work together, officers must use more detailed instructions or messages to be sure they are clearly received. For example, a new recruit might feel insecure unless all the necessary details about how to enter a building are spelled out. A seasoned, well-organized fire fighter may not need details, however, and might even resent step-by-step instructions. Exhibit 3.3 gives commands in a ladder of abstractions.

Using specialized symbols is an excellent way to eliminate excess words from messages. Fire departments often develop sets of symbols, such as numbers, codes, or terminology, to communicate clear and precise information and quickly translate it into action.

Most organizations develop specialized terminology or jargon, which is often unintelligible to all but those who use it. Such terminology enables people to communicate with one another in terms that have a narrow and specific application. The terms *BLEVE* (boiling liquid-expanding vapor explosion) and *fire loading* (the weight of combustibles per square foot of floor area), for example, mean little or nothing to people outside the fire service. Some departments use the term *working fire* or *all hands* to indicate a serious fire that might require a multiple-alarm response.

Exhibit 3.3 Communications in the Ladder of Abstractions

- **For the least experienced, least trained fire fighter:** Get into the building; take a power saw with a carbide-tipped blade so you could cut a window even if it is not glass. Cut the lower left window so you can reach in and open the door from the inside. If that doesn't work, cut all of the panes so that you can climb in.

- **For a somewhat experienced, trained fire fighter:** Get into the building through that door; take the right cutting tool because those could be plastic windows.

- **For a more experienced, more thoroughly trained fire fighter:** Get into the building through that door.

- **For a highly experienced, highly trained fire fighter:** Get into the building.

A fire fighter who can implement a command without requiring detailed instructions saves time and avoids misunderstandings. In addition, the officer in charge is free to concentrate on other problems by knowing that the fire fighter does not need step-by-step directions. A good training program results in more rapid development and implementation of strategies, swifter decisions about tactics to be used, and improved communications of those tactics. Fire fighters benefit from the greater self-confidence and closer working relationships that come from better understanding and from being a member of an efficient, well-functioning, competent team.

THE 3CS, LINKING ELEMENTS, AND AN OFFICER'S ROLE AT AN EMERGENCY

Several important considerations related to the 3Cs and their guidelines directly affect the role of the company officer on the fireground, including goals and objectives, leadership style, enforcement of rules, and postincident analysis.

Goals and Leadership Style

The company officer must be able to use a wide range of leadership styles. The style required at the beginning and through the climax of the emergency is different from the style required during overhaul and salvage operations. For example, the IC would not hold a conference on the fireground immediately after sizeup to discuss what steps to take next. Instead, the IC must be decisive and quick to lay out the attack strategy and take the necessary steps so it will be implemented immediately. Participative decision making and goal setting are not totally inappropriate, however; a great deal of participative goal setting takes place in a smoothly coordinated team, although at first this might not be apparent. (See Additional Insight 1.1, Participation in Decision Making, and Additional Insight 3.1, Goal Setting and Implementation.)

The meaning of a command is influenced by the context in which it is given. When an officer in charge and the individual line officers have the kind of understanding characteristic of a well-trained team, commands have deeper meanings. Brief instructions or orders from the officer in charge at the fireground to a line officer might really be highly complex. By accepting the instructions or orders, the lower-level officer assumes these responsibilities:

- To lay out the tactical steps and set additional objectives to achieve the goal
- To change the tactical steps as necessary to achieve the objectives toward the goal
- To immediately notify the IC or other officer in charge if the resources are inadequate to meet the objective or the goal (for instance, to give notice if a team is reaching the point of exhaustion or if any other matter related to fire fighters' safety comes up)

Thus, even though the officer in charge determines most of the goals and objectives, an effective team uses participative decision making on the fireground.

Adherence to standard operating procedures or guidelines and to other rules on the fireground is crucial. Little time is available to explain procedures and SOPs/SOGs. Fire fighters must have a thorough understanding of the reasoning behind the rules before an emergency. When an officer gives instructions to withdraw or accept relief, for example, a fire fighter who resists must be aware that resistance is a major breach of the command authority, even though such devotion to duty might seem exemplary.

An officer should not take disciplinary steps on the emergency incident scene to achieve compliance with an order. Through prior training, personnel should understand that disciplinary consequences result from deliberate failure to follow instructions.

Postincident Analysis

Most fire-fighting teams conduct a postincident analysis after each response (except the most routine ones) to review strategy and tactics and to think about possible improvements. Conclusions have limited application if they concentrate, as they often do, only on fire-fighting strategy and tactics and do not take into consideration all elements that contribute to an effective operation.

For example, any indication that the goal-setting process is not under-stood by everyone can become the basis for team training and practice ses-sions. Fire fighters understand their responsibilities better when objectives are delegated to them. Delegation of responsibility is an important part of effective team work, as distinguished from a work group that depends on detailed instructions from its leader.

In analyzing what occurred during a run, the fire-fighting team has many opportunities to suggest methods for improving effectiveness. An of-ficer could follow these suggestions:

- Encourage members of the team to express their views.
- Thoroughly explore these views and give credit for any ideas that will be beneficial in the future.
- Plan, jointly with the team, training and drills that will lead to improvement.
- Agree on goals and objectives that should be set for activities other than emergencies and on revisions to SOPs or SOGs.

The team may also decide on methods that can be used to enhance coor-dination, cooperation, or rule adherence. The group may look at whether an emergency has helped cement and increase team spirit or has done damage to the cohesiveness of the team. The extent to which the goal-setting process has worked is another important subject for review during the analysis of every emergency. Finally, the discussion may include how an emergency might have added to the satisfaction that team members gain from their work to see what can be done to improve work-related satisfaction in a more general way.

Every emergency provides many opportunities for the company officer to analyze, either alone or jointly with the team, what impact the emergency has had on the effectiveness of that team. From this analysis, changes can be made that will help the team perform its mission more effectively in the fu-ture. The more the officer thinks of the relevant functional guidelines and the 3Cs guidelines, the more effective the critique will be. A comprehensive listing of all guidelines is given in Appendix A, Decision Guidelines.

CONCLUDING REMARKS

Detailed guidelines provide a useful structure for the decisions an officer must make at an emergency incident. With practice, an officer automatically recalls the guideline, which brings up consideration of the issues. As we have pointed out, an officer is free to adapt the guidelines to particular needs, situ-ations, and styles. Using the guidelines is important, not their specific ideas.

CHAPTER 3
STUDY QUESTIONS AND ACTIVITIES

If you are working alone, prepare your own written responses to these questions. If you are studying with a team or working in class, discuss the questions with the group and write a consensus answer.

1. What elements should be included in a strategic plan for a large fire?
2. List and explain the three questions that officers should consider during sizeup as a basis for determining strategy.
3. When formulating rescue strategies at the fireground, what factors should the officer in charge consider?
4. How would the ICS system be used if a tornado were expected?
5. For assignments, instructions, and orders on or off the fireground, why it is important to organize the information to be communicated?
6. Explain how the ladder of abstractions can help clarify messages and instructions.
7. What distinguishes a goal or objective from an action step?
8. Except during an emergency, what types of situations require that goals and/or objectives be set (think in terms of importance and urgency)?
9. In the analysis of the factory fire scenario, what questions about interunit communications might be addressed at the time that instructions are confirmed? What questions or issues should be addressed at that time to ensure the best possible communications?
10. How does ensuring coordination differ from gaining cooperation?

CHAPTER 3
REFERENCES

Brunacini, A. V. 1985. *Fire Command*. Quincy, MA: National Fire Protection Association.

Cote, A. E., ed. 1997. *Fire Protection Handbook*, 18th ed. Quincy, MA: National Fire Protection Association.

Kimball, W. Y. 1966. *Fire Attack 1: Command Decisions and Company Operations*. Quincy, MA: National Fire Protection Association.

Kipp, J. D, and Lofflin, M. E. 1996. *Emergency Incident Risk Management*. New York: Van Nostrand Reinhold.

NFPA 1500, *Standard on Fire Department Occupational Safety and Health Program*, National Fire Protection Association, Quincy, MA.

NFPA 1561, *Standard on Fire Department Incident Management System*, National Fire Protection Association, Quincy, MA.

Rausch, Erwin. 1978. *Balancing Needs of People and Organizations —The Linking Elements Concept*. Washington, DC: Bureau of National Affairs (Cranford, NJ: Didactic Systems, 1985).

Rausch, Erwin, and Washbush, John B. 1998. *High Quality Leadership: Practical Guidelines to Becoming a More Effective Manager*. Milwaukee: ASQ Quality Press.

GOAL SETTING AND IMPLEMENTATION

Goals and objectives exist at all levels of an organization, whether all, some, or none of the staff members know them, and whether or not they are part of the management and leadership system. In some fire departments, goals and objectives, other than those at the emergency scene, are only ideas in the minds of some officers. In others, they are paper tigers, used primarily to communicate lofty ideals and dreams. Many organizations, both in the fire service and elsewhere, have formal or informal goals that limp along, partly ignored and partly used to satisfy procedural requirements.*

Setting goals and objectives at the emergency incident scene is entirely different from setting them for other fire department activities. At the emergency scene, the officer in charge sets the goals and objectives. The participation of other personnel is limited to communicating significant potential problems. In a nonemergency situation, the officer solicits appropriate participation (with the right people, with appropriate authority, at the right time) to set goals and objectives. (See Additional Insight 1.1, Participation in Decision Making and Planning.) The successful use of goals and objectives in nonemergency situations depends on how well they are set, communicated, understood, and respected.

At the highest organizational level, the longest-range goals are often referred to as "vision," an important element of effective leadership. Vision is meaningless, however, unless it is shared and accepted. Vision for the future of the services the township provides to citizens is primarily determined by the highest level in the respective government. The portion of this vision that pertains to the fire department is significantly influenced by the vision that the chief of the department develops and communicates.

Vision is not confined to the top of a department, however. Although the chief's vision may set the tone for the entire department, one can talk about vision at the level of a shift or a company. The vision is likely to be the goal of the respective officer to make the unit highly effective. This vision is manifested in specific long-term and short-term goals and objectives, that conform to the goals of the entire fire department.

Whether originated by the officer or based on suggestions by others, nonemergency goals and objectives address functional issues that pertain to fire service activities such as fire prevention, fire safety education, and management of physical and financial resources. *How* the specifics are determined, *what* subsidiary goals and

---

*For a thorough discussion of the issues that make a goals program successful, you may refer to Rausch (1978 and 1985), and for scenarios that illustrate this discussion, see Rausch and Washbush (1998).

objectives are set, and *how* they are implemented are the issues from the management/leadership perspective.

Goals and objectives are decisions. It is therefore important to keep in mind the 3Cs management/leadership issues: control, competence, and climate. The functional aspects are usually dominant and have a tendency to overshadow all other considerations. That is why the officer has to bring balance by considering them.

The idea of working toward goals seems deceptively simple, yet much is involved. Specifically, if a goals program is to be successful, an officer should ask the following questions. The questions relate to goals and objectives that are different from most of those at the emergency scene.

1. Are the goals and objectives of high quality?
    1(a). Are the unit's goals and objectives in line with the larger organization's goals?
    1(b). Do they address matters that are important rather than those that are urgent?
    1(c). Are they both challenging and realistic (achievable)?
    1(d). Is it possible to determine whether or not they are achieved?
    1(e). Are they "true" goals and objectives, or are they action steps?
    1(f). Are they for a meaningful time span?
2. Are the goals and objectives communicated effectively to all stakeholders?
3. Is the number of goals or objectives appropriate for the organizational unit and for each of its members, considering their abilities and work load?
4. Have stakeholders had appropriate participation in setting the goals and objectives?
5. Have you, the manager/leader, accepted your share of the responsibility for achieving the goals and objectives?
6. Do goals and objectives address not only the fire service functions but also the management/leadership aspects of control, competence, and climate?
7. Is the award/reward system of the organization coordinated with performance on the achievement of goals and objectives?

Let's look at each of these recommendations for successfully managing with goals and objectives.

1(a). *Are the goals and objectives in line with the larger organization's goals?* The goals and objectives of an organizational unit are of high quality only if they are in line with and contribute to the larger organization's goals.

1(b). *Do they address matters that are important rather than those that are urgent?* At the incident scene, all goals and objectives that pertain to the emergency itself, though not all those that pertain to overhaul and salvage, are both important and urgent. As a management tool in nonemergency functions, goals and objectives are useful only if they focus on achievements related to a major project or to some significant improvement. If goals or objectives are set on trivial matters, they are not likely to earn the respect of the staff. Goals and objectives help to separate important matters from unimportant ones. The relative importance and urgency of an issue determine whether it deserves a goal.

Except for goals and objectives at the emergency incident scene, it is not the important and urgent matters that deserve to be considered for goals and objectives. Those matters automatically receive maximum immediate attention and, because they are already urgent, it is too late to set goals and objectives for them.

Matters that are important and not urgent are the best candidates for goals and objectives. Goals and objectives bring with them timelines for action steps and thus can ensure that these matters do not get pushed aside for other urgent ones, until they reach crisis stage. At that point it may be too late to deal with them most effectively. (For more information on the relationship between importance and urgency, see Additional Insight 12.1, Time Management and Delegation.)

1(c). *Are the goals and objectives both challenging and realistic (achievable)?* According to much of what has been published on the subject, goals and objectives can motivate both organizations and individuals if they are challenging and realistic. Goals are realistic if they are believed to be achievable. In practice, a contradiction exists here. On the one hand, *realistic* and *achievable* mean that the goal or objective can be reached. On the other hand, *challenging* means that, even with maximum effort, matters beyond the control of the individual and of the organization may prevent the attainment of the goal or objective.

All goals and objectives predict what can be achieved with diligence and maximum reasonable effort. To determine what the goal or objective will require in the way of budget, time, and other inputs requires forecasting. Therefore, if a goal or objective is to be both challenging and fully realistic, and if those who set the goal are good at forecasting, then the chances are greater than 50/50 that slightly less than the goal will be achieved (because of the "challenging" aspect) and less than 50/50 that the actual outcome will be better than the goal. Perfect achievement of a goal, in the sense of meeting the exact budget, quantity, quality, completion time, and other conditions, occurs rarely.

From the perspective of the control guideline, the best goals and objectives are indeed those that are challenging and realistic. To set a realistic goal or objective, again in nonemergency situations, managers must lay out the steps that may possibly be taken to achieve the goal. A goal or objective that is set without a review of the alternative steps needed to reach it (plan alternatives) is more of a guess than a serious effort to provide meaningful direction for the team or staff.

Managers may sometimes find it impossible to be specific on how much is to be accomplished by when. In such situations, they can temporarily define a goal or objective by the action steps that are planned to achieve it. Often this alternative is more useful than specifics, such as quantities, quality, and timelines, at least in the interim, until the view is clearer and specific attributes of the goal or objective can be set. If the action steps are fully appropriate and competently executed, they will achieve the best possible outcome.

1(d). *Is it possible to determine whether or not a goal or objective is achieved?* Often goals or objectives are stated too vaguely to be useful. A goals program that provides long-term goals for general guidance and supports them with short-term goals and objectives needs to have the specifics (dates, budgets, quantities, quality specifications, and so on) incorporated only in the short-term goals and objectives.

At the emergency scene, many goals and objectives are set without the specifics that nonemergency situations require. An objective such as "search and rescue inhabitants at risk" or "attack with interior hose lines" does not contain the specifics that permit a clear determination of whether or not it was achieved. If search and rescue is able to rescue only three people out of four in the structure, was the objective achieved? What if a very valiant effort did not rescue a single person? If two lines were used and it would have been possible, though not efficient, to use three, how should achievement of the objective be evaluated? Because the rules for goals and objective at the emergency have somewhat different meaning, the need for specificity with which to evaluate achievement does not necessarily detract from their usefulness.

1(e). *Are the goals and objectives "true" goals and objectives or are they action steps?* An important consideration is whether the steps that have to be taken to achieve the "goal" are totally under the control of the person, team, or organization that has accepted responsibility for achieving it. The question you have to ask is: If that person/team/organization wants to achieve the goal, can anything other than an emergency or a major, totally unforeseeable event stand in the way? If the desired result can definitely be achieved with competent action and adequate effort, possibly with some extra effort, then it is not a goal. It is merely a task or project that requires one or a series of actions steps, possibly directed toward a goal. An end result is a goal or objective only if circumstances beyond the control of the individual/team/organization may hinder its achievement.

Goals and objectives involve entirely different tasks and responsibilities from those of action steps. A goal or objective requires consideration of all the issues involved in making a goals program successful. A task or project requires only what you do with other day-to-day activities; you agree on a completion date and then follow up. This criterion applies to the emergency scene as well. Although the strategic goals are indeed goals, some of the expected results that some departments may consider to be goals or objectives are really only action steps.

Sometimes an expected result is not obviously a goal, objective, or action step. In questionable cases, the particular performance element can be labeled as an action step, especially if it does not involve numbers to be achieved. Then, if the action step is not completed or if major obstacles develop, the action step can be changed to a goal or an objective and treated accordingly, possibly by developing an action plan. The distinction between action steps and goals is of utmost importance to control and climate. As is discussed later, the officer (or higher-level officer) has to retain the major share of responsibility for the acheivement of a goal. The person (officer or fire fighter) accepting an action step can usually be held fully accountable for its accomplishment—except, of course, if it turns out that it was a goal or objective but not recognized as such.

1(f). *Are the goals and objectives for a meaningful time span?* Setting goals and objectives properly for nonemergency functions requires considerable time and effort and involves timelines for the supporting action steps. Setting goals and objectives is impractical for urgent matters that have to be done in a few days. These matters require immediate attention, and milestones are usually not practical.

2. *Are the goals and objectives communicated effectively to all stakeholders?* Timely communications in appropriate forms with all stakeholders are of utmost importance to ensure that everyone knows what is to be achieved, when, and what respective roles staff members play. Only with that knowledge can there be full coordination and cooperation. Sound communications contribute to a motivational climate for goal achievement and are important to hold the stress related to the "challenging" aspect of goals and objectives within reasonable limits.

3. *Is the number of goals and objectives appropriate for the organizational unit and for each of its members, considering their abilities and work load?* An individual and a company can work on only a limited number of goals and objectives. When they are set on only those end results that are most important to the organizational unit, the special attention they deserve will focus on a limited number of matters. The current and anticipated work load as well as the individual staff member or company abilities also affect the number of goals and objectives that can be handled. Of course, all work that is not covered by a goal must be done at least as well as it always has been. Quality performance demands full, normal attention to all work that is part of one's job and additional extra attention to matters that will ensure the achievement of goals and objectives.

To keep the number of goals as small as possible, managers should set goals primarily on issues that contribute to improving the operation of the department or company. One goal that should always be on the list, however, is that all procedures and policies are adhered to at all times and that performance is maintained at the level as it would be if there were no other goals.

4. *Have stakeholders had appropriate participation in setting the goals and objectives?* The importance of participation in decision making should be kept in mind throughout the goal-setting process and even during the assignment of tasks or projects. Without appropriate participation, the climate is less motivational and the man-

ager is seen as a less effective leader. Participation has many facets, however, and appropriate participation can take place even at the emergency incident. (See Additional Insight 1.1, Participation in Decision Making.)

5. *Have you, the manager/leader, accepted your share of the responsibility for achieving the goals and objectives?* The setting of goals and objectives clearly has a great impact on the climate as well as on control. It can easily involve political games. In an organization that has a climate favorable to achievement and open communications, an organizational unit or an individual is likely to feel safe to set and to shoot for challenging, ambitious goals and objectives. However, where the climate is such that negative consequences can result from failure to achieve a goal, then strong incentive exists to accept conservative, less challenging goals or objectives.

To ensure that the goals program contributes to a positive climate, you, as the manager/leader, have to be involved and be satisfied with the methods your staff uses to accomplish their goals and objectives. You have to provide any support that may be needed and, most important, you have to be prepared to accept your full share of responsibility for the outcome of the strategies.

6. *Do goals and objectives address not only fire service functions but also the management/leadership aspects of control, competence, and climate?* To help ensure that goals and objectives will be achieved, staff members must be competent in all aspects of their respective responsibilities. Goals and objectives should therefore consider any competence needs that may exist and include whatever competence development may be necessary.

Staff members should feel satisfied with the goals program. A manager who accepts his or her share of the responsibilities and makes the most effective use of the organization's reward system can help ensure that the climate guideline is satisfied.

7. *Is the award/reward system of the organization coordinated with performance on the achievement of goals and objectives?* All people, of course, desire to be rewarded fairly, if not generously. Psychological rewards play as important a role as tangible rewards do. [See Additional Insight 5.1, Enhancing Work Satisfaction (Providing Recognition) and Performance Evaluation.]

ADDITIONAL INSIGHT 3.1

Fire Prevention and Code Enforcement

INTRODUCTION

Most of a fire department's resources, including personnel, equipment, facilities, and support services, are committed to fire suppression efforts. Increasingly, however, fire departments are recognizing that fire prevention activities deserve almost equal, if not fully equal, attention and greater resources than they have received. Including loss reduction with the fire prevention function makes the case for adequate resources even stronger.

As used in this book, *fire prevention* concerns those policies, strategies, and activities that are intended to prevent fires from starting. In contrast, *loss reduction* concerns those efforts that are designed to reduce the cost of fires in terms of property damage and human casualties (deaths and injuries).

Fire investigation, public fire and life safety education, and construction plan reviews to ensure good engineering and construction practices are usually part of the fire prevention section or bureau of fire departments. These activities not only contribute to the prevention effort but also reduce loss (i.e., property damage and human casualties and injuries) when a fire does occur.

In the past, fire departments have often considered the role of fire prevention to be secondary to the suppression of fires. Fire prevention assignments were for the "sick, lame, and lazy." Real fire fighters went to fires. "Over the last 20 years, the fire inspector's job has evolved from that of a hard-nosed code enforcer to a highly skilled professional well versed in all aspects of fire prevention....The days of placing a fire inspector into a fire prevention bureau to bide his time until retirement are long gone," observed Chief William Peterson of the Plano, Texas, Fire Department (Scott 1997).

Many communities have broadened the focus of fire prevention to include communitywide risk management. Fire prevention is also the starting point for other areas of risk analysis. The modern fire chief staffs the fire prevention bureau to manage a wide range of code enforcement and preincident planning.

For fire inspection to bring full benefits, information should be funneled directly into the database that suppression personnel use for their prefire planning data. (See Chapter 6 for more information on the inspection/prefire planning interface.)

In this chapter we discuss fire prevention and the code enforcement aspect of fire prevention. Chapter 5 is on public fire and life safety education, and Chapter 6 deals with prefire planning, related loss reduction activities, and fire investigation as it contributes data for both fire prevention and loss reduction. The topic of sound engineering and construction practices is part of all three functions. The training of fire department personnel for fire prevention is covered in Chapter 11. Water supply issues, which play a major role in loss reduction, are discussed in Chapter 7.

SCENARIO
THE NEW INSPECTOR

The residential fire was over; the mopping up had begun. As fire fighter Albert surveyed the scene in his part-time role as investigator, he saw little that was unusual. It was obvious that the toaster oven had been the problem.

When Albert questioned the distraught homeowner, she told him that she had started to cook the pizza when she remembered that she had to get something from the store. She had to wait at the supermarket checkout, and then the five-minute trip home took longer than ten minutes because of heavy traffic. By the time she arrived, the fire had engulfed the cabinets over the under-the-cabinet toaster oven that had apparently malfunctioned. A neighbor had noticed the smoke and called the fire department, which had reached the scene only moments earlier.

As Albert drove back to headquarters, he noticed the new hydrant just around the corner, opposite the new supermarket, that had been used to fight the fire. Something bothered him about that hydrant. Then he remembered. The building department had sent the supermarket construction drawings to his captain, the fire official, for fire department approval. The captain was preparing to leave for a vacation and asked Albert to review the drawings and to return them with his comments. At that time Albert had just begun to serve as one of the four part-time shift investigators after almost ten years as a fire fighter. He had seen nothing wrong with the drawings, after noticing the conveniently located hydrant, and he had left them on the captain's desk with a note saying that they seemed OK to him. Apparently the captain, having returned from vacation with a lot of work waiting for him, had approved the drawings and sent them in.

When Albert returned to his office, he looked at the hydrant data in the computer file. Sure enough, the hydrant was not listed. Albert pulled up the prefire plan for the supermarket and found only two hydrants listed: the one he had seen and another one about 500 feet closer to the fire station. No information was provided about the water mains that supplied the hydrants.

Over the next few days, he checked the hydrant and found that it supplied about 400 gallons per minute. Records of the water company showed that both hydrants were supplied by the same 6-inch main. The nearest hydrant on another main, also on a 6-inch line, was more than 1500 feet from the supermarket. It was time to get the captain involved.

The captain was not happy with the news. His first impulse was to blame Albert, but he realized that would only lead to a confrontation over an event that had happened a long time ago, possibly, even before Albert became the part-time investigator. The captain also became aware that he was partly to blame because he should have taken the time to check the plans more carefully. He knew that Albert was still inexperienced when the drawings were submitted.

What should be done? The information had to be fed into both the prefire plan and the hydrant records. Something also needed to be done about the water supply so that it would be able to meet the needs of the sprinklers and multiple lines in case of a serious fire at the supermarket.

The captain told Albert that he would take the problem to the fire chief. They would decide who should negotiate with the water company and what steps should be taken to attempt to force the company to provide a larger line if negotiations failed. In preparation, the captain asked Albert to check whether any ISO or local fire code provisions require the water company to provide larger mains when greater risks developed with construction of industrial sites, high-rises, or multiple residential occupancies.

Scenario Analysis: Fire Service Function Perspective

Before reading the analysis, you may want to give some thought to what the characters in the scenario did well, in your opinion, and what they could have handled more effectively.

This scenario involves fire prevention and code enforcement as well as prefire planning and loss reduction issues. The following goals are relevant:

> The primary goals of fire prevention and code enforcement are to create a community safe from fire, through adherence to codes, construction plan reviews, and field inspections.

foot notes

> The primary goals of prefire planning and related loss reduction functions are to ensure that department members have thorough plans for attacking fires most effectively, with the available water supply and other extinguishing agents, and that the members are knowledgeable, skilled, and equipped to implement the plans.

These are the guidelines for fire prevention and code enforcement:

> What else needs to be done to ensure thorough adherence to codes—specifically, what should be changed with respect to relations with architects, enforcement, competence development of inspectors, and communications with interested parties (architects, engineers, property owners, fire fighters, and contractors)?

The guidelines for prefire planning and loss reduction are:

> What else needs to be done to ensure that adequate information is available to responding companies, appropriately analyzed and formulated into plans; that these plans are used in staff development, fire investigations, and water supply review and testing; and that the information management systems are as effective as possible?

We will address the prefire planning guidelines in greater detail in Chapter 6. Here we discuss the specific segments for fire prevention and code enforcement by considering the following questions:

- What should be changed with respect to relations with architects to enhance code compliance?
- What should be changed with respect to enforcement to enhance code compliance?
- What should be changed with respect to the competence development of inspectors?
- What should be changed with respect to communications with interested parties (architects, engineers, property owners, fire fighters, and contractors) to enhance code compliance?

These guidelines raise many issues for consideration. The captain, the investigator, and possibly the fire chief might ask these questions to prevent problems like the one that occurred in the new-inspector scenario:

- What procedures should we change in approving construction plans? Should we require double signatures? Should we keep a log of all requests for approval, showing the individuals who signed off?
- How should we keep the water supply and hydrant information? What safeguards should we establish to ensure that this information is updated regularly?
- What water supply conditions should we consider before approving water company plans to add new mains or improve existing mains?

- What else should we include in the training of new inspectors?
- What additional regular communications with the various stakeholders and others should we consider?

For prefire planning, they should raise at least one question:

- What information about hydrant locations, main connections, and flow capacities should be available on the prefire plans?

If the department reaches sound answers to these questions and implements them, then the likelihood decreases that a problem similar to the one in the scenario will develop in the future.

The problem in the scenario should trigger a broad review of the fire prevention and prefire planning practices and comparisons with other well-run departments. Such reviews and comparisons are called *benchmarking* in industry. The practice ensures that a department's procedures are brought up to date and sharpened to match the most effective practices that are used by other fire departments.

Scenario Analysis: Management/Leadership Perspective

Before reading the analysis, you may want to give some thought again to what the characters in the scenario did well and what they could have handled more effectively.

The Competence Guideline.   The discussion in Chapter 2 emphasized the control guideline of the 3Cs. This chapter's emphasis is on the competence guideline.

> What else needs to be done so that all those who will be involved in implementing the decision and those who will otherwise be affected (all the stakeholders) have the necessary competencies to ensure effective progress toward excellence in fire department operations and service to the community?

Specific questions follow:

1. Are changes needed in recruiting and selecting to fill vacancies?
2. Is the management of learning concepts applied effectively?
3. Are coaching and counseling on self-development used to their best advantage?

If you refer back to this guideline in Chapter 1, you can see that the first part of the guideline is the same as the one presented there. The list of specific issues is new here. We will review the scenario from the perspective of these specific questions, taking them one at a time.

1. *Are changes needed in recruiting and selecting to fill vacancies?*   Choosing fire fighter Albert to serve as part-time investigator was apparently a good decision. He is thoughtful and conscientious, and he appears to accept responsibility for self-development. These important characteristics deserve special consideration in the selection of candidates for positions of responsibility.

2. *Is the management of learning concepts applied effectively?*   The scenario did not indicate what training Albert actually received. He undoubtedly received some schooling, possibly at a local fire academy or college, and was certified by the regulatory body in the state. However, the management of learning concepts requires that KSA (knowledge, skills, and abilities) levels be identified carefully whenever someone assumes a new position or responsibility, and that gaps between high-level competencies and current levels be worked on until full competence has been achieved. (See Additional Insight 11.1, Management of Learning and Coaching.) That clearly did not happen in the scenario. Albert's KSAs were not evaluated when he assumed the position. Otherwise, his lack of knowledge about water supply and other record-keeping requirements would have been detected and remedied before he was asked to review construction plans.

Officers, and especially higher-level officers who think of guidelines when making decisions, might ask themselves these questions when placing someone in a new responsibility:

- Which KSAs are necessary and which are desirable for the effective performance of new responsibilities?
- How can the person's KSA levels best be determined?
- What steps can (should) be taken to remedy any weaknesses?
- In what ways can advantage be taken of any KSA strengths?

3. *Are coaching and counseling on self-development used to their best advantage?* Coaching is an effective technique for remedying KSA weaknesses, in conjunction with the management of learning steps (identification of KSA levels, determination of appropriate methods for development to eliminate weaknesses, and reviews to ensure that progress continues to full competence). Coaching is especially needed in situations where only one person needs to gain new KSAs.

The captain (and to some extent the captain's supervisor) appears to have dropped the ball in the scenario. Effective coaching would have revealed Albert's lack of knowledge about hydrant locations and flow rates.

Albert also needed coaching on the department's policies and procedures on record keeping. The captain's somewhat careless handling of the drawing approval indicates that he needed to develop competence in attention to detail and in helping fire fighters develop full competence in nonroutine and nonemergency duties.

Albert is a willing learner. Some fire fighters or officers are reluctant to devote the effort to learning, especially if they think they already know enough about a topic or if they just do not see the need to learn more. Then counseling is recommended. (See Additional Insight 9.1, Positive Discipline and Counseling.)

The Control and Climate Guidelines. A fire department that manages learning and coaching effectively will have good control over its activities. If gaps in competence exist, some functions will not be performed well, and other functions may be missed entirely. Planning for improvement is likely to be inadequate. Quite simply, without high levels of competence, the department's other steps to achieve meaningful control over its direction can be frustrated by lack of competence. As was pointed out previously and is stressed again below, the 3Cs are extensively interlinked, supporting each other in numerous ways. Progress toward long-range and short-range progress is highly dependent on attention to all three.

Cooperation is one aspect of control and it depends on smooth resolution of conflicts. In the scenario, the effective handling of emerging and existing conflicts was of considerable importance. Both the captain and the fire chief displayed intuitive or learned competence in conflict prevention—the captain when he did not blame Albert. (See Additional Insight 4.1, Management of Potentially Damaging Conflict.)

Climate also is tied to competence. People feel much better about their work when they know that they are highly competent in their respective functions and when they work with others who are equally professional.

We cannot overemphasize the interdependence of the 3Cs. Although we present the three components of the 3Cs model—control, competence, and climate—as three distinct entities, actions taken to achieve each one of them often influence one or both of the others. Control involves coordination, which requires competence in all the tasks for which coordination is needed. Both control and competence have a powerful impact on climate. Job satisfaction decreases when a lack of coordination leads to frustration. Lack of confidence may also result from inadequate competence. Steps taken to achieve a high level of competence bring better control and climate. Anything done to enhance the climate, in turn, positively affects both control and competence. The enhanced climate fosters higher levels of motivation for the work and for learning.

FIRE PREVENTION AND CODE ENFORCEMENT: FUNCTIONAL ISSUES

Function of Fire Prevention

The *Fire Protection Handbook* describes fire prevention as follows (Cote 1997):

> Fire prevention includes all fire service activity that decreases the incidence of uncontrolled fire. Usually, fire prevention methods utilized by the fire service focus on inspection, which includes engineering and code enforcement, public fire safety education, and fire investigation. Inspection, including enforcement, is the legal means of discovering and correcting deficiencies that pose a threat to life and property from fire. Enforcement is implemented when other methods fail. Education informs and instructs the general public about the dangers of fire and about fire-safe behavior. Fire investigation aids fire prevention efforts by indicating problem areas that may require corrective educational efforts or legislation.
>
> Good engineering practices, including plans review—another fire prevention method—can provide built-in safeguards that help prevent fires from starting and limit the spread of fire should it occur. [p. 10–168]

The participation of fire suppression personnel in fire prevention activities is as necessary as their participation in tactical operations. The majority of the fire department's resources are committed to fire suppression and are distributed systematically throughout the protected area. Fire department personnel are the logical resources for fire prevention efforts, when they are not occupied with responses to incidents and related activities.

Fire suppression personnel can perform routine inspections on a regular basis within their first-due response area, and designated fire prevention personnel, where available, can perform follow-up inspections, enforcement, and special technical inspections. The total involvement of all personnel, particularly those assigned to suppression activities, will not only decrease the incidence of fire but also demonstrate the highly effective use of personnel and competent management.

History of Fire Prevention: A Brief Overview

One of the first tasks of the early settlers in the Boston area was to build shelters against the harsh New England winters. Using local materials, they constructed wood houses with thatched roofs similar to the ones they were accustomed to in Europe. The chimneys were made from wood frames covered with mud or clay. Exposure to the elements dried the thatch and washed or blew away the mud or clay that protected the wood frames of the chimney stacks. These structures invited catastrophe from fire; burning embers, drawn up the chimneys, ignited the roofs and set the houses ablaze.

Early Fire Laws. Recognizing these construction hazards, the leaders of the Bay Colony outlawed thatched roofs and wood chimneys. A fine of ten shillings (a large sum in those days) was levied on any homeowner who had a chimney fire. People were thus encouraged to keep their chimneys free from soot and creosote. In effect, the first fire law was thus established and enforced.

As the town of Boston grew, the need increased for new laws to protect it from the ravages of fire. These new laws outlined the joint responsibilities of the homeowner and the fire protection authorities. They required every homeowner to have a ladder long enough to reach the ridgepole of the roof. They also required that homeowners have in their possession poles with swabs on the ends of them. When soaked in water, these poles were used to help extinguish roof fires.

The modern fire department was most likely started when Boston provided centrally located equipment and supplies to help residents extinguish fires. Attached to the outside of the town meeting house were several ladders and poles with hooks. The purpose of the hook-ended poles was to tear away neighboring structures and thus stop a fire from spreading. A cistern was created that provided a readily available water supply, and night patrols were formed to sound fire alarms.

The town of Boston enacted laws to punish people who exposed themselves and others to fire risk. No person was allowed to build a fire within "three rods" (49.5 feet) of any building or in ships tied up at the docks. It was illegal to carry "burning brands" for lighting fires (there were no matches in those days), except in covered containers. The penalty for arson was death.

Thus, several of the important elements of organized fire prevention and control existed in these early days: codes and enforcement for fire prevention, quick alarm, water supply, and readily available implements for control of fire. Despite such precautions, conflagrations were commonplace in Boston and other municipalities. More laws were needed to govern building construction and to provide for public fire protection. The result was a growing body of rules and regulations concerning fire prevention and control, which also strengthened the ability of fire departments to enforce the codes when necessary.

The Beginnings of Fire Insurance.   In early cities in the United States, building codes were primitive and buildings were usually constructed close to one another. The year before the great Chicago fire of 1871, the London insurance company, Lloyd's, stopped writing policies in Chicago because Lloyd's officials were horrified at the haphazard way construction was proceeding. Other insurance companies, too, had difficulty selling policies at the high premiums they had to charge because of the poor construction. Even with these high rates, the companies often suffered great losses when fires spread out of control. As often happens, even today, many of the fire laws and insurance policies were written as the result of tragedies.

The National Board of Fire Underwriters* realized that adjusting and standardizing rates were merely a paper solution to an essentially technical problem. It began to emphasize safe building construction, control of fire hazards, and improvements in both water supplies and fire departments. New, tall buildings constructed of steel and concrete adhered to controlled specifications that helped to limit the risk of fire. These were called Class A buildings. Today we refer to these sorts of structures as only fire resistive; however, in their day, they were a tremendous advancement.

Although some new Class A steel and concrete structures had been built in downtown San Francisco in 1906, much of the city was still composed of flimsily built, fire-prone wood shanties. The National Board of Fire Underwriters was so alarmed by these hazardous conditions that it predicted a major disaster. ("San Francisco has violated all underwriting traditions and precedents by not burning up.)" That same year, after a devastating earthquake, the city of San Francisco did indeed burn.

Even though the contents of the new buildings were destroyed, the steel and concrete walls, frames, and floors remained intact and could be renovated. After analyzing of the fire damage, fire protection engineers realized that further improvements were necessary. For example, glass needed to be reinforced to prevent shattering and deformation under the intense heat of fire, and auxiliary water towers on roofs were needed to supplement the regular local water supplies. Furthermore, vertical spaces, especially stairways and elevator shafts in tall buildings, should be enclosed to stop the vertical spread of fire.

Increasing awareness of the importance of fire prevention brought additional knowledge about the subject. Engineers began to accumulate informa-

---

* In 1965 the National Board of Fire Underwriters (organized in 1866) merged with the Association of Casualty and Surety Companies (organized in 1926) and the former American Insurance Association (founded in 1953) to become the American Insurance Association. The basic objective of the association is to promote the economic, legislative, and public standing of its participating insurance companies.

tion about fire hazards in building construction and in manufacturing processes. A new science was developed to meet newly perceived needs.

Principles of Fire Prevention

From its studies of the San Francisco disaster and other major fires, the National Board of Fire Underwriters became convinced of the need for more detailed, comprehensive standards and codes relating to the construction, design, and maintenance of buildings. Regulations based on such codes could undoubtedly prevent most fires and reduce losses in those that did occur. We will use the following definitions in our discussion:

- A **code** is a standard that is an extensive compilation of provisions covering broad subject matter or that is suitable for adoption into law independently of other codes and standards.
- A **standard** is a document, the main text of which contains only mandatory provisions using the word *shall* to indicate requirements and which is in a form generally suitable for mandatory reference by another standard or code or for adoption into law. Nonmandatory provisions shall be located in an appendix, footnote, or fine-print note and are not to be considered a part of the requirements of a standard.
- A **recommended practice** is a document that is similar in content and structure to a code or standard but that contains only nonmandatory provisions using the word *should* to indicate recommendations in the body of the text. If a recommended practice is adopted in a regulation, a local government and/or a fire department can decide whether to enforce it.

The *National Fire Codes* from NFPA include a wide range of codes, standards, and recommended practices. Codes alone are only guidelines unless they are adopted by some level of government. If they are to be meaningful and fulfill the purpose for which they were created, regulations must be enacted to cover their enforcement. (See the section Legal Foundations for Activities to Enforce Fire Code Regulations later in this chapter.) Thus, the fire department and the local authority have the responsibility to identify and order the correction of potential fire hazards. The local government has the power to do this through the enforcement of state regulations in support of codes where they exist and through the enactment of its own ordinances. If changes within the district have made the present codes inadequate, such as the development of mobile or trailer parks, the fire departments must voice the need for modification and help develop new codes and regulations where they are needed.

To be effective, regulations must be supported with inspections. The fire department performs these inspections as the local government's arm for the enforcement of the regulations and ordinances.

In most states, all buildings (except one- and two-family homes), and especially those in which many people work, live, or meet, must be inspected on a predetermined schedule. The purpose of these inspections is to ensure that the buildings are free from any known hazards and that they do indeed conform to the standards and codes specified in the regulations and ordinances.

Identification alone does not always bring compliance. However, if the owner of a building refuses to remove identified fire hazards or to renovate a building so that it conforms to the standards, violations can be enforced. Fire safety ordinances and regulations built around a model fire code not only outline inspection procedures but also carry penalties. Violators can be fined, and certificates of occupancy or permits for specific businesses or manufacturing processes can be withheld until code compliance is reached.

In the United States, the full value of fire prevention was not realized until fire departments and agencies began to compile meaningful information about the causes and circumstances of fires. Progressive departments began to initiate more effective fire prevention efforts in addition to maintaining their fire-fighting forces. The results of such efforts are defined more clearly every year. Fire prevention received its greatest endorsement in 1973 when the National Commission of Fire Prevention and Control reported on the fire problem in America. Throughout the report, *America Burning*, the commission gave top priority to the necessity for increased fire prevention activities in reducing fire loss (NCFPC 1973).

ORGANIZATION FOR FIRE PREVENTION

In Canada, a Dominion Fire Commission with a staff of inspectors and personnel supports local fire operations in the provinces, which follow very similar practices. The various provinces also have provincial fire commissioners. The United States has no similar uniformity. Although certain branches of the federal government conduct research and gather data concerning fire problems, government support is far less comprehensive. The National Fire Academy does train selected fire prevention personnel, and the U.S. Fire Administration provides materials for training, logistics, and occasional grant money for local programs.

Most states have state-level offices to oversee certain phases of fire prevention. The chief administrator at the state level is usually called the state fire marshal (discussed later on in this chapter).

Chief of Fire Prevention or Local Fire Marshal

Various state, local, and fire district regulations delegate the responsibility and authority of fire prevention to the fire chief or fire department head. That person may then delegate the authority to an individual or a division, depending on the size of the department. The individual or head of the division should be a high-ranking chief officer and should also function as a staff officer to the fire chief. This division of the fire service usually is called the fire prevention bureau, and its top officer is designated as the chief of fire prevention, the fire marshal, or the fire official, depending on local practice, custom, and regulations. Where size permits, a bureau is divided into subdepartments of inspections, investigations, and public education. These subdepartments are then headed by subordinate chiefs or officers.

Fire Inspector or Fire Prevention Officer

The positions of fire inspector and fire prevention officer usually have different meanings in different departments. Sometimes both titles denote the position responsible for conducting fire inspections assigned to the fire prevention bureau. In bureaus that are not large enough for multiple subdepartments, the fire inspector may also be responsible for both conducting fire investigations and performing public education duties. NFPA 1031, *Standard for Professional Qualifications for Fire Inspector and Plan Examiner,* provides complete details about fire inspector qualifications.

Fire Protection Engineer

The complexity and magnitude of fire protection problems create demand for the services of fire protection engineers. Although most of their work is done on a consulting basis, some public fire protection agencies have recognized the need for full-time staff engineers to conduct in-depth plan reviews, perform engineering calculations, and conduct comprehensive field inspections.

The Society of Fire Protection Engineers defines fire protection engineering as follows (Custer and Mechamp 1997):

> The application of science and engineering principles to protect people and their environment from destructive fire includes: analysis of fire hazards; mitigation of fire damage by proper design, construction, arrangement, and use of buildings, materials, structures, industrial processes, and transportation systems; the design, installation, and maintenance of fire detection and suppression and communication systems; and post-fire investigation and analysis.

Assignment of Responsibility for Fire Prevention

How are departments organized to carry out all the fire prevention tasks? In small departments, the chief might conduct inspections with the assistance of fire fighters who are specially interested, trained, and certified in this facet of fire department work. In a large city, the fire department might assign these functions to two or more distinct sections: one responsible for fire prevention and education and the other in charge of inspection, enforcement, and investigations. Medium-sized departments might concentrate all of the fire prevention duties in one centralized bureau.

Generally, fire prevention personnel do not work in shifts. However, in every department, whatever its size, at least one trained person should always be on call to carry out immediate investigations of fires that have resulted in loss of life, serious injury, or severe property damage, or that are considered suspicious in origin. Sometimes regular-shift personnel perform this function on a part-time basis when needed.

A fire fighter who has assumed fire prevention duties might receive compensation for the extra responsibilities and training that are required. In addition, the assignment may lead to greater competence and understanding of fire department activities. This broader perspective may enhance the likelihood of promotion.

Alternative Organizational Patterns

Instead of keeping fire prevention solely as a staff function, some departments have modified their organizational framework by assigning full- or part-time fire prevention inspectors to shifts. This method has two advantages. First, someone with special knowledge about specific buildings and their hazards is always on hand to provide additional information to fire fighters. Second, an immediate investigation into the cause of a fire can proceed. A major disadvantage to shift assignments, however, is that very few technical inspections can be conducted at night. Occasional delays in technical inspections may result.

Fire Prevention Activities in Volunteer Departments

Volunteer departments can organize themselves to carry out fire prevention activities in various ways. Here are a few possibilities:
- Offer special training for those fire fighters who wish to assume fire prevention duties.
- Hire paid staff on a full- or part-time basis. Sometimes retired fire department personnel can be hired for these positions.

- Have chief, deputy chiefs, or both assume fire prevention duties, sometimes in rotation.

STANDARDS AND CODES

Fire prevention programs and activities are based on standards and codes. Unless some level of government creates enabling legislation, however, the codes and standards are not likely to be used. Fire prevention programs are still operating in some places without the necessary enabling legislation, but then a great deal depends on the personal influence of the individual charged with administering the program. In the absence of law, the risks and potential liabilities are great.

Fire departments can work best with fire safety regulations that have been promulgated and subsequently enforced by the different levels of government. Although some of these functions overlap, federal and state laws generally govern those areas that cannot be regulated at the local level.

The major goal of any successful code or standard is to provide a reasonable degree of safety to life and property from fire, and accordingly, most well-developed fire prevention recommendations receive public acceptance and compliance. As important a role as the codes and standards play, they provide only the *minimum* requirements for any given situation. They are not intended to prevent steps to bring adherence to more stringent requirements.

The Federal Government

Under the U.S. Constitution, the legal authority of the federal government is limited to enforcing laws of an interstate or international character. Thus, in fire matters, federal laws mainly relate to transportation, including the shipment of hazardous substances by road or rail across state lines and the enactment and enforcement of fire protection regulations aboard planes and ships. In addition, fire prevention programs in national parks and forests are under the jurisdiction of the Forest Fire Service of the U.S. Department of Agriculture. The U.S. Department of Defense also delivers a wide range of fire protection services in each of the armed services.

The U.S. federal government contributes to research on fire prevention and protection through various agencies. A wide range of fire research is conducted by the Building and Fire Research Laboratory of the National Institute of Standards and Technology within the Department of Commerce. The U.S. Fire Administration of the Federal Emergency Management Agency (FEMA) conducts programs in such areas as public fire education, arson, and fire data analysis. Also under the FEMA umbrella are the National Fire Acad-emy (NFA) and the Emergency Management Institute (EMI), both located in Emmitsburg, Maryland. A wide range of resident and field programs are available to members of the emergency response community.

In Canada, assistance is available from Fire Prevention Canada, located in Ottawa. The agency's mission is "to achieve a fire-safe environment for Canadians through education and to provide Canada with dynamic leader-ship and a national focus in the field of fire prevention" (Cote 1997, p. C–17).

National Fire Protection Association (NFPA)

Nongovernment organizations develop most of the knowledge and stan-dards relating to fire prevention. The most important of these organizations in the United States is the National Fire Protection Association. Based in Quincy, Massachusetts, the NFPA was organized in 1896 "to promote the science and improve the methods of fire protection and prevention, to ob-tain and circulate information on these subjects, and to secure the coopera-tion of its members in establishing proper safeguards against loss of life and property by fire" (Tryon 1969, pp. 3–7, 3–8). The NFPA was organized by 18 men drawn primarily from the insurance industry. Today, many different in-terests are represented in the NFPA membership, including the fire service itself, the scientific community, educators, government officials, and private industry.

One of NFPA's most important functions is to develop fire safety codes and standards for processes, materials, and operations that may present po-tential fire hazards or are important for fire safety. It also prepares codes and standards for the professional qualifications (i.e., competency requirements) of positions in the fire service. Although NFPA documents often are adopted and incorporated into state and local ordinances, the NFPA considers its sta-tus to be solely advisory. The codes and standards are published in reference volumes and cover a wide range of subjects, including flammable liquids and gases, electrical equipment and installations, building construction, sprinkler systems, and fire alarm systems. Two of the better-known codes are the National Electrical Code® and the Life Safety Code®. The codes are revised periodically to include updated construction techniques, processes, materials, and uses. The NFPA is also at the forefront of developing a single national fire prevention code. NFPA 1, *Fire Prevention Code,* is widely used both in the United States and in other regions of the world.

Preparation of NFPA Codes and Standards.   NFPA technical committees that administer codes and standards for fire safety are charged to ensure that they do not bring prohibitive expense, interference with established processes and methods, or undue inconvenience. Each committee is a balanced working group made up of all the interests concerned with a particular document. In general, committees include appropriate manufacturers, users, installers and maintainers, labor and insurance representatives, researchers and testers, enforcing authorities, consumers, manufacturers, and special experts.

NFPA Codes and Standards. NFPA codes and standards are adopted more widely each year as increasing numbers of government, insurance, and industry officials and others recognize the scope of the world's tragic fire problem. Today, millions throughout the world are protected from fire by these codes and standards.

NFPA codes and standards are used in many contexts. For example, NFPA aviation documents are referenced by airports throughout the world. In the United States, scores of NFPA codes and standards are referenced by the federal government's Occupational Safety and Health Administration, the Veterans Administration, the Department of Health and Human Services, the Department of Defense, and other federal agencies.

NFPA develops "full consensus" codes and standards—codes and standards built on a foundation of maximum participation and substantial agreement by a broad variety of interests. This philosophy has led to reasonable codes and standards that provide adequate protection to the public yet do not stifle design or development. The result is technical codes and standards that truly represent society's acceptance of risk toward fire.

NFPA prides itself in supporting a flexible system that depends largely on volunteers and therefore produces fire safety codes and standards at no cost to taxpayers. But the process doesn't stop with the completion of a code or a standard. The rapid pace of technology creates a need for frequent updating of information, and NFPA's system has a built-in mechanism for regular updating.

Fire safety is everybody's business; everyone deserves to be heard when it comes to fire safety. That's why, after more than 100 years, the NFPA's codes and standards process has evolved into one of the fairest and most effective technical document development systems in the world.

American Insurance Association (AIA)

A series of studies conducted by the American Insurance Association isolated the factors that contributed to the major fires in U.S. cities in the late 1800s and early 1900s. The AIA used this information and the early NFPA standards to establish levels of adequacy for fire prevention in cities. AIA activities have led to the development of the *National Building Code*, a model code that has been adopted by many municipalities across the United States. The AIA has also suggested a fire prevention code for cities. Both the *National Building Code* and the *AIA Fire Prevention Code* are based largely on NFPA standards as well as on recommendations relating to various problems encountered by industries in their manufacturing processes.

Underwriters Laboratories

At one time the AIA also sponsored Underwriters Laboratories (UL), a testing laboratory originally organized to investigate electrical hazards. UL is now an independent, nonprofit membership organization supported by fees from manufacturers who want their products tested both prior to marketing and periodically for continuing compliance with nationally recognized safety standards. The official UL label is issued only after testing and follow-up inspections.

UL's current corporate membership is drawn from consumer interest groups, public safety bodies or agencies (responsible primarily for enforcement in the field of public safety), government bodies or agencies, the insurance industry, safety experts, standardization experts, public utilities, educators, and corporations (usually at the officer level). UL is managed by a board of trustees drawn from the aforementioned categories, plus an additional "at large" position. Only one officer of the corporation is included on its board of trustees.

Factory Mutual Research Corporation

The Engineering Division of the Factory Mutual Research Corporation (FM) maintains laboratories for testing building materials and fire equipment, Like UL, it issues labels to indicate that certain products have passed its tests. Factory Mutual provides the protected companies with support, such as risk surveys, testing, and studies.

The FM research staff includes standards, research, and approvals groups. The standards group is made up of engineers in many fields who develop information and recommendations based on research and loss experience. They are also available to offer advice to Factory Mutual members on specific loss reduction matters.

The research group consists of two subgroups of scientists: One does basic research and the other assists in the development of standards for direct application in industry (Cote 1997, p. C–11). The goal of the first subgroup is to secure information pertaining to the initial phases of fire, its detection, and growth patterns. Theories developed by this subgroup are expected to lead to new methods of loss reduction and control. The subgroup that works on applications is concerned with the ignition and flammability of materials, improvements in the effectiveness of fire protection systems, and new suppression agents and systems. It performs fire modeling studies, rack storage and plastics storage fire tests, and design and cost evaluation of effective fire protection systems.

The approvals group subjects equipment and materials submitted by manufacturers to stringent tests to determine whether the devices will operate dependably and to ensure that materials have an acceptable low flammability rating when subjected to fire tests. It issues an approval guide annually.

Other Groups

Other technical groups prepare standards for specific manufacturing processes or for potential fire risks. Just as the AIA uses the NFPA standards to prepare its own codes and grading schedules, these groups, which represent various occupations or industries with fire protection interests, such as the Chlorine Institute and the American Petroleum Institute, prepare even more stringent codes, using NFPA and other standards as a base. Their purpose is to obtain lower insurance rates for those industries that comply with stricter requirements and allow regular inspections by the group's inspectors.

State Regulatory Offices

The principal regulatory authority for implementing fire laws at the state level is often the state fire marshal's office. Almost all states have a state fire marshal's office. In most states, enabling legislation gives the fire marshal the authority to draw up regulations covering various hazards, and the regulations usually have the effect of law.

The makeup of state fire marshal offices differs from state to state. Most receive their authority from the state legislature and are answerable to the governor, a high state officer, or a commission created for that purpose. Few fire marshal offices are separate state agencies, but rather a division of the state insurance department, the state police, state building department, state commerce division, or some other state agency.

The precise responsibilities and organization of the state fire marshal's office vary from state to state. The authority of a state fire marshal's office usually extends to the following general areas:

- Prevention of fires
- Storage, sale, and use of combustibles and explosives, including fireworks
- Installation and maintenance of automatic fire and smoke alarms and sprinkler systems
- Construction, maintenance, and regulation of fire escapes
- Means and adequacy of exits, in case of fire, in public places or buildings where many people live, work, or congregate (such as schools, hospitals, and large industrial complexes)
- Suppression of arson and investigation of the cause, origin, and circumstances of fire incidents

In discharging its responsibility, the state fire marshal's office usually maintains statewide fire records. As a state agency, the fire marshal's office normally is concerned with functions that are outside the scope of municipal, county, or fire district organizations. For instance, the state fire marshal's office may oversee district, municipal, and county fire prevention groups. The office frequently provides technical assistance and other services to unincorporated areas. In addition, the state fire marshal is responsible for the inspection and code enforcement of state facilities.

Although the state fire marshal's office has legal authority for fire prevention, much of this power is delegated to the local fire departments and local government. Fire departments are responsible for inspecting private properties to check for fire hazards or code violations, and local authorities are given the power through enabling acts to adopt their own regulations related to fire prevention.

Local Codes and Ordinances

Local codes and ordinances hold the greatest interest for fire fighters and officers. Many local codes and ordinances incorporate the standards and codes set by state and private organizations. Some states have adopted uniform codes in areas such as building construction; these uniform codes might supersede any existing local ordinance.

Laws for local fire safety generally fall into two categories: (1) those related to buildings and (2) those related to hazardous materials, processes, and machinery that might be used in buildings. In general, requirements relating to construction are given in the building code and are enforced by the building inspector and the building inspector's department. The requirements relating to hazardous materials, hazardous processes, and the safe operation of machinery or equipment are the responsibility of the fire department and thus are covered by the fire code.

The following items are usually covered by municipal building codes:

- Administration, which spells out the powers and duties of the building official
- Classification of buildings by occupancy
- Establishment of fire limits or fire zones
- Establishment of height and area limits
- Establishment of restrictions as to type of construction and use of buildings
- Special occupancy provisions that stipulate construction requirements for various occupancies such as theaters, piers and wharves, and garages
- Requirements for light and ventilation
- Exit requirements
- Materials, loads, and stresses
- Construction requirements
- Precautions during building construction
- Requirements for fire resistance, including materials, protection of structural members, fire walls, partitions, enclosure of stairs and shafts, roof structures, and roof coverings
- Chimneys and heating appliances
- Elevators
- Plumbing
- Electrical installations
- Gas piping and appliances
- Signs and billboards
- Fire extinguishing equipment

Fire prevention codes cover:

- Administration, which includes the organization of the bureau of fire prevention and defines its powers and duties
- Explosives, ammunition, and blasting agents
- Flammable and combustible liquids
- Liquefied petroleum gases and compressed gases
- Lumberyards and woodworking plants
- Dry cleaning establishments
- Garages
- Application of flammable finishes
- Cellulose nitrate motion picture film
- Combustible metals
- Fireworks
- Fumigation and thermal insecticidal fogging

- Fruit-ripening processes
- Combustible fibers
- Hazardous chemicals
- Hazardous occupancies
- Maintenance of fire equipment
- Maintenance of exit ways
- Oil-burning equipment
- Welding and cutting
- Dust explosion prevention
- Bowling establishments
- Automobile tire-rebuilding plants
- Automobile wrecking yards, junkyards, and waste material handling plants
- Manufacture of organic coatings
- Ovens and furnaces
- Tents
- General precautions against fire

Although these coverages may seem to overlap, close scrutiny will show that the inclusion of the original fire prevention item (for example, duct, vent, exit, or sprinkler system) is supervised by the building department but its continuing adequacy is the responsibility of the fire department. Regardless of whether a particular provision is in a building code or a fire prevention code, however, the fire official has input and jurisdictional authority for all fire protection features in new and existing construction.

Jurisdictional problems with fire and building codes can usually be lessened by a frank approach and open communication. When high-level officials from both departments meet on a continuing basis to discuss their responsibilities and mutual concerns, they can usually resolve the problems. Some European countries have eliminated jurisdiction disputes. In Germany, for example, fire department officials are responsible for both building and fire codes. Some local governments in the United States have also elected to assign fire fighters the job of inspecting for building and fire code violations at the same time.

Model Building Codes

In an effort to lessen some of the confusion about enforcement of fire codes, some states depend on model building codes, such as (1) the *Basic Building Code* of the Building Officials and Code Administrators International, Inc. (BOCA); (2) the *Uniform Building Code* of the International Conference of Building Officials; and (3) the *Standard Building Code* of the Southern Building Code Congress. Each model building code has an accompanying fire prevention code. The states that have adopted one of the model building codes might recommend that local governments also adopt the companion fire prevention code to provide a uniform functional separation of the building and fire departments. Because the codes are to be adopted in their entirety, the duties of each department are clarified in relation to one another.

Code Modification

What can a fire prevention officer do to change or modify an inadequate code? The best way is to work within the various systems to support the codes that are currently in use. Each has a mechanism for changing the respective code document.

Good Engineering and Construction Practices

Sound practices in building design, construction, and renovation are important for making buildings strong enough to support their intended uses and for ensuring that they are as resistant to fire as possible. Overseeing building construction is primarily the responsibility of the local building department, because fire department personnel are generally not expert in the relevant issues. However, ensuring appropriate fire prevention features of new construction requires interaction among building officials, engineers, architects, and the designated fire department official.

Fire department surveillance is particularly important for renovations, in which, as a rule, the building department's role is minimal. Therefore, in most jurisdictions, the building department has primary responsibility for new construction, and the fire department is the main agency for seeing to it that all structures are maintained in fire-safe conditions.

INSPECTIONS

Inspections are the first step in ensuring adherence to the regulations that are based on codes. To be fully qualified to conduct fire inspections, department members should be thoroughly familiar with codes at all levels and be competent in accordance with NFPA 1031. Enforcement is the second step.

Enforcement requires support of the judiciary for the regulations and for enforcement actions. (We will discuss legal foundations for enforcement activities later in this chapter.)

Local fire codes call for the inspection of several categories of hazards within the district served by a fire department. The frequency of inspections varies from state to state and depends on the type of occupancy and level of hazard. The hazard levels are listed here in order:

1. One- and two-family dwellings (usually not inspected except at the request of owner/occupant)
2. Three- or more family dwellings
3. Commercial office buildings
4. Industrial (high, moderate, and low hazard)
5. Mercantile (high, moderate, and low hazard)
6. Public assembly complexes
7. Institutions (hospitals, nursing homes, group homes, and so forth)

Institutions, public assembly complexes, and high-hazard mercantile and industrial plants frequently are designated as "target properties" for inspections.

Types of Inspections and Personnel

The two major types of inspections are company (regular) and technical. In some departments fire fighters perform company inspections (or field inspections as they sometimes are called). These routine inspections check for compliance with the general regulations concerning access to standpipes and sprinkler valves, adequacy of fire extinguishers, lack of obstructions to emergency evacuation exits, and the more obvious safety problems such as multiple connections from electrical outlets (circuit overloads) (see Figure 4.1). When inspectors find code violations, they issue violation notices or report the violations to the fire prevention staff, who issue notices. Company fire fighters or inspectors make follow-up visits to ensure that deficiencies are corrected.

Because the complexities of many modern industrial processes and operations are beyond the scope of standard fire department training, specially trained fire prevention inspectors perform technical inspections. Such inspectors may be available in the private sector or from county and state agencies. Detailed inspections ensure that hazardous materials and processes are subject to definite safety procedures and regulations. Businesses and industries that use hazardous materials are required to obtain permits from the fire department and cannot start hazardous activities (or continue them) without such permits. The technical inspections must be carried out before a permit is issued or renewed.

Inspections of occupancies other than the interiors of single-family and two-family residences are usually required by law. These inspections identify those conditions that violate the fire code and that have the potential to cause fire or endanger life and property. Conditions of interest to fire officers for prefire planning and training purposes are emphasized.

Other inspections are conducted by fire company personnel to supplement inspections of the fire prevention bureau. These inspections normally are conducted in the fire company's first-due area. Before fire fighters perform such inspection work, they must receive proper training and be granted the authority to conduct inspections as fire prevention officers. The fire prevention bureau provides assistance where needed in obtaining compliance to company recommendations.

Some inspections are made by fire company personnel for prefire planning and training purposes. They emphasize conditions that violate the fire code and that are likely to cause fire or endanger life. Conditions that require more than on-the-spot correction usually are referred to the fire prevention bureau for review and issuance of violation notices.

Figure 4.1 Inspection reveals a parked vehicle blocking the sprinkler Siamese connection, which can cause a problem during an emergency. (Courtesy of Harry Carter)

Fire company personnel may conduct inspections in private homes when requested by the owner or occupants. Recommendations that result from these inspections are not mandatory; however, if inspectors find definite code violations, an effort is made to have the hazard corrected through proper department channels of authority (see Figure 4.2).

Objectives of Inspections

The functions that are performed during an inspection and the compilation of the report that results from the inspection are covered in detail in the *NFPA Inspection Manual* (NFPA 1994). These are the major inspection objectives of an inspection:

- Uncovering code violations and potential fire hazards.

- Acquainting fire fighters with fire codes, SARA (Superfund Amendments and Reauthorization Act), Title III, and OSHA (Occupational Safety and Health Act) building and safety requirements. Knowledge of these codes often helps in actual fire-fighting operations.

Figure 4.2 Inspection of the house in the foreground revealed unsafe fire escape stairs. Most of the wood in the fire escape is rotted, and the handrails have collapsed. (Courtesy of Amy E. Dean)

- Familiarizing fire fighters with contents and construction hazards, thus making fire fighting operationally more efficient and safer. Appropriate risk management is a critical, and frequently overlooked, element in personnel safety at the operational level. For example, fire fighters gain a better understanding of how a fire can spread through vertical openings (such as elevator shafts, stairwells, and the stairways of buildings) and gain awareness of where the primary potential sources of fire hazard are located.

- Using personnel more efficiently for assigned fire prevention duties. Most fire departments inspect the most important properties (according to local regulations) about four times a year, depending on the model fire code used by the department. Moderate- and low-hazard occupancies are likely to be inspected at least twice a year. This schedule sometimes is met by combining technical inspections with company inspections. Still, the workload is often heavy, and some inspections are scheduled specifically to develop methods that are most efficient.

Inspectors from the Fire Prevention Bureau might inspect higher-risk properties twice a year, especially when permits are due or when changes of occupancy or process indicate a more detailed inspection. That leaves fewer inspections to be completed by the fire suppression personnel.

The Inspection Process

Whether inspections are carried out by fire fighters or by fire prevention bureau personnel, the inspectors follow the same general steps.

1. Before conducting an inspection, the inspectors review all records related to the specific area, premises, or occupancy. An inspector needs to know which violations have been found previously, what steps the owner has taken to correct them, and the causes of any fire that might have occurred in the building.

2. The inspectors make a comprehensive list of items to look for in each occupancy. Although some large fire departments might produce lists for each type of occupancy, fire fighters must sometimes compile their own. Exhibit 4.1 is a sample fire inspection report for fire prevention inspectors. Sample lists and inspection forms for the 14 occupancies identified in the *Life Safety Code* are given in the *NFPA Inspection Manual* (NFPA 1994). Forms for these occupancies can also be found in NFPA's *Fire Protection Handbook* (Cote 1997).

Exhibit 4.1 Sample Fire Inspection Report (*Source:* Courtesy of Plainfield Fire Department, Plainfield, New Jersey)

FIRE INSPECTION REPORT FOR			DISTRICT	
Street			Number	
Owner/Agent/Superintendent				
Address of above				
Class Construction	Roof		Stories	
Occupancy			Fl.	No. of Tenants
			DATES OF INSPECTIONS	
FIRE HAZARDS				
Heating System				
Clearances				
Heat Deflectors				
Condition of Flue				
Flue Pipe Fit				
Burner Controls				
Storage of Explosives				
Storage of Flammables				
CONDITION OF				
Fuses/Breakers				
Electrical Wiring				
Electrical Appliances				
Chimneys				
Vent Ducts				
Rubbish				
Storage of Ashes				
Air Conditioning				
Gas Appliances				
Miscellaneous				
FIRE PROTECTION				
Sprinklers	Wet	Dry		
Standpipes				
Second Egress				
Extinguishers				
Fire Doors				
Fire Escapes	W	M		
Aisles				
Halls				
Chutes				
STRUCTURAL DEFECTS				
Roots				
Walls				
Floors				
Foundations				
Stairs				
Elevators				
Enclosures				
Stairway Enclosures				
LOCATION OF CUT OFFS				
Sprinkler				
Standpipe				
Gas				
Electricity				
Water				
LOCATION OF SIAMESE CONNECTIONS				
Sprinkler	Standpipe			
Inspector's Initials				
Violation Notice Issued				
Date Corrected				
REMARKS				

3. Before entering the building, the inspectors examine the property to see where apparatus might be stationed for prefire plans, check the location of fire escapes and their condition, and survey for obstructions of standpipes or emergency exits.

4. Upon entering the building, the inspectors inform the person in charge that an inspection is about to be made (see Figure 4.3).

5. The actual inspection often begins at the roof, where the inspector can see exposures and adjacent rooftops and parapets. Wherever it begins, however, the inspection should proceed in a systematic fashion (e.g., from attic to basement, from floor to floor). Each item or measurement should be checked off or recorded on the inspector's reporting form. The inspector must check certain specific points: The panic hardware on exit doors could have been removed inadvertently and replaced with regular fixtures. The door itself could be jammed or blocked. No-smoking or exit signs might be unreadable or blocked. There might be a need to designate a smoking-permitted area. All exit signs must be visible from all approaches.

6. During a company inspection, the inspector should give special attention to items of interest for prefire planning, such as the way in which

Figure 4.3 The fire inspector should always make contact with building management before beginning an inspection. (Courtesy of Harry Carter)

the building could be ventilated, fire protection equipment, stairways and corridors, doorways and exits, heating and exhaust ducts, insulation and wiring, storage areas, facilities for disposal of refuse, and general maintenance procedures.

Many fire departments have fully computerized their fire inspection operations. Inspectors can also use handheld computers to record the inspection data. When the inspectors return to the department at the end of the day, they transfer the data from the handheld device into the department's computer system and automatically add it to the ongoing database.

All conditions that are not in compliance with regulations should be included in a report. (see Figure 4.4). Violation notices should be issued for all

Figure 4.4 A lightly supported television antenna that is too close to power lines should be noted in an inspection report.

violations noted and certainly for all those that might contribute to the start or significant spread of fire. The owner or person in charge should get a copy of the inspection report in addition to any notice of violation.

An inspector who has a well-groomed appearance, including a clean and pressed uniform, presents a positive and professional image for the department. As the inspection proceeds, the inspector (whether a fire fighter or a technical inspector from the fire prevention bureau) has the opportunity to emphasize the preventive aspects of the inspection work. The inspector can explain the implications of violations to the owner or employee of the premises being inspected and thereby enhance the image of the department while also educating the public. The inspector can also distribute fire safety education material.

Private Dwelling Inspections

Inspectors should examine residential areas in much the same manner as they conduct other inspections. The standard break point for inspections is at the three-family or higher level. Fire departments may offer courtesy inspections to the one- and two-family units, but they cannot enter apartments without the permission of the occupants. The inspection of private dwellings is not required by regulation, and it is often specifically prohibited except when requested by the owner or occupant.

Apartment buildings are subject to inspections in both common and private areas. In practice, however, access to individual apartments may not be possible because the occupants are not at home.

Fire departments should attempt to inspect individual apartments only if they have sufficient personnel. When inspecting private dwellings, inspectors should point out possible escape routes in the event of fire, provide information about fire detection equipment and portable fire extinguishers, and distribute fire and life safety education materials such as NFPA's E.D.I.T.H. (Exit Drills in the Home) flyers or other available brochures.

Code Violations

Reasons for Violations. People sometimes violate codes because compliance costs money to buy sufficient fire extinguishers, to put in additional electrical circuits, or to add extra storage space. Sometimes people simply forget to be careful about fire hazards. They forget to check whether the fire escape is in good condition and whether the sprinklers work. People who do not have enough storage space for boxes may pile them in a corridor for a few days until they find more room. It often takes a fire inspection to call attention to the fact that obstructions in corridors and doorways are serious fire code violations. Most inspections find some code violations.

Correction of Violations.   If corrections of code violations involve a substantial amount of money, the inspector may help the owner determine how to rectify the situation. Of course, the owner may resist. Sometimes the fire prevention bureau can provide special technical advice. At other times the company officers must do the best they can to resolve the conflict. (See Additional Insight 4.1, Management of Potentially Damaging Conflict.)

Fire departments allow the owner a grace period for correcting violations. The technical inspector or company officer determines how much time to allow. The time usually depends on whether the violation is a simple one, like removing obstructions in doorways, or a comparatively difficult one, like installing a fire escape.

The inspector usually schedules follow-up visits to determine whether the owner has corrected the violations. In the case of voluntary inspections of private dwellings, follow-up visits are made only if requested by the owner or occupant.

Role of the Company Officer in Inspections

The company officer has the responsibility to integrate all aspects of company inspections into the general purpose and goals of fire prevention. This task is not easy because many fire fighters have negative feelings about conducting inspections.

Fire Fighter Attitudes Toward Inspections.   Negative feelings are usually rather fundamental.

- Most fire fighters join the fire service because they are interested in fighting fires. Fire prevention is a different kind of work that does not give them the same level of satisfaction.
- Some fire fighters do not have outgoing personalities. They cannot relate easily to the people with whom they must deal when doing inspection work.
- To carry out inspections, fire fighters must wear clean, pressed uniforms, which either cost extra money or require extra effort.
- Inspections often occur at inconvenient times for a building owner—for example, when an owner is busy or the premises to be inspected are in full use. A fire fighter may feel uncomfortable intruding.

It is not easy to confront these issues. Still, competent officers will not ignore them. The following suggestions for officers can help, although they are not likely to completely overcome the resistance to inspection work.

- Do not ignore the negative feelings. Hold meetings from time, especially before inspection runs, to discuss them and explore ways to reduce them.
- Arrange the inspection procedure so that contacts with the occupants are handled primarily by those fire fighters who are most comfortable with strangers.
- Show as much appreciation as possible for little things that fire fighters do, prior to and during inspections, to make the work more pleasant or effective. [See Additional Insight 5.1, Enhancing Work Satisfaction (Providing Recognition) and Performance Evaluation.]

In addition to the fundamental reasons, there are other reasons fire fighters do not like inspection work:

- Fire fighters often have difficulty accepting that inspections have to be part of *their* job rather than that of inspectors. Although they may understand the reasons, emotionally they are committed to responding to emergencies, and they don't like to be assigned other duties.
- When asked why they are conducting an inspection, fire fighters who have not been trained properly might have no response other than to say that they were told to do it.
- When asked other questions, some fire fighters feel insecure about the accuracy of their responses and become frustrated when confronted with a questioner who assumes that all fire fighters know all there is to know about fire prevention.
- To many fire fighters, it is frustrating to impose on people and to carry out some safety procedures that are not well understood by the public.
- The attitudes and reactions of the people whose premises are being inspected often make fire fighters feel uncomfortable.
- Inspections sometimes make the workday longer.

Good communications and expressions of appreciation can alleviate these problems, too. In addition, a company officer has three specific ways for dealing with dissatisfaction with inspection duties: (1) careful planning and scheduling, (2) better training and (3) appropriately supportive supervision.

Careful Planning and Scheduling.   A company officer may be responsible for planning company inspections so that the entire area assigned to the jurisdiction is covered within a prescribed period of time. To do this, the company

officer must estimate how long it will take the available fire fighters to conduct each inspection as well as the number of hours needed to cover the entire area undergoing inspection.

Because inspections often lengthen the workday, the officer who wants
to reduce negative attitudes must carefully plan and schedule inspections so
that they can be finished within the allotted time. The company officer
should ensure that all of the materials needed to carry out inspections are
available, such as reporting sheets, checklists, records, maps, flashlights,
tape measures, pencils, and special equipment like hard hats. A book of
codes and regulations can provide easy reference when questions arise. Another good resource is the *NFPA Inspection Manual* (NFPA 1994).

Training in Understanding and Application of Codes.   Fire fighters will
better understand the importance of inspections and how they relate to the
full range of fire service work if they understand codes and the purposes of
inspections. This awareness makes fire fighters better able to impart knowledge to the people whose premises they inspect and promotes a more comfortable atmosphere for both parties involved. Most people are comfortable
answering questions only in areas where they are able to provide adequate
responses. Thus, fire fighters feel more confident if they are able to accurately answer many of the questions they are asked during inspections.
They should receive training in the appropriate codes and in answering
common questions.

However, it is not sufficient for fire fighters merely to be aware of and
understand codes and the purpose of inspections. It is also necessary that
fire fighters be trained to apply this knowledge to identify code violations
(see Figure 4.5). They need skill in analyzing the present conditions in a
building and in determining how a fire might begin and spread. Competent
fire fighters also must be able to determine whether escape routes are adequate for quick evacuation in case of fire or other emergencies. They must
know more than simply that electrical circuits cannot be overloaded or that
flammables must be stored in approved, self-closing containers, although
these precautions are important parts of the codes.

Fire fighters must apply their knowledge of fire safety and codes to each
situation. For example, if a floor or space is very tall, an adequately trained
fire fighter will recognize the need for fire curtains so that when heat from
fire rises, it does not spread but is contained. Fire fighters should either
know or quickly find the answers to specific questions: How should fire
doors be hung, and when do they meet specifications? When are the windows an approved design? What are the characteristics of a fire escape that
is in compliance with the codes?

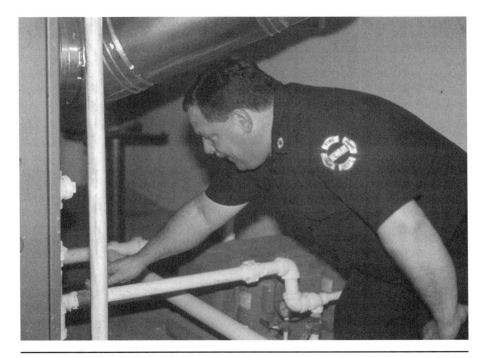

Figure 4.5 Understanding fire prevention codes is essential for fire fighters to be able to identify violations during inspections. (Courtesy of Harry Carter)

Beyond providing fire fighters with technical training, company officers should help them understand the importance of good public relations to gain cooperation from the community. For example, although regulations generally give an inspector the right of entry, advance notification of an inspection is good practice. With large-scale facilities, the company officer usually makes that call. When a company plans several simultaneous inspections during one stop, such as at an office building or strip mall, the fire fighters may make their own notification calls. In certain controversial situations, an appointment or a warrant might even be necessary. Upon arrival at the inspection site, an inspector should seek out the person in charge and request that person's permission to make the inspection. In addition, the inspector should suggest that the owner, the manager, or a representative go along on the inspection.

Appropriate and Supportive Supervision.   During fire prevention inspections, company officers should provide guidance appropriate to the respective fire fighter's level of competence and work maturity. Company officers should maintain more frequent contact with the less knowledgeable or seasoned fire fighters than with those who are more competent. (See Additional Insight 1.1, Participation in Decision Making and Planning.) The officer should provide follow-up analysis of knowledge, skills, and abilities (KSA) for coaching and training, and for performance evaluation. Often the officer may have an opportunity to commend a fire fighter in the presence of members of the public and other fire fighters, thus adding satisfying moments to field inspections. [See Additional Insight 5.1, Enhancing Work Satisfaction (Providing Recognition) and Performance Evaluation.]

OPERATIONAL TASKS

Operational tasks, in addition to inspections, and issues related to effective fire prevention include:

1. Construction plans review
2. Consultation
3. Prefire plans
4. Public fire and life safety education
5. Records and reports
6. Fire investigation

We will introduce each operational task as it relates to the fire service and the officer's responsibilities.

1. Construction Plans Review.   The review of construction plans for different classes of buildings is now legally mandated in many localities. The fire service can thereby see that fire protection standards are met before construction begins. Reviews must be followed with on-site inspections to ensure that the fire protection provided for in the plan is not overlooked or compromised during construction.

2. Consultation.   The general public looks to the public fire service for answers to its fire problems. Because fire prevention covers such a broad area and reaches so many people, consultation services are necessary. Fire prevention officers are sometimes called on to explain the fine points of fire codes to professionals, such as architects and engineers, who might be dealing with fire codes for the first time when they prepare to file their fire plans for review. Officers also give information to owners of small structures who need clarifications on code-related matters, such as storing combustibles, and on voluntary requests for inspections.

3. Prefire Plans.   Prefire plans may include detailed layouts of properties (except for one- and two-family dwellings) in the fire district, showing entrances, exits, stairs, firewalls, standpipes, areas covered by sprinkler systems, and other information pertinent to a fire attack. Sometimes plans also show outlines for attack preparation, such as the positioning of apparatus and initial hose layouts. Prefire plans are drawn up by fire fighters and their company officers. After they have been approved by the department, the plans usually are carried on the apparatus in hard copy or are available in electronic format, so that they can be consulted on the way to a fire. Prefire plans can also serve as a basis for simulation in fire company training drills.

Prefire plans and inspections overlap in two ways. First, they familiarize personnel with buildings where other than routine fires could occur. Second, prefire plan surveys can sometimes uncover code violations, thus helping to support inspection work. We will discuss prefire planning in more detail in Chapter 6, Prefire Planning and Related Loss Reduction Activities.

4. Public Fire and Life Safety Education.   Educational programs help the fire department gain the cooperation of the citizens it serves. The department programs can use media publicity, flyers, and educational and informational programs aimed at audiences of all ages.

Public fire and life safety education is a vital tool in fire prevention. If people are to take the initiative in helping to prevent fires, they must learn how to do that. Three examples of public education programs are Fire Prevention Week (conducted annually in October), NFPA's Operation E.D.I.T.H. (Exit Drills in the Home), and NFPA's *Learn Not to Burn Curriculum*. Since its introduction in 1979, the curriculum has been used in more than 43,000 U.S. elementary school classrooms and is endorsed by the National Education Association. In 1998 NFPA released *Risk Watch*™, an injury prevention program for children in preschool through grade 8.

In addition to NFPA, many public and private organizations provide materials for use in public fire and life safety education programs. Chapter 5 is devoted almost exclusively to public fire and life safety education.

5. Records and Reports.   Records are essential for effective fire prevention, and they form the basis for studying trends to develop new programs. The National Fire Incident Reporting System (NFIRS) developed by the U.S. Fire

Administration, based on NFPA 901, *Standard Classifications for Incident Reporting and Fire Protection Data,* has improved the availability of fire loss data. Records that should be maintained include

- Copies of violations, inspection reports, and follow-ups
- Prefire plan maps and attack practices
- Statistical information organized into maps, charts, and other diagrams for use in fire prevention planning
- Recommendations dealing with specific problems in certain occupancies or areas, or for decisions about future needs

Records and reports of fire prevention activities should be clear and concise. Every time inspectors or fire prevention officers visit a site, they should include information about that location in a report that is kept in the occupancy file of that location. The file should contain the following data:

- A complete history of the building site
- Building plans
- Specifications (when possible)
- Permits issued for the use, storage, and handling of hazardous materials
- Inspection reports
- Fire incident reports

Photographs are invaluable in records and reports. A photograph that is properly taken and correctly identified is excellent evidence for a city attorney, an owner of the building, a chief officer, a judge, or members of a jury. Photographs and detailed reports can resolve arguments about actual conditions at the time of fire exposure. Photographs are also useful for educational purposes, such as training programs. Many larger fire departments have full-time photographers with complete camera and laboratory facilities. Even the smallest department can use instantly developing cameras at inspections as long as there are SOPs/SOGs and the inspectors have some training in them.

6. Fire Investigation.   Company officers play a critical role in fire investigations because they are usually the first on the scene to observe what is occurring. They must be trained to understand and perform this role.

Fire departments endeavor to investigate all fires to determine the first ignition sequence. Data compiled from investigations are useful in determining future fire prevention strategies. In cases where arson is suspected, an officer should summon appropriate investigative assistance if it is not available at the local level. Such assistance can come from local, county, or state agencies.

Most fire departments were organized for fighting fires. Few departments were set up to develop and compile comprehensive, in-depth information on the number of fires that occur by location and occupancy, the fire ignition sequence or causative factors, the time of day or week of occurrence, the room or floor in which the fire occurred, and similar information that is basic to any effective evaluation of a fire incident.

Obviously, fire prevention is a major concern of all fire department personnel. However, a point that has not been sufficiently recognized is that the comprehensive investigation of fires and all the factors that influence or contribute to their ignition sequence, spread, or extension is the very foundation on which fire prevention is built. Without the extensive and detailed information obtained from investigations, it is not possible to develop the most effective regulatory codes, standards, and inspection and suppression procedures designed to prevent or control fires. Two NFPA documents are available to assist fire departments with fire investigations. NFPA 921, *Guide for Fire and Explosion Investigations,* and NFPA 1033, *Standard for Professional Qualifications for Fire Investigator.*

Water Supply

We will present water supply issues in more detail in Chapter 6 because they are relevant to the loss reduction function of a fire department. Obviously, water availability will not prevent a fire. An adequate and conveniently located water supply will, however, reduce the damage that a fire inflicts.

The fire department usually is responsible for making recommendations to the local government about the adequacy of water supplies for fire fighting, especially when plans are considered for the development of new industrial, commercial, or residential areas in the community. As a general rule, the Fire Prevention Bureau works with a local authority and the water company to make surveys that ensure sufficient water supply to extinguish fires in the district. They examine the size of water mains in relation to the size and population density of buildings and in relation to sprinkler systems and water towers (where they exist). They also check the number, location, and maintenance of hydrants.

Personnel Assignment and Fire Prevention Priorities

To a certain extent, the number of personnel assigned to fire prevention duties reflects not only the size of the department but also the department's financial resources and the priority assigned to fire prevention by the fire chief, local government officials, and taxpayers. The 1973 report of the

National Commission on Fire Prevention and Control recommended that local governments make fire prevention at least equal to fire suppression in fire department priorities. Few departments (if any), however, have committed 50% of their resources to prevention. That kind of attention would require massive changes in the way fire departments are organized. Many of the jobs that currently are done during the day, such as checking of apparatus, housekeeping chores, and even some in-service inspections, would have to be performed at night to free personnel for additional daytime fire prevention activities. Everyone from the taxpayer to the fire fighter would have to be willing to accept changes. Because society generally is unaware of the need to emphasize fire prevention, the change is not likely to occur in the near future.

Barriers to Expansion of Fire Prevention Activities.   Inspections make up the bulk of fire prevention activities. Most departments do not have enough inspectors to cope effectively with the present workload. If a greater proportion of available personnel were assigned to carry out more inspections during daytime shifts, then a fire department might be able to complete its inspection schedule. One way to overcome the staff shortage for adequate inspections is to shift some fire fighters from other shifts to business-hour duty. That might deplete the night shifts, however, so a department would have to place some of the business-hour inspection personnel on rotating on-call status to respond in the case of a serious fire. Hiring and training additional personnel would require that more resources be allocated at a time when local authorities are looking for ways to trim their budgets. Some existing funds might have to be reallocated from fire fighting activities to fire prevention.

Ways to Expand Fire Prevention Activities.   Fire fighters conduct inspections not only to carry out the required number of inspections per occupancy but also to familiarize themselves with the district and thus lay the foundation for more effective fighting in fires. Although it might be difficult to commit a greater percentage of available working hours to fire prevention, inspections do assist fire-fighting activities. In addition to justifying a larger proportion of time for inspections, fire departments could take steps to conduct more inspections without relying on extra time from fire fighters, including:

- Inspecting some restaurants, bars, schools, and other establishments in the evening.
- Overcoming some of the obstacles involved in inspections to produce greater efficiency (as discussed earlier in this chapter).

- Conducting public education programs to reduce property damage and human casualties and the need for any expansion of fire-fighting capabilities. Such programs also can help change the attitudes of public and government officials regarding current financial allocations for fire prevention.

Role of the Company Officer in Fire Prevention

From the viewpoint of the company officer, fire prevention can provide meaning and greater job satisfaction to a company, especially one that fights fires only infrequently. Fire prevention activities provide officers with an opportunity to stimulate job enthusiasm that can spread to other activities as well. To do this is not easy. Such enthusiasm can, however, be instilled through creative leadership. Company officers should consider the following suggestions:

- Use prefire plans to simulate fighting fires at various locations, and then discuss what fire prevention activities might have prevented the fire or reduced losses.
- Rotate command of the simulations among the fire fighters so they take an even greater interest in the simulation, and then discuss the benefits of fire prevention efforts.
- Discuss the manufacturing and other processes used at various locations, the fire hazards they contain, and the implications for fire fighting after inspections are conducted.
- Make training sessions more interesting and realistic so that fire fighters will better appreciate the sessions conducted on codes, inspection procedures, and fire investigation.

Fire Prevention Goals and Objectives for Company Officer and Chief

As in other fire service activities, improvement in fire prevention starts with sound goals and objectives. For example, an officer might have these objectives:

- Prepare an inspection schedule for the entire year, so that all properties are inspected as frequently as required, allowing for an appropriate number of repeat inspections.

- Adhere to the prepared inspection schedule.

- Develop jointly with the training officer a KSA list for inspection procedures and related tasks, by (date).

- By (date), set individual learning plans for each fire fighter, based on the KSA list and an analysis of each fire fighter's learning needs.

- Review all prefire plans and update as needed within three days after the property has been inspected.

- By (date), add simulations based on prefire plans to at least half of all training sessions.

The chief might have the following objectives:

- Obtain funds for electronic equipment to enter inspection data in the field, by (date).
- By (date), prepare a plan for making all necessary and desirable inspections, including clear definitions of "necessary" and "desirable" inspections.
- Ensure that all district commanders have Fire Prevention Week plans ready and approved by June 1.

LEGAL FOUNDATIONS FOR ACTIVITIES TO ENFORCE FIRE CODE REGULATIONS

The importance placed on fire prevention activities in the United States is indicated by the wide range of powers given most fire marshals in rights of entry for fire inspection and investigation, fire marshal's hearings, rights of subpoena of any records or persons who might have information concerning fire ignition sequence, and other provisions. These powers have been upheld by most courts of law, and they are likely to be honored when members of the fire service use them in the performance of their duties.

Individual Rights Versus Public Protection

Ordinances and regulations confer the authority for inspection of premises on fire department officials, subject to certain safeguards. The safeguards protect the rights of the individual against unreasonable search and seizure as guaranteed by the U.S. Constitution. When a conflict arises between the rights of the individual and the fire department's police (inspection) powers, the issue is subject to rulings by the courts. Two sample cases—*Camara v Municipal Court of the City and County of San Francisco* (387 U.S. 523 [1967]) and *See v City of Seattle* (387 U.S. 541 [1967])—involved the right of individuals to refuse admission to the fire department to inspect commercial occupancies, as provided for in fire code safety regulations. In both cases, the courts maintained that a person cannot be prosecuted for resisting inspection unless a search warrant has been issued by a legally appointed judicial officer.

In the Camara case, the U.S. Supreme Court ruled in favor of the appellant; the right of entry without a warrant involved a housing code inspector. In the See case (which involved a warehouse), the fire department wanted to inspect under the authority of a City of Seattle ordinance granting the fire chief the right "to enter all buildings and premises except the interior of dwellings as may often be necessary." The owner refused to permit an inspection on the grounds that the ordinance was invalid. He said the fire chief had neither a search warrant nor any probable cause to believe that a violation of any law existed on the premises. In upholding the constitutionality of the ordinance, the Washington Supreme Court (later overruled by the U.S. Supreme Court) said:

> The purpose of the fire code inspection is to correct conditions hazardous to life and property. The problem of keeping cities and their inhabitants free from explosions and fires is a serious task facing all fire departments. It is obvious that routine inspections are necessary to ensure the safeguarding of life and property. The need to conduct routine inspections of commercial premises, in regard to which probable cause for the issuance of a warrant could not ordinarily be established, outweighs the interest in privacy with respect to such premises. The purpose of the inspection contemplated by the code is not unreasonable. (*See v. City of Seattle*, 387 U.S. 541, 87 S. Ct. 1737 [1967])

The U.S. Supreme Court reversed the conviction of the warehouse owner on the ground that the Seattle ordinance, which authorized a warrantless inspection of his warehouse, was an unconstitutional violation of his rights under the Fourth and Fourteenth Amendments. The court cited its decision in *Camara v San Francisco* and declared that "the businessman, like the occupant of a residence, has the constitutional right to go about his business free from unreasonable entries upon his private commercial property. "The court felt that its decisions restricting administrative agencies in their attempts to subpoena corporate books and records supported the court's view that any agency's particular demand for access should be measured in terms of probable cause to issue a warrant against a flexible standard of reasonableness that takes into account the public need for effective enforcement of the particular regulation involved. "But the decision to enter and inspect will not be the product of unreviewed discretion of the enforcement officer in the field."

The court concluded:

> We therefore conclude that administrative entry, without consent, upon
> the portions of commercial premises which are not open to the public,
> may only be compelled through prosecution or physical force within the
> framework of a warrant procedure. We do not in any way imply that busi-
> ness premises may not reasonably be inspected in many more situations
> than private homes, nor do we question such accepted regulatory tech-
> niques as licensing programs which require inspections prior to operating
> a business or marketing a product. Any constitutional challenge to such
> programs can only be resolved, as many have been in the past, on a case-
> by-case basis under the general Fourth Amendment standard of reason-
> ableness. We hold only that the basic component of a reasonable search
> under the Fourth Amendment—that it not be enforced without a suitable
> warrant procedure—is applicable in this context, as in others, to business
> as well as to residential premises. Therefore, appellant may not be prose-
> cuted for exercising his constitutional right to insist that the fire inspector
> obtain a warrant authorizing entry upon appellant's locked warehouse.
> (*Camera v. Municipal Court of the City and County of San Francisco*, 387 U.S.
> 523, 87 S. Ct. 1727 [1967])

Although in theory the courts might ultimately decide that certain inspections constitute an invasion of privacy, in practice a search warrant is
often not needed because property owners are aware that the fire department can usually obtain one. When a fire department does seek a warrant,
the department must show that probable cause exists, based on the length
of time since the last inspection, the nature of the occupancy, and the condition of the entire area (but not necessarily specific knowledge of a
violation).

Fire Code Retroactivity in Life Safety Considerations

The courts have attempted to maintain a balance between an individual's
right to privacy and the need to protect against potential fire hazards. However, they have consistently recognized the need to protect life by upholding
the right of the fire department to enforce regulations retroactively when life
safety is involved. When life safety is involved, a local government sometimes adds items to the fire safety code regulations, such as the installation
of sprinkler systems in nursing homes and high-rise apartment buildings.
Owners then have to install sprinkler systems in all buildings in both occupancy categories, whether the structure is already in existence, is in the
process of being built, or is planned for the future.

Role of the Company Officer in Legal Matters

Company officers must know their legal responsibilities as specified in the fire codes and interpreted by the courts. In the performance of duty, the company officer must keep these guidelines in mind:

- One- and two-family units usually can be inspected only at the request of the owner or occupant. An officer should know, however, whether local regulations require or permit an inspection. If a citizen complains about another person's violation of the safety ordinances, the department may have the responsibility to investigate and inspect the premises.
- When company fire fighters conduct inspections, the officer must be satisfied that they have received adequate training, including the basics of legal foundations for fire prevention activities.
- Officers and inspectors must be aware that they have to inform an owner, or person in charge, of noncompliance with fire codes before a hazard can be considered a violation.
- A reasonable period of time must be allowed for an owner to correct code violations. Although unreasonableness in itself is not illegal, an officer must be aware that it may invalidate a department's police power to enforce compliance. Judges tend to dismiss cases brought by the department for failure to comply with ordinances unless a reasonable amount of time for correction has been allowed. Furthermore, harsh interpretation of codes and ordinances could eventually lead to poor public relations.
- An inspector is legally responsible for recording all violations; nothing can be disregarded intentionally.
- An officer or inspector cannot recommend a specific company or individual as the contractor to correct deficiencies; only the procedures to be carried out can be suggested.

CONCLUDING REMARKS

This chapter, like the preceding one, has provided evidence of the value of using fire service function guidelines—the ones suggested here or modifications. They provide focus for the work in the function, emphasizing what is important. In fire prevention, guidelines, such as those derived from the 3Cs, are exceptionally useful in decision making. A comprehensive listing of all guidelines is given in Appendix A, Decision Guidelines.

Every fire fighter and officer recognizes the importance of fire prevention, and all are aware that a reduction in fire emergencies can be traced to
the increased emphasis on prevention activities. Still, fire prevention is not
the most exciting work. It can be boring or unpleasant, at least for those
fire fighters and officers who are uncomfortable when confronted by the
public or asked questions during inspections. The 3C guidelines can be of
help.

1. The control guideline helps the fire officer plan the inspection program
   so that even the work of a fairly demanding district can be completed as
   expected. Appropriate participation and effective coordination with
   other functions such as prefire planning and training can substantially
   enhance a company's and a department's productivity.
2. Competence enhancement in prevention is of great importance, not only
   to achieve the highest possible productivity without stress or strain but
   also to provide fire fighters with satisfaction in their work. The fewest
   disturbing moments occur for those who are most knowledgeable.
3. The officer who considers the climate guideline with all decisions ensures that the positives dominate the negativs in fire prevention.

An officer who develops the habit of thinking of the 3Cs or similar
guidelines with every decision will probably also review the functional
guidelines when working in a specific function. There are far more functional guidelines, so it is difficult to remember them all. For that reason, Appendix A is a source for a quick review of those that are applicable to the
specific activity of the moment.

CHAPTER 4
STUDY QUESTIONS AND ACTIVITIES

If you are working alone, prepare your own written responses to these questions. If you are studying with a team or working in class, discuss the questions with the group and write a consensus answer.

1. Describe the purpose of fire prevention and loss reduction activities.
   Then list the activities that are part of fire prevention, and explain what
   role each plays in preventing fires.
2. How do fire prevention activities satisfy the goals and guidelines of fire
   prevention?

3. How is your department organized for fire prevention? What could be done to make this function more effective?

4. What is the role of each organization in fire prevention and in formulating standards and codes?
   a. National Fire Academy (NFA)/Emergency Management Institute (EMI)
   b. U.S. Fire Administration
   c. National Board of Fire Underwriters
   d. National Fire Protection Association
   e. American Insurance Association
   f. Underwriters Laboratories
   g. Factory Mutual Research Corporation

5. Explain fire prevention regulations and the organizations that promulgate them. Discuss what a fire prevention officer can do to bring about changes when a local regulation is not in the best interest of the community.

6. What are the major responsibilities of each of these officials in fire prevention?
   a. State/provincial fire marshal
   b. Chief of fire prevention or local fire marshal
   c. Fire inspector or fire prevention officer
   d. Fire protection engineer

7. How can a fire department influence architects and contractors to be conscious of fire safety in their work?

8. Why is it important that all fire department personnel participate in fire prevention activities as well as in fire suppression operations?

9. Assume that you have been asked to discuss the inspection process during a local fire department training seminar. Prepare an outline of your presentation.

10. If you are a new company officer and the members of your company have a negative attitude toward fire inspections, how can you eliminate some of the negative feelings?

11. List the tasks and concerns that are important to effective fire prevention. Write a summary of the tasks and concerns, including the items listed here. (If you are working in a group, divide the items on the list equally among the group members.)
   a. The importance of each task and concern in the overall program of fire prevention

b. The duties that must be performed in each task

c. The role of fire department members in each task

CHAPTER 4 ....

REFERENCES

Cote, A. E., ed. 1997. *Fire Protection Handbook*, 18th ed. Quincy, MA: National Fire Protection Association.

Custer, R. L. P., and Mechamp, B. J. 1997. *Introduction to Performance-Based Fire Safety*, National Fire Protection Association, Quincy, MA, p.vi.

Fisher, Roger, and Ury, William. 1981. *Getting to Yes.* Boston: Houghton Mifflin.

National Commission on Fire Prevention and Control. 1973. *America Burning, The Report of the National Commission on Fire Prevention and Control.* Washington, DC: U.S. Government Printing Office.

National Fire Protection Association. 1994. E.D.I.T.H. (Exit Drills in the Home). Quincy, MA: National Fire Protection Association.

National Fire Protection Association. 1994. *NFPA Inspection Manual*, 7th ed. Quincy, MA: National Fire Protection Association.

National Fire Protection Association. *Risk Watch*™. Quincy, MA: National Fire Protection Association.

National Fire Protection Association. *Learn Not to Burn Curriculum*, 3rd ed. Quincy, MA: National Fire Protection Association.

NFPA 1, *Fire Prevention Code*, National Fire Protection Association, Quincy, MA.

NFPA 101®, Life Safety Code®, National Fire Protection Association, Quincy, MA.

NFPA 901, *Standard Classifications for Incident Reporting and Fire Protection Data*, National Fire Protection Association, Quincy, MA.

NFPA 921, *Guide for Fire and Explosion Investigations*, National Fire Protection Association, Quincy, MA.

NFPA 1031, *Standard for Professional Qualifications for Fire Inspector and Plan Examiner*, National Fire Protection Association, Quincy, MA.

NFPA 1033, *Standard for Professional Qualifications for Fire Investigator*, National Fire Protection Association, Quincy, MA

Scott, David. 1997. "The New Inspector: Everywhere at Once." *NFPA Journal* (January/February): 84.

Tryon G. H., ed. 1969. *Fire Protection Handbook*, 13th ed. Quincy, MA: National Fire Protection Association.

Ury, William. 1991. *Getting Past No.* New York: Bantam Books.

CHAPTER 4
ADDITIONAL READINGS

Bass, Bernard M., and Vaughan, James A. 1966. *Training in Industry; The Management of Learning.* Belmont, CA: Wadsworth.

Berne, Eric. 1967. *Games People Play—The Psychology of Human Relationships.* New York: Grove Press.

Bramson, Robert M. 1981. *Coping with Difficult People.* Garden City, NY: Anchor Press.

Fast, Julius. 1970. *Body Language.* New York: M. Evans and Company.

Gordon, Myron. 1981. *Making Meetings More Productive.* New York: Sterling Publishing.

Gray, John. 1992. *Men Are from Mars, Women Are from Venus: A Practical Guide for Improving Communications and Getting What You Want in Your Relationship.* New York: Harper Collins.

Harris, Thomas A. 1976. *I'm OK, You're OK—A Practical Guide to Transactional Analysis.* New York: Avon.

Knowles, Malcolm. 1990. *The Adult Learner, A Neglected Species,* 4th ed. Houston: Gulf.

Lefrancois, Guy R. 1995. *Theories of Human Learning,* 3rd ed. Pacific Grove, CA: Brooks/Cole Publishing.

Luft, Joseph. 1970. *Group Processes: An Introduction to Group Dynamics.* Mountain View, CA: Mayfield.

Mager, Robert F. 1968. *Developing Attitude Toward Learning.* Palo Alto, CA: Fearon.

Rausch, Erwin. 1978. *Balancing Needs of People and Organizations—The Linking Elements Concept.* Washington, DC: Bureau of National Affairs (Cranford, NJ: Didactic Systems, 1985).

Rausch, Erwin, and Rausch, George. 1971. *Leading Groups to Better Decisions; A Business Game.* Cranford, NJ: Didactic Systems.

Rausch, Erwin, and Wohlking, Wallace. 1969. *Handling Conflict in Management I, II and III—Business Games.* Cranford, NJ: Didactic Systems.

Senge, Peter M. 1990. *The Fifth Discipline—The Art and Practice of the Learning Organization.* New York: Currency and Doubleday.

Skinner, B. F. 1968. *The Technology of Teaching.* New York: Appleton Century Crofts.

Tannen, Deborah. 1990. *You Just Don't Understand.* New York: Ballantine Books.

MANAGEMENT OF POTENTIALLY DAMAGING CONFLICT

In my controversial moments my perception's rather fine,
I can see both points of view, the one that's wrong, and mine.
                                        —Anonymous

Preventing and resolving potentially damaging conflicts can be beneficial in ways beyond the issues involved. Organizations and individuals who are able to prevent or settle conflicts also gain increased trust and confidence as well as loyalty and respect.

Not all conflict is damaging, of course. Healthy competition is conflict. Up to the limit that an individual can tolerate it effectively, competition can be beneficial. The presence of conflict beyond competition, however, whether it is based on conflict of interest, conflict of goals, dissatisfaction (often not recognized as conflict), or personality clashes, affects control, competence, and climate. Control benefits from the best possible cooperation, enhanced by the absence of conflict. Competence increases because everyone learns, from the resolution of conflicts, about the many different perspectives on the issues that are in dispute. Climate, of course, is much better when no conflicts remain simmering or unresolved. A personal goal for fire officers is to ensure that conflicts do not find fertile ground to grow in their organization.

Unresolved disagreement about what to do is likely to be damaging. At the least it takes time and attention away from more important things. At the worst, it creates an escalating spiral of frustration, mistrust, serious discord, refusal to cooperate, punitive actions, and retribution. To manage discord, this spiral must be broken or prevented—the earlier the better. Of course, an officer cannot prevent or resolve all damaging conflict. The toughest types of disputes to resolve are usually based on complex issues, including irreconcilable conflict of interest between the parties and psychological factors such as pride, feelings of security, self-esteem, and hate. When even compromise is not in the cards, conflict sometimes cannot be resolved. It must be played out to the end, where one side wins and the other loses. The major obstacle often lies in the commitment of at least one party to "winning" on his or her own terms. That type of conflict (except when it is part of a game or sport) requires early intervention because it can set the escalating spiral in motion.

Conflict Manager.   For fire officers, the most productive strategy for managing conflict is to ensure that every fire fighter and officer knows how to be a "conflict manager" and is willing to accept that role whenever a conflict threatens or arises. To implement such a strategy will require initial meetings to explain the concept and refresher meetings to develop habits.

Anyone and everyone can be a manager of conflict; you do not need to be in a position of authority. You need to understand conflict and you can then just appoint yourself. A dispute can involve as many managers of conflict as there are people involved in it. If everyone in the organizational unit takes this role in every conflict, all will have their eyes on winning the war, not winning sectional battles. They will be aware that the enemy is not the antagonist in a specific disagreement. The enemy is the stress, trouble, strained relationships, unsatisfactory climate, and losses that follow in the wake of poorly resolved conflicts.

In this sense, a war is completely over only if or when a dispute is resolved in such a way that both sides "win." Ideally that happens when all parties are as satisfied as they would have been if they had "won" the way they originally wanted to win. Often it doesn't take a truly creative solution to end a war. A satisfactory compromise usually is adequate.

Officers and fire fighters can appoint themselves as managers of conflicts in which they are involved. They need not, and probably should not, tell anyone that they have assumed that role. Every conflict can use at least one manager on each side. The more conflict managers, the better. Those involved in the dispute who understand conflict well enough to have appointed themselves conflict managers will welcome the tacit decisions of others to quietly join them as additional conflict managers.

Managers of conflict provide leadership to help achieve a rational, unemotional search for the "best" outcome, one that satisfies the perceptions of both parties that it is the best solution, in their respective short-term and long-term interests. The emphasis in the search is on perceptions because people and groups often do not know or may even misunderstand their own best interests. Resolving conflict must expose the reality, so that each party can gain a realistic perspective of its best interest.

Keeping an eye on both the issues on the table and the process of conflict resolution is not necessarily easy to do, especially in the heat of battle. Short sections from two poems and a limerick are so much to the point here that they deserve quoting:

If you can keep your head when all about you
Are losing theirs and blaming it on you,
If you can trust yourself, when all men doubt you,
But make allowance for their doubting too;

—Rudyard Kipling, If—

In the world's broad field of battle,
In the bivouac of Life,
Be not like dumb, driven cattle!
Be a hero in the strife!

—Henry W. Longfellow, A Psalm of Life

As you ramble on through Life, brother,

Whatever be your Goal,

Keep your Eye upon the Donut

And not upon the Hole

—Cover, Mayfair, New York, Coffee Shops Menu

Resolving Conflict.   To resolve disagreements, conflict managers help the parties proceed through several steps.

1. Prevent or reduce undesirable emotions.
2. Identify the central issues in the conflict.
3. Identify possible solutions.
4. Choose the best solution, with consideration for the views of the parties and others who are seriously affected.
5. Implement the best solution, including communicating it to those who are affected by but not parties to the dispute.
6. Monitor how well the outcome has helped to end the dispute completely.

Whether the conflict is between two individuals, a group and an individual, or two groups, the steps are the same. Communications skills are central to the success of these steps.

This approach is, of course, not the only one that can be used. In their best-selling book, *Getting to Yes*, Roger Fisher and William Ury (1981), of the Program on Negotiations at Harvard Law School, suggest a different, somewhat more elaborate approach. That book and a follow-up volume by William Ury (1991), entitled *Getting Past No*, present another very useful perspective.

The third step—identifying possible solutions (ways to settle the conflict)—deserves more explanation. In this step conflict managers suggest, or respond to, specific possible solutions offered by the parties. If these suggestions do not close the gap between the parties, competent conflict managers may suggest other alternatives based on one of the five groups of solutions listed next. Outcomes based on four of these solution groups are always available. The first one, the creative win-win settlement, is not always within reach, but it is the one that conflict managers should focus on throughout. One can never tell when such a solution might pop up unexpectedly.

1. *Creative solution* is a win-win outcome that leaves both parties satisfied, or almost as satisfied as if they had "won" the way they wanted to win.
2. *Compromise* gives each party somewhat less than it had hoped to get but enough so it considers the outcome fair and reasonable.
3. *Postponement of the discussion*, depending on the situation, sometimes brings a temporary win for one party if time is in its favor. Frequently, however, postponement is neutral and brings cooler heads and solutions that are better all around.

4. *Concession by one party* is usually a loss for that party in the short term. It can turn into a win-win over the long run, however, especially if the goodwill it generates brings reciprocal actions that increase mutual respect and support.

5. *Use of authority* may feel like a win at least for the short run, whether the party using it is in a position of authority or not. Over the long run, the benefits of using authority are not clear; that's why it should be used cautiously.

In some instances, a party to a conflict or a conflict manager has to use position authority or authority as a parent. In the case of parents, authority often works to the long-run benefit of both parties. The same can happen in a work environment. Using authority is also the best thing to do in the rare instances when one side wants the responsibility for the outcome to rest solely with the other side.

When either party to a conflict is not in a position of authority over the other, authority can still be used in a conflict. One side may refuse to listen or agree to a course of action, rightfully or wrongfully withhold an object or service wanted by the other party, demand a postponement of the discussion when the postponement benefits one side, or just refuse to cooperate with further attempts to resolve the problem. The party that uses authority like this can "win," for the moment at least, but the temporary advantage can also trigger retaliation from the adversary.

If fire officers have the authority to do so, they may want to consider bringing in a third party when a dispute is not resolving or is festering. Exhibit 4.2 lists the roles a third party can play in resolving conflict.

An officer should ensure that all the members of the department understand how to be managers of conflict and create a climate in which they want to apply this skill.

Exhibit 4.2 Third-Party Roles in Resolving Conflict

The role of a third party in resolving conflict may be assumed by a volunteer neutral person or group, a person with higher-level responsibilities for all or some of the issues, or a professional mediator or arbitrator.

- **Mediators** are impartial persons, or panels, without authority to impose a settlement. The function of mediators is to help the parties reach a mutually satisfactory outcome of the dispute. Sometimes mediators are given the authority to recommend a resolution, as is the case with fact-finding panels that are sometimes appointed by municipal, state, and even federal authorities, primarily in labor disputes.

- **Arbitrators,** by contrast, specify a settlement, like judges, or set specific limitations. Arbitrators may sit alone or as panels. Their authority derives from a contract, from other agreement between the parties, from policies of the organization to which the parties belong, or from governmental regulation. Arbitration eliminates the need to take a dispute to court. A special case of arbitration is one in which an officer makes the decision on an issue on which other department members at lower levels disagree.

ADDITIONAL INSIGHT 4.1

Then most conflicts will have a happy ending. Conflict managers are wise to also consider issues that might bring new conflicts in the wake of resolved ones:

- The long range impact of the solution (Bitter experiences have shown, time and again, that short-term gain may lead to long-term loss, and vice versa.)
- The effect of the settlement on preventing future conflict
- The outcome's likely impact on individuals and groups who are not parties to the dispute

ADDITIONAL INSIGHT 4.1

Fire and Life Safety Education

INTRODUCTION

>   Contemporary fire departments provide a rich variety of public services: emergency medical services, plans review, and inspection services, among others. In many career and volunteer fire departments, this list of services includes fire and life safety education. (Cote 1997, p. 2-3)

Fire and life safety education, as pointed out in Chapter 5, is part of a fire department's fire prevention effort. It is designed to (1) increase awareness of human actions that can lead to fires, burns, scalding, and fire-related injuries or fatalities, and (2) develop skills and knowledge to prevent fires or to minimize injuries from fire and heat when a fire does occur. Safety education should reach children as well as adults to change their attitudes and behaviors about the prevention of fire. It encompasses a wide spectrum of programs and activities directed toward such diverse audiences as preschoolers, schoolchildren, senior citizens, homeowners, apartment dwellers, physically challenged people, hospital and nursing home staffs, and church, service, and civic organizations.

SCENARIO

RECOGNITION FOR FIRE AND LIFE SAFETY EDUCATION EFFORTS

Friday, the last day of the Fire Prevention Week (FPW) program, was a festive affair at the largest of the three grade schools in town. Pictures and stories about fire and life safety, prepared by the children, were posted in the hallways and classrooms. Three of the fire department's six-member volunteer fire and life safety education team were in the school auditorium with others who had supported FPW. The volunteers had spent many hours on the program, most on their own time.

The mayor was at the school along with several members of the town council, the principals of the three schools, the superintendent of schools, and, of course, Pat Lynn, the fire chief. Mrs. Tatre, the school superintendent, opened the brief program by introducing the dignitaries. She gave an overview of the FPW program in all schools. At the close of her talk, she stressed how much the schools appreciate the help they receive from the fire department, not only with educational programs and materials for the children but also with fire safety reviews for the teachers and the maintenance staff. Then she turned the program over to the mayor.

The mayor pointed "with pride" to the safety record of the schools: no burn or scalding accidents in more than three years and only a minor one in the preceding three years. He gave credit to the fire department for its efforts and the educational materials it made available. He expressed his gratitude for Chief Lynn's devotion to fire prevention and the fire and life safety education program and asked him to say a few words.

After thanking the mayor, the council members, and the superintendent, Chief Lynn spoke: "We are delighted that there haven't been any burn injuries in the schools, and we want to be sure that the children know how much they have to thank their teachers for being careful and also themselves for remembering the rules." The chief continued, "Now, I can't step down without letting you meet the man who is in charge of this successful fire and life safety education program, Captain Drake. Come up here, Captain, and let them see you."

Captain Drake thanked everyone. He said he wanted to thank his team for putting together a program that taught everyone.

One of the parents asked whether she could ask a question. "Certainly," the captain replied.

"Speaking for myself, I want to first state that I am delighted with the fire department's efforts at fire prevention. I understand that you are planning a program for some of the children for next month. Can you tell us a little about that?"

"We are constantly planning new programs, so I really don't know which one you are referring to. Can you tell me a little more about the specific program? Where will it be held?" answered the captain, a little flustered.

"I am talking about the program for this school."

"Oh, that program. I am not too familiar with it, but one member of my staff, who has been working on it can tell you about it. Let me ask the staff members to stand up. They are Inspector Jackson and fire fighters Clark and Morales. Please give them a hand."

As the team members stood up, the captain introduced them individually and asked Inspector Jackson to answer the question. Jackson described a new injury prevention program called *Risk Watch*™ that would be implemented on a pilot basis after a teacher in-service program.

When Jackson finished, the school superintendent asked for questions or comments. The remaining questions related to other school matters. Finally, the dignitaries and fire department personnel left.

Back at headquarters, Chief Lynn asked the captain to his office. I'd like to talk to you about two things. "It seems to me that we are doing something worthwhile with this education program. The schools and the parents seem to have appreciated this year's program even more than those in past years. Did you notice that, too?"

Captain Drake thought for a moment. "Yes, I had the same impression. Maybe that's because Jackson and the five fire fighters dug in harder. They really seemed to enjoy it this year. And the students scored much better on the posttest!"

"That may be it. But, you know, the way that the people reacted, I think we ought to consider doing more of this public education. We really haven't done much except for the grade schools and Fire Prevention Week. We haven't had many fires, but let's take a look at what we did have, and see whether that tells us something about education opportunities. Why don't you get the staff together, look at the records, and come up with a few ideas on what we could and possibly should do. Now, to the other matter. That was quite embarrassing for both of us, when you had to call Jackson to explain next month's program at the school. Shouldn't you have known about that?"

"Yeah. You're right, but you know how busy I've been. Still, I've got to tell the guys to keep me in the loop, even when things are hectic. And before another public program, I'll make sure I get a briefing on all that's going on. That should prevent something like this from happening again."

Chief Lynn didn't look happy, but he accepted the explanation and promise with just a brief: "Well, try to keep abreast on a continuing basis, if you can."

"OK, Chief." With that Captain Jackson left.

When the captain spoke with the team, they did not seem as enthusiastic as they had been in the recent past. They felt that it wasn't easy to come up with something new for each grade. The captain wasn't sure, but he had the feeling the team members were not happy that the chief and he, too, had not given them recognition by name and voluntarily. After all, they had neighbors and friends in the audience.

The team thought that the existing program was adequate because it covered, in addition to grade school pupils, their teachers, school maintenance people, nursing home residents, and hospital and nursing home maintenance staff. They also expressed concern that any expansion of the programs would require still more of their personal time. Nevertheless, they did come up with three suggestions. They stressed, however, that they could not commit themselves to implementing the ideas.

- Send a mailing to local clubs and associations, offering to speak at their meetings and distribute fire and life safety materials.
- Have an open house at the two fire stations, with tours for children and adults, brief talks on fire and life safety at home, and maybe a demonstration or film.
- Recruit volunteer speakers to augment the fire department team with presentations, articles for the local paper, and other steps so that more organizations will request presentations on fire and life safety.

Scenario Analysis: Fire Service Function Perspective

As in earlier chapters, this analysis is based on the goal and reasonable guidelines for the chapter's topic, which is fire and life safety education. Before reading the analysis, you may want to give some thought to the strengths and the weaknesses of the way the fire department discharged its responsibility for fire and life safety education.

The fire and life safety education goals and guidelines are relevant:

> The primary goals of fire and life safety education are to provide comprehensive fire and injury prevention programs designed to eliminate or mitigate situations that endanger lives, health, property, and the environment, and to motivate changes in behavior by members of the public, through various programs and delivery methods, using all mobilizable community and department resources.

The guideline follows:

> What else needs to be done to increase public awareness of fire and injury prevention issues and to motivate more widespread and effective fire and life safety behavior?

The following specific questions are based on this guideline:

1. What can be done to enlist more community resources to reach more people in the community effectively with fire and life safety messages?
2. What arrangements can be made with nonfire resources (including the print, radio, and TV media) to get more fire and life safety messages to more people?
3. What can be done to motivate those who have been exposed to fire and life safety messages to act on them?
4. What can be done to evaluate to what extent fire and life safety is being practiced in residences, businesses, and institutional occupancies?

We will consider these guidelines one at a time.

1. *What else can be done to enlist more community resources to reach more people in the community effectively with fire and life safety messages?* The suggestions from the team members are a good start in enlisting community resources. Suggestions from others involved in fire and life safety—classroom teachers, school administrators, health educators, and community organizers—would also be helpful. These suggestions cannot be implemented, however, without the active support of the chief and the captain. They can take the next step that is most helpful in bringing significant results: developing a formal plan with goals, objectives, and action steps (tactics).

Questions that this guideline segment might trigger if the fire and life safety education team used guidelines when making decisions include the following:

- Which of our suggestions can we implement even if no additional department time is provided for the effort?
- What is a meaningful plan for those action steps that can be implemented?
- What is a meaningful plan for carrying out all the action steps if the department provides support?

2. *What arrangements can be made with nonfire resources (including the print, radio, and TV media) to get more fire and life safety messages to more people?* None of the suggestions from the team pertained to the media. Questions for the fire and life safety education team on this issue include the following:

- What newspapers and radio and TV stations cover our community, and how can we encourage them to include fire and life safety education messages in their programs or publications?
- What else can we do to obtain media coverage of our fire and life safety education events and programs?

3. *What else can be done to motivate those who have been exposed to fire and life safety messages to act on them?* Motivating people to take action on the basis of an educational experience is always a difficult task. Reminders and reinforcing experiences are the most effective. The fire and life safety education team might ask these questions on this guideline segment:

- How can we present the fire and life safety messages so that they will be most motivational?
- What can we do to provide reminders?
- What additional fire and life safety education activities can we combine with inspections?
- Can we make up a mailing list of people who have attended our programs and send them short reminders on smoke detectors, escape routes, stop-drop-and-roll, and other safety issues?

4. *What else can be done to evaluate to what extent fire and life safety is being practiced in residences, businesses, and institutional occupancies?* The questions for the fire and life safety education team could include the following:

- What techniques can we use during a presentation to determine how well the audience has learned fire and life safety behaviors, information, and attitudes?
- What techniques can we use to measure the short- and long-term impact of fire and life safety education?
- Could/should we distribute a questionnaire at inspections and fires asking whether the people involved have attended fire and life safety education programs and what is of interest in such programs?

Scenario Analysis: Management/Leadership Perspective

Before reading the analysis, you may want to give some thought to the strengths and weaknesses of the way Chief Lynn and Captain Drake handled the situation from the management/leadership perspective. In addition to the fire-relevant service functional guidelines, officer decisions involve all three management/leadership considerations: control, command, and climate (the 3Cs). In the scenario analysis in this chapter, we will concentrate on the climate guideline.

The Climate Guideline. A more detailed version of the climate guideline asks:

> What else needs to be done so that the various groups and individuals who have to implement the decision or plan and those who will be affected by it (all the stakeholders) will be in favor of it or at least have as positive a view as possible, so there will be a favorable climate?

The specific questions for this guideline follow:

1. Are appropriate psychological and tangible rewards offered and provided effectively and efficiently to bring about the highest possible level of satisfaction from the creation and use of the product/service?*
2. Are policies in place to help reduce work-related stress?

We will review the scenario from the perspective of the specific questions of the guideline, which lists specific issues guideline segments, taking them one at a time.

1. *Are appropriate psychological and tangible rewards offered and provided effectively and efficiently to bring about the highest possible level of satisfaction from the creation and use of the product/service?* The scenario revealed many opportunities to provide greater job satisfaction to the staff members and presumably also to others. Obviously, Captain Jackson missed an opportunity to provide recognition to the members of the team. Expressing appreciation for the efforts of department members is one major management/leadership

---

* *Product/service* is a generic term that applies in every industry and government activity. The fire service has many services and even a few products. Beyond fire suppression, fire prevention activities, EMS, and the nontraditional services, a fire department provides a psychological service in reassuring citizens that there is an agency they can call for many serious and even minor emergencies. The products of a fire department are mostly in the fire and life safety education area; some are related to EMS, and some are in the nontraditional services, such as some meals and lodging-related materials in major disaster situations.

behavior that brings higher levels of satisfaction. The others are open communications, mutual respect, and participation in decisions.

The practices in the scenario's department seemed to encourage participation in decisions. The scenario provided little evidence of appreciation for effort and effective, two-way, open communications. [See Additional Insight 5.1, Enhancing Work Satisfaction (Providing Recognition) and Performance Evaluation.]

The chief and the captain might have asked themselves the following questions:

- What opportunities do I have at school ceremonies or other public and private events to show how much the efforts of the fire safety education team members are appreciated by the public and the department's leadership?
- What should I do about department or team communications so that the officers and the team members stay abreast of plans?
- To what extent do I keep fire fighters informed about developments pertaining to fire and life safety education that affect the department?
- What can I do to ensure that the team will find greater satisfaction from devoting effort to fire and life safety education?

2. *Are policies in place to help reduce work-related stress?* The scenario did not describe any work-related stress for the fire department staff, except possibly stress that resulted from delivering programs on their own time or at the expense of other fire department work. We can assume that there were no unusual stresses beyond those that are "normal" for fire department work. Still, if the captain were aware of this aspect of the guideline, he might have asked himself an additional question:

- What support should I provide to ensure that the team members are aware of my willingness to pitch in when needed so they feel more secure that help will be available if the fire and life safety education function adds excessively to other deadline pressures?

The Control and Competence Guidelines. From the control perspective, in his discussion with Captain Drake, the chief could have focused more on what the captain would do to keep abreast of fire and life safety education activities and not treat them as secondary responsibilities. If the chief were aware of the control guideline, he might have asked himself these questions:

- What else should I do to find out whether Captain Drake gives adequate priority to fire and life safety education?

- What help, if any, should I provide Captain Drake so that other responsibilities do not suffer if he devotes more attention to fire and life safety education?

The captain's failure to reward the fire fighters appropriately for their efforts appears to have made it more difficult for him to obtain their full cooperation with expanding the program. Thus, the team has less motivation to set challenging goals and to devote serious effort to achieving them. If the captain were aware of the control guideline, he might have asked himself questions such as these:

- What goals, if any, should I ask for?
- What support should I provide so that the team members understand my interest in fire and life safety education and know that I give it high priority? (See Additional Insight 3.1, Goal Setting and Implementation.)

If the chief were aware of the competence guideline, he might have asked himself what developmental actions would help Captain Drake show more appreciation for the efforts of his staff and other members of the department. The situation at the school event should have reminded the chief to question the extent to which all officers in the department have the desirable level of competence for providing such rewards.

This analysis shows that thinking along guidelines can improve not only individual decisions but also entire functions. Officers who develop the habits of considering the 3Cs guidelines with every decision and of reviewing the fire service function guidelines from time to time can significantly enhance their personal competence. At the same time, they can make suggestions for department policies and procedures to improve its productivity along all the dimensions that count—all aspects of fire prevention and suppression, including fire and life safety education.

OVERVIEW OF FIRE AND LIFE SAFETY EDUCATION

Historically people have become conscious of the need for fire safety whenever a spectacular fire grabbed the headlines and made them aware of the cost of fire in terms of both human suffering and dollars. For a long time, however, the general public was exposed to fire-related safety messages only during Fire Prevention Week, which was established by President Warren G. Harding in 1922. (See Figure 5.1 for a listing of the important milestones in fire and life safety education.)

1909	NFPA's Franklin Wentworth begins sending fire prevention bulletins to correspondents in 70 cities, with the hope that local newspapers will publish the bulletins as news articles.
1911	Fire Marshals Association of North America proposes the October 9 anniversary of the Great Chicago Fire as a day to observe fire prevention.
1912	NFPA publishes *Syllabus for Public Instruction in Fire Prevention*—fire safety topics for teachers to use in the classroom.
1916	NFPA and the National Safety Council establish a Committee on Fire and Accident Prevention. Communities nationwide organize Fire Prevention Day activities.
1920	President Woodrow Wilson signs first presidential proclamation for Fire Prevention Day.
1922	President Warren G. Harding signs first Fire Prevention Week proclamation.
1923	23 states have legislation requiring fire safety education in schools.
1927	NFPA begins sponsoring national Fire Prevention Contest.
1942	New York University publishes *Fire Prevention Education*.
1946	U.S. government publishes *Curriculum Guide for Fire Safety*.
1947	Hartford Insurance Group begins the Junior Fire Marshal Program, perhaps the first nationally distributed fire safety program for children.
1948	American Mutual Insurance Alliance publishes first edition of *Tested Activities for Fire Prevention Committees*, based on Fire Prevention Contest entries.
1950	In October, 7,000 newspapers receive the ad, "Don't Gamble with Fire—The Odds Are Against You," developed by the Advertising Council and NFPA.
1954	Ted Royal of the Advertising Council creates Sparky® the Fire Dog.
1965	*Fire Journal* begins a regular column on "Reaching the Public."
1966	"Wingspread Conference" highlights the need for public education.
1973	• The National Commission on Fire Prevention and Control publishes its report, *America Burning*. • The Fire Department Instructors' Conference offers its first presentation on fire and life safety education, delivered by Cathy Lohr of North Carolina.
1974	• NFPA and the Public Service Council release the first television *Learn Not to Burn*® public service announcements, starring the actor Dick Van Dyke. • *The Fire Prevention and Control Act* establishes the National Fire Prevention and Control Administration.
1975	The National Fire Prevention and Control Administration holds its first national fire safety education conference.
1977	• NFPA 1031, *Standard for Professional Qualifications for Fire Inspector, Fire Investigator, and Fire Prevention Education Officer*, is published. • National Fire Prevention and Control Administration releases *Public Fire Education Planning: A Five-Step Process*. • National Fire Prevention and Control Administration launches national smoke detector campaign.
1979	• J.C. Robertson's *Introduction to Fire Prevention* is published by Glencoe Press. • The *Learn Not to Burn Curriculum* is published by NFPA. • International Fire Service Training Association (IFSTA) releases IFSTA 606, *Public Fire Education*.
1981	NFPA establishes its Education Section.
1985	• The National Education Association recommends the *Learn Not to Burn Curriculum*. • NFPA publishes *Firesafety Educator's Handbook*
1986	Learn Not to Burn Foundation incorporated.
1987	• The first edition of NFPA 1035, *Standard for Professional Qualifications for Public Fire Educator*, encourages civilians to become public fire educators in the fire department. • TriData Corporation publishes *Overcoming Barriers to Public Fire Education*.
1990	Oklahoma State University publishes the first issue of the *Public Fire Education Digest*.
1994	TriData Corporation releases *Proving Public Fire Education Works*.

*This timeline was originally prepared for IFSTA's *Fire and Life Safety Educator*, based on information from Pam Powell's "Firesafety Education: It's Older Than You Think. (*Fire Journal*, May 1986) and information provided by Nancy Trench, Fire Service Training, Oklahoma State University.

Figure 5.1 The timeline lists the milestones in the development of fire and life safety education. (*Source*: Cote, *Fire Protection Handbook*, 1997, p. 2–5)

The publication of the report of the President's Commission for Fire Prevention and Control in 1973 increased fire departments' interest in fire and life safety education. "In the commission's poll of those who live daily with destructive fires—fire service personnel—98 percent of those who replied agreed that there is (was) a need for greater education of the public in fire and life safety" (NCFPC 1973, p. 105). The members of the commission felt that public education could have an impact in two areas. First, education could change people's behavior toward a safer direction. Second, education could train people to spot faulty equipment and unsafe acts.

This twofold approach, it was felt, could improve the general level of fire and life safety in America. As one writer noted, "a significant factor contributing to the cause and spread of fire is human failure—failure to recognize hazards and take adequate preventive measures, failure to act intelligently at the outbreak of fire, failure to take action which would limit damage" (Cote 1986, p. 3-3).

Although direct action by government is limited to requirements such as mandating smoke detectors, educational efforts can lead indirectly to more appropriate fire and life safety behavior. That is how fire departments acquired the role to educate their "customers."

By the 1980s, the need for what was known as "public fire education" or "public fire safety education" was well established. An increasing number of fire departments began to provide fire safety education in their communities, but progress was slow. Gus Welter of the National Volunteer Fire Council summed up the problem by asking, "Why is there always time to put out fires, but not to teach fire prevention?" (Shaenman et al. 1987, 1).

In 1981 the National Fire Protection Association established an Education Section to help meet the need for fire prevention education. The association's *Learn Not to Burn*® programs were developed and field tested in consultation with educators, fire and burn prevention experts, and curriculum specialists. The program was recommended by the National Education Association in 1985 and has since been used successfully by many schools and fire departments. The program consists of a structured series of fire and life safety lessons that includes a curriculum for students in kindergarten through grade 8, resource material for teachers of preschool students, and *Learn Not to Burn Resource Books* for children in kindergarten through grade 3.

Beginning in the 1980s and with more intensity in the 1990s, "public fire safety education" began to expand into what became known as "fire and life safety education." In addition to traditional fire and burn safety, the field grew to include a variety of injury prevention messages, including topics such as electrical safety, pedestrian safety, water safety, and poison prevention.

The following factors contributed to the emergence of fire and life safety education from the foundation of public fire education:

- The publication of the influential book *Injury in America* by the National Research Council in 1985
- The growing awareness of injury, including injury to children, as a widespread public health problem in the United States
- The fire service role as first responders to a wide variety of nonfire emergencies, thus easing the way for fire and life safety educators to serve as the education component of fire department first responder programs

The National Fire Protection Association also contributed to the growth of comprehensive injury prevention education with the release of the *Risk Watch™* curriculum in 1998. With funding from Lowe's Home Safety Council and the assistance of a broad-based technical advisory group, NFPA developed and pilot tested *Risk Watch™*. The curriculum presents injury prevention lessons for children in preschool through grade 8 in such areas as motor vehicle safety; fire and burn prevention; choking, suffocation, and strangulation prevention; poisoning prevention; fall prevention; firearms injury prevention; bike and pedestrian safety; and water safety. Fire departments were among the first to adopt *Risk Watch™*, often using it in conjunction with *Learn Not to Burn* materials or other existing fire safety curricula.

ORGANIZING FOR FIRE AND LIFE SAFETY EDUCATION PROGRAMS

Successful fire and life safety education begins with good decisions by the managers of fire and life safety programs—decisions that identify the major problems and target audiences, design a program best able to address those problems with that audience, and deliver the right program to the right people. [Cote 1986, p. 3-5]

Although many states offer programs for fire safety education, they are not likely to be used without active leadership from local fire departments. The responsibility for organizing the fire department for fire and life safety education programs lies with the chief. The chief obtains funding and other resources for the program.* Usually the responsibility for planning and delivering the services lies with officers and teams selected by the chief. NFPA Standard 1035, *Standard on Professional Qualifications for Public Fire and Life Safety Educator,* lists the qualifications for three levels of educators (see Exhibit 5.1).

---

*Both public and private funding sources may be approached to help finance expanded programs. Local service organizations are the sources to consider first. Public libraries have books and periodicals on this topic, which is beyond the scope of this book.

Exhibit 5.1 Levels of Fire and Life Safety Educators

- **Level I** is for those who coordinate and deliver existing programs.
- **Level II** is for individuals who prepare educational programs and information.
- **Level III** is for educators who create, administer, and evaluate educational programs and information.

Source: NFPA 1035.

In many departments, the responsibility for fire and life safety education rests with the officer in charge of fire prevention. Personnel are usually assigned from the fire prevention division or the community relations division. In some smaller departments, the function is assigned as an additional duty for shift personnel or to a part-time specialist who is hired to develop and deliver these educational programs. Sometimes specialists are public educators who work on an as-needed basis in their spare time. Some excellent existing programs started out as part-time and grew into full-time operations.

Developing or enhancing an effective fire and life safety education program in a community involves the same steps that are taken for any other program:

1. Identifying (new or changed) needs
2. Setting goals and objectives
3. Developing and implementing a comprehensive program
4. Monitoring, evaluating, and correcting the program

Identifying Needs

Before a fire department develops a new or revised program to address local fire and life safety education needs, it should review local, regional, and national fire and burn injury data so that it can target the limited resources available for fire and life safety education at those audiences that will benefit most. The department should ask these questions:

- What types of fires occurred and how frequently, locally and in broader areas?
- What were the causes?
- Are there specific patterns for the community that can be identified?

Answers to these and other questions can be found in fire department records as well as records from hospitals, insurance agencies, and state agencies. The U.S. Fire Administration, which maintains the National Fire Incident Reporting System (NFIRS), and the National Fire Protection Association are excellent sources of supplementary information. For procedural questions, the *Fire Safety Educator's Handbook* is a useful reference (Adams 1983, p. 180).

Fire and life safety education programs should focus on generating public interest and cooperation in the areas of need. Each program component should be designed to be "functional" (to reduce potential fire risks) as well as appealing and motivational to target audiences. Field research in the community can provide information on the learning capabilities, attitudes and behaviors, accessibility, and current knowledge of people in the target audiences. The needs assessment should also include research on the programs of other fire departments for ideas about how to develop new programs or adapt existing ones to meet target audiences' needs.

Many well-intentioned programs are not successful because the message and the medium are not appropriate for the target audiences. For example, children in higher grades may not pay attention to messages they perceive as repeating programs presented in earlier grades. Effective programs develop messages that address the target audience with appropriate tone and language and with attention-getting techniques (see Figure 5.2).

Figure 5.2 A fire officer teaches a lesson on fire and life safety to grade-school children.
(*Source*: Courtesy of Harry Carter, Newark, New Jersey, Fire Department)

Setting Goals and Objectives

Fire safety education is one area in which setting goals and objectives can bring significant benefits. When something is important yet not urgent, it is not likely to receive the attention it deserves until a crisis occurs. Setting goals and objectives can prevent such crises by making sure that appropriate action is taken before conditions precipitate a crisis. The action plans that support goals and the timeposts they require see to it that the necessary steps are taken on the matter for which goals and objectives were set.

Exhibit 5.2 lists sample goals and objectives a fire department might set for its fire and life safety education program. So that those who will be involved in implementing the program segments have the necessary competencies, objectives should include the KSAs (knowledge, skills, and abilities) for the various elements that will be delivered (presentations in the various levels of schools, equipment and fire demonstrations, presentations to adult and senior groups, etc.). Some staff members, especially those members of the fire department who lead the program, should take a course in fire and life safety education at the National Fire Academy or at a state or local school where such courses are taught by professionals.

Exhibit 5.2 Sample Goals and Objectives for a Fire and Life Safety Education Program

- To provide coordination with schools, and any requested support, so that at least one fire and life safety lesson is given during October/November to all children enrolled in local preschool and day care centers and a second one during April/May
- To provide coordination with schools, and any requested support, so that four fire and life safety lessons are given to all fifth-graders in local public schools beginning in September and ending in June
- To plan and implement the annual fire and life safety assembly programs for grades K–4 in each of the local public elementary schools during January, February, and March*
- To ensure that fire and life safety messages for children are included in ongoing programs conducted by the Parks and Recreation Department and the public libraries
- To identify levels of relevant KSAs for all department members and community volunteers who will be involved in delivering fire and life safety education presentations and to develop individual development plans that will assure full competence (as defined by the plans), no later than January (year)
- To conduct three lessons in fire prevention and fire survival (emergency fire response) for all employees, on all three shifts, at the four local nursing homes within the next month

*Note that developing programs for children that age can be difficult without specific early childhood education experience. It is therefore advisable to plan such programs in conjunction with specially trained and experienced educators in that field or to use materials from established programs such as *Learn Not to Burn*™.

Setting goals and objectives requires appropriate participation by both those who will be involved and those who will be affected. (See Additional Insight 1.1, Participation in Decision Making and Planning, and Additional Insight 3.1, Goal Setting and Implementation.) In addition to the department members who are on the team, other individuals might be invited to help with planning and delivering the program for the community. The others may include individuals who can contribute ideas and possibly also services, such as educators, community leaders, representatives of the medical community, fire investigators, social workers, and representatives from the media, the building industry, and civic associations. If the group exceeds six or seven people, then smaller subgroups or committees may be charged with recommending goals and objectives for various aspects of the program. After the group has set goals and objectives, members who will not have an active role in implementation, either as members of the department or as volunteers, can sometimes be invited to serve in an advisory capacity.

In addition to program goals and objectives, specific programs have learning objectives. The National Fire Protection Association's *Learn Not to Burn Curriculum* and *Risk Watch*™ have learning objectives, and so do state programs. For example, the New Jersey Division of Fire Safety's *N.J. Fire Safety Skills Curriculum*, which is modeled on a program developed in Oregon, sets goals and objectives for kindergarten and grades 3, 5, 8, and 11. Kindergarten is included, though not for all goals and objectives, because it has been determined that students at this age can begin to understand the dangers of uncontrolled fire, how to escape from it, and how to treat simple injuries caused by fire. The learning goals and objectives of the New Jersey program include those listed in Exhibit 5.3.

Exhibit 5.3 Learning Objectives for Students in the New Jersey Fire Safety Skills Program

- Recognizing the basic components and hazards of heat, smoke, gases, and flame
- Recognizing the importance of early detection, quick reporting, and rapid suppression of fire
- Demonstrating ability to escape life-threatening fire environments
- Demonstrating ability to survive clothing fires
- Demonstrating basic first aid skills for minor burn injuries
- Knowing how to practice home and outdoor fire prevention
- Recognizing the causes of arson and its impact on the community
- Recognizing the role of the fire department in the community

Source: New Jersey Dept. of Community Affairs, undated.

Developing and Implementing a Comprehensive Program

Materials. When developing fire and life safety education programs, a fire department may use materials, including posters, flyers, and other handouts, published by the NFPA and other organizations with fire protection interests (see Figure 5.3). Many of these materials are for children in schools; others are for adults and can be left with local businesses and in public places. When inspectors check private dwellings, they may leave flyers with information about specific hazards. School systems often distribute fire prevention flyers and posters to students to take home to their parents.

Many organizations distribute fire safety materials, including the U.S. Fire Administration, the International Association of Fire Chiefs, some state divisions of fire and life safety, local fire prevention associations, the Red Cross, and such private organizations as the Shriners, the Association of Home Appliance Manufacturers, the Gas Appliance Manufacturers Associa-

Figure 5.3 An NFPA poster emphasizes public participation in fire prevention.

tion, Allstate Insurance Company, First Alert, BIC, and Energizer Batteries. Some county fire marshals' offices and possibly other state agencies provide materials for programs directed at juvenile fire-setters and other young people who have a special fascination with fire.

Pilot Programs. While a fire department is developing a program, pilot tests are a useful way to validate the elements and teaching aids. The results of pilot tests indicate the effectiveness, accuracy, and appropriateness of the program before a department commits large amounts of time and resources to the project (Cote 1986, p. 3-7). There is no prescribed length for pilot tests. They can range from a few deliveries to a small number of groups to a town-wide program with several months of presentations and alterations to fine-tune a large-scale county or state program.

Preliminary materials should be used in pilot tests whenever possible, so that they can be revised easily after the results of the pilot have been compiled. Audience reaction must be monitored carefully to ensure that the material is readable, understandable, appealing, and relevant to their actual and perceived needs. Pilot programs also provide opportunities for instructors to practice their delivery skills and enhance their knowledge of the subject. When everything is ready, the pilot program can be delivered to the target audiences on a wider scale. It can also serve as a foundation for programs directed at specific segments of a target audience.

A comprehensive fire and life safety education program should reach all members of the community, sometimes more than once, by targeting school-children who will bring the message home and by delivering the same message to targeted adult groups. Often a target audience already has a communication network of some kind. There are places where an audience meets or publications from which it already obtains information (Cote 1986, p. 2-12). Tying into any such identifiable, existing networks can help a program achieve practical impact.

Schoolchildren. School programs for children and young people, like the Oregon program mentioned earlier, usually reach grade-school students several times as they progress toward graduation. School programs can be built around NFPA's *Learn Not to Burn* program. Alternatively, the fire and life safety skills curricula and supporting materials that many states offer to local school systems and fire departments may be used (see Figure 5.4). In addition, to support in-class programs taught by the regular teachers and sometimes augmented and assisted by fire department personnel, fire departments often offer programs such as fire drills, in which fire fighters or cartoon characters block exits, and "Smoke Houses," sometimes called "Fire

Figure 5.4 The captain from the Toronto Fire Department receives a certificate from children who had learned about fire and life safety from him. (*Source*: *Toronto Sun*, Toronto, Ontario, Canada)

Safety Houses" (trailers resembling houses, possibly with more than one low-ceiling floors, where hazards are simulated and students can practice window emergency exits to escape simulated smoke conditions).

To make the best use of educational materials, some fire department fire and life safety education teams arrange with local school administrators or curriculum coordinators to schedule in-service training for teachers. Fire department personnel or specially trained educators provide the instruction.

Instructors for in-service programs or other fire and life safety programs do not have to be certified teachers, but they should have demonstrated competence in teaching the target audience. Nothing is more important to the success of educational efforts than competent and dedicated instructors. Two ways to ensure effectiveness are to provide training for the instructors and to give them the necessary support to implement quality programs.

Staff members can get initial training from the National Fire Academy in Emmitsburg, Maryland, which offers residential courses in public fire education. The academy also offers public fire and life safety education training through field courses. Printed materials, such as the *Fire Safety Educator's Handbook,* are available from NFPA to assist with the training of instructors (Adams 1983). The International Fire Service Training Association (IFSTA) publishes a manual, *Public Fire and Life Safety Educator*, that is used by many fire departments for instructor self-development (IFSTA 1997).

In addition to programs aimed at students and in-service programs for teachers, instruction should also reach school administrators and maintenance personnel. Many fire departments conduct such programs in coordination with school authorities. The programs cover housekeeping to avoid the accumulation of combustible material, smoke and heat detection and alarm systems, monitoring of suppression systems, compartmentalization, deadend corridors, and other topics of local significance. Instructors often review fire drill schedules and procedures with the administrators, either in conjunction with these programs or as part of the fire inspections.

Children in Alternative Education. A relatively new development, children taught at home by their parents, requires a different approach for reaching the children and their families. Some fire departments offer and advertise special sessions for these families at the fire station. Others encourage parents to bring the children to school when fire and life safety education events or classes are scheduled.

Boy Scouts, Girl Scouts, and Other Activity Groups. Programs for scout troops and other activity groups for school-age children are similar to those offered in the grade schools. They may also be variations of fire station open house programs or simply opportunities for children to see, touch, and feel fire apparatus and equipment. Apparatus and equipment demonstrations, though valuable public relations and "attention getters," are certainly no substitute for teaching fire and life safety behaviors, however.

Adults. Because the adults in a community are never all available at the same time, a comprehensive program must try to reach them in as many ways as the program staff and budget allow. Special messages may be given to specific groups, such as parents of small children and parents of adolescents considered to be possible fire-setter risks.

Adults can be exposed to fire and life safety messages in a variety of ways. During Fire Prevention Week, which is observed in a large number of communities, the fire department can stage events such as parades, fire station open houses, demonstrations of fire suppression, equipment and apparatus demonstrations, and window decorating contests with holiday or seasonal themes (see Figure 5.5).

Fire department staff can distribute fire and life safety literature to adults during building inspections, at club and association meetings, and at special events such as store opening celebrations and fairs. They can also reach adults in church groups, parent/teacher organizations, fraternal associations, or other citizens' groups. Some adults may be exposed to programs presented at local colleges and in adult and continuing education settings.

Before groups agree to host a fire and life safety education program—whether a single program or a series of lectures or training sessions—many

Figure 5.5 Fire departments can reach many segments of the community by sponsoring window display contests. (*Source*: Courtesy of Roselle, New Jersey, Fire Department)

require that a proposal be given to their decision-making committee or person. The proposal should make its value stand out over other proposals that are competing for the group's limited time or resources. Sample handouts should be included with the proposal. If the program consists of a series of training sessions, the proposal should include the following detailed sections:

1. Statement of needs
2. Behavioral objectives
3. Format description, including the number and length of lessons, the topics to be covered, and audiovisual materials to be used
4. Estimates of income from requests for funding, to corporate or private sponsors, for the costs of developing the program
5. Description of evaluation tools

Another way to reach adults is through special interest groups that spring up spontaneously or that are identified by the program goals to provide channels of communication between various segments of the community and the fire department. The information needs of special interest groups are similar to those of the general public, but different enough to require individual programs. Groups form around educational, industrial, institutional, residential, high-rise dwellers, civic, service, professional, and commercial interests. Fire department personnel can reach members of these groups through their own newsletters, through public service columns in print media and on radio or television, or through speakers bureaus.

Building inspections, whether required or requested, are an opportunity to reach adults with messages about fire and other life safety hazards in the home, plant, or office. When carried out in a competent and professional manner, inspections can make the public receptive to safety information. Although, single-family home inspections may not seem to be a very efficient use of department personnel time, they can include fire and life safety messages and handouts and encourage homeowners to spread the word about the benefits of greater fire and life safety awareness. Other homeowners may then call the local fire department and ask to have their homes inspected, or they may request literature on fire and life safety. A number of years ago, Howard Boyd, the former fire marshal of Metropolitan Nashville and Davidson County, Tennessee, stressed that fire prevention bureau education programs need to concentrate on people who live in one- and two-family dwellings, the location of most deaths due to fire.

Older Adults. Older adults in private residences are no different from other adults, except that they have more discretionary time and therefore may be more easily motivated to obtain and read literature and act on the suggestions it contains. Many older adults reside in multiapartment senior housing facilities and in Continuing Care Retirement Communities (CCRCs), which provide assisted living services. The administrators of these facilities offer many social and educational programs to their residents, and most welcome sessions on fire and life safety education. NFPA has developed a new fire safety education program for the senior citizen community. This resource should be explored for its local applicability.

Residents of Hospitals, Nursing Homes, and Other Group Homes. In most states, hospitals, nursing homes, and group homes are inspected several times each year. These inspections are opportunities to deliver fire and life safety messages to the management, staff (with emphasis on engineering and maintenance), and even those residents who can participate in identifying hazards, practicing fire-safe behavior, and evacuating in emergencies.

Motels and Hotels. Most of the people in these facilities are transient and do not attend any functions that might be appropriate for delivering fire and life safety education messages. There are many places where literature may be left, however—on the counters and in the rooms. Many facilities are required to display "In case of fire emergency" instructions on the inside of room doors. Some managers might be willing to augment these notices with "Fire Safety Suggestions for the Home" that remind the occupant of important hazards and safety procedures when they return to their permanent residences, such as smoke alarms and carbon monoxide detectors. Even the required messages have fallout benefits in reminding travelers of hazards and emergency evacuation procedures that might lead them to review the safety in their homes. Here, too, as in nursing homes and other health care institutions, staff training in conjunction with inspections can help increase fire and life safety hazard awareness and lead to safer practices.

The Media. The media can play two distinct roles in support of local fire safety education. They can actually deliver public fire safety education messages, and they can provide publicity to keep the fire department's mission in the public eye. Assistance in developing media relations programs can be found in NFPA's *Fire Protection Handbook* (Cote 1997, Section 2-6).

The media reaches audiences of all ages and in all situations with articles by reporters and editors that provide publicity and also with public education messages that they might publish as a public service or for minimal fees. Local daily and weekly publications and local radio and TV stations provide many opportunities to spread fire and life safety messages widely. Some local "shopper" tabloids may even be willing to distribute free flyers with the paper as a public service. For more information, refer to the NFPA's *Fire Protection Handbook* (Cote 1997, Section 2-6). Every article and every news item that offers further information brings inquiries and requests for the literature that is offered. All these publicity programs help the public understand what is important in preventing fires and fire-related injuries.

To be most effective, publicity should adhere to the following guidelines:

- Information should be disseminated in a steady and continual flow over a period of time. (Isolated messages are not as likely to be remembered.)
- Publicity should be geared to events and seasons by emphasizing the current seasonal hazards. A fire and life safety message can be offered to the newspaper or broadcaster as an addition to announcements of Fire Prevention Week, delivery of new apparatus, plans to build a new fire station, and so on.
- Sometimes news items can be used to pave the way for more specific educational information. For example, an article that describes how the fire department brought a specific house fire under control with minimal damage and no loss of life might mention that the occupants had recently installed a smoke detector. A follow-up article on smoke detection

equipment and carbon monoxide alarms reinforces the message that working alarms save lives and give further motivating detail.

Timing is important for programs as well as for media publicity. A fire and life safety educational message is best received when it refers to a prominent recent fire event and when it is appropriate for the target audience. The message should not, for example, be delivered to a senior citizen audience on the same evening as the annual banquet, and a lecture on the chemistry of fire is far less appropriate for a first-grade class than is a discussion of behaviors from the NFPA's *Learn Not to Burn Curriculum.*

A fire department can gain the attention of the public through the media more easily if the department's personnel are organized to deal with the media. Many departments have specially trained public information officers. In the absence of such officers, the Fire Prevention Bureau usually handles publicity. The chief or a designated officer should work with editors of local papers and the managers of local radio/TV stations to arrange comprehensive public education programs that avoid the limitations of single, uncoordinated articles. A member of the department can then consult directly with the newspaper or radio/TV station staff in the development of specific segments of the program. This type of planning can help bring higher priority at times when the media might not normally consider items on fire prevention to be of foremost importance.

Television or radio interviews and prepared articles for the newspapers (especially during Fire Prevention Week in October) draw attention to particular hazards and escape techniques, explain critical elements of the fire codes, and solicit compliance. Some departments enlist volunteers to write general-interest articles on specific aspects of fire and life safety. An article could discuss how to deal with grease fires in the kitchen, how to plan escape routes, how to conduct exit drills, stop-drop-and-roll, Operation E.D.I.T.H. (Exit Drills in the Home), how to protect children against accidents causing burns or scalding, and what to do in case of a fire.

Every large fire presents an opportunity to publicize the importance of fire prevention, especially when it involves loss of life, serious injury, or large property loss. At such times, both the press and the public are more receptive to ways to reduce the risk of fire. People are then more likely to take steps to protect themselves. Another publicity opportunity comes when something of importance or of special interest to the public occurs that is related to fire and life safety and that must be communicated immediately via the media. In addition to a large fire that is in progress and of concern to the public, such an issue could be a particular potential fire problem (such as hazardous toys or garments) or the need for special care in forests during a prolonged dry spell.

Monitoring, Evaluating, and Correcting the Program

Monitoring has two aspects: (1) checking on how well a program adheres to its goals and objectives and the delivery timetable and (2) determining a program's effectiveness. For guidance in implementing a monitoring system, refer to Section 2-7 of the NFPA's *Fire Protection Handbook* (Cote 1997).

Tracking adherence is quite easy. All it takes is the maintenance of accurate records on publications, program deliveries, dates, and attendance. More complicated is the monitoring of effectiveness. Documentation is also an essential element of this aspect of monitoring. To ensure that a program fully serves its intended purpose, its results should be compared with its objectives. In large communities where one objective may be to lower the number of various types of fire incidents, a department can evaluate the effectiveness of the program by seeing whether the number of such incidents has decreased since the program started. Information compiled when local fire problems were identified can be used as baseline data to evaluate the effectiveness of the fire and life safety education program. In small communities where very few fires occur, a department may have to take a program's effectiveness on faith. Monitors may compare the number of programs conducted, the people actually attending, and the personnel participating with what was planned. The measure of effectiveness is based on this comparison. Information from pre- or posttesting or from any surveys or questionnaires completed by the target audience can also be helpful (Cote 1986, p. 3-8).

Not every program results in a lower incidence of fires, but documenting the benefits that a program attains will ensure continued acceptance and support for fire and life safety programs. If a fire and life safety education program can demonstrate that it brought about a change in behavior, then the probability is good that fires will be reduced and lives saved in the future. That is the goal. The NFPA's *Learn Not to Burn Curriculum* is an excellent example of this. From 1975 to 1988, NFPA records show that more than 250 lives have been saved by the use of the key fire safety behaviors taught in this curriculum. "The critical success factors incorporated in NFPA's *Learn Not to Burn* Champion Award Program are reflected in other national initiatives, as well. A strong coalition effort between the United States Fire Administration and the American Red Cross in 1994 used market research and demographic analysis to develop media materials . . . in both Spanish and English" (Cote 1997, p. 2-7). The coalition campaign succeeded in delivering a much-needed fire safety message to Spanish-speaking communities.

The purpose of evaluation is to provide the information a department needs for the changes that will make the program more effective. That is true of all aspects of fire and life safety education. Sometimes evaluations of past programs do not, by themselves, provide adequate and sufficiently detailed information for effective corrections. That's when the cycle starts again from the beginning, with revised programs and new pilots and then evaluation of the pilots to identify the critical elements that will make the revised program more effective than the one it replaces.

CONCLUDING REMARKS

The fire and life safety education function offers many opportunities for fire fighters and officers to perform interesting and beneficial tasks, such as speaking, visiting schools, or writing, all of which can have a positive impact on the community. At the same time, these activities are rewarding to those fire fighters who are involved.

Reviewing the fire service functional guidelines can occasionally help officers find interesting challenges and opportunities and useful ideas for meeting them. Such a review would, of course, be in addition to considering the relevant fire service functional guidelines as well as the 3Cs guidelines whenever a decision has to be made. The guidelines raise practical questions.

- What resources (department member time, apparatus, special equipment, other resources) should be devoted to the fire and life safety education effort?
- What kinds of plans should be prepared to contact groups for which fire and life safety presentations can be made or other educational events or demonstrations be scheduled?
- What schedules should be developed for presentations or other events?
- How can more fire fighters become interested and involved in the fire and life safety education effort?
- What training do fire fighters and others need for the type of involvement they would like and for which they can be scheduled?
- What nontangible rewards can be planned for department members who devote effort (including personal time) to fire and life safety education?

These questions are not an exhaustive list, and they are not necessarily the most appropriate ones for a given department. However, the questions do emphasize the way the functional guidelines and the 3Cs or similar guidelines interact for greater effectiveness of the function.

All of the preceding questions, in one way or another, relate to the functional guidelines because plans, schedules, human resources, training, and even rewards are likely to lead to effective educational programs and public involvement. More specifically, with respect to the 3Cs guidelines, the first three questions concern control, the fifth involves competence, and the fourth and sixth address climate.

CHAPTER 5
STUDY QUESTIONS AND ACTIVITIES

If you are working alone, prepare your own written responses to these questions. If you are studying with a team or working in class, discuss the questions with the group and write a consensus answer.

1. Discuss the main purposes of fire and life safety education.
2. Develop a series of goals and objectives for implementing a community fire and life safety program by (a) a small local fire department and (b) a large metropolitan fire department.
3. Who are the major audiences for fire and life safety education messages in a community?
4. Think about the program in effect in your department and list additional steps that would be beneficial, in two separate groups: (a) those that would require significant time and money resources and (b) those that could be implemented without any resources or with insignificant expenditures of time and money.
5. What are the different ways in which the adults in a community can be reached with fire and life safety messages?
6. Why might a program proposal be required before a fire and life safety education session can be scheduled? What should the proposal cover?
7. Discuss the benefits of pilot testing a new or revised community fire and life safety education program.
8. How can the effectiveness of fire and life safety education programs be evaluated?
9. What are the different methods a fire department can use for public education in fire prevention?
10. What are the specific responsibilities of a company officer with respect to fire and life safety education, considering the objectives and guidelines.
11. In what ways can a company officer apply the 3Cs guidelines to fire and life safety education and thereby improve the company's performance?

CHAPTER 5

REFERENCES

Adams, R. C., ed. 1983. *Fire Safety Educator's Handbook: A Comprehensive Guide to Planning, Designing, and Implementing Fire Safety Programs*. Quincy, MA: National Fire Protection Association.

Cote, A. E., ed. 1986. *Fire Protection Handbook*, 16th ed. Quincy, MA : National Fire Protection Association.

Cote, Arthur E., ed. 1997. *Fire Protection Handbook,* 18th ed. Quincy, MA : National Fire Protection Association.

Deming, W. Edwards. 1986. *Out of the Crises*. Cambridge, MA: MIT Center for Advanced Engineering Study.

Didactic Systems. 1977. *Providing Recognition: A Handbook of Ideas*. Cranford, NJ: Didactic Systems.

Didactic Systems. 1996. *ASP—The Achievement Stimulating Process, A Handbook for Managers/Leaders*. Cranford, NJ: Didactic Systems.

International Fire Service Training Association. 1997. *Public Fire and Life Safety Educator*, 2nd ed. Stillwater, OK: International Fire Service Training Association.

Meyer, Herbert H., Kay, Emanuel, and French, John R. P. Jr. 1965. "Split Roles in Performance Appraisal." *Harvard Business Review* (January/February).

National Commission on Fire Prevention and Control. 1973. *America Burning: The Report of the National Commission on Fire Prevention and Control*. Washington, DC: U.S. Government Printing Office.

National Fire Protection Association. *Learn Not to Burn Curriculum*, 3rd ed. 1987. Quincy, MA: National Fire Protection Association.

National Fire Protection Association. 1987. *Learn Not to Burn Resource Books*. Quincy, MA: National Fire Protection Association.

National Research Council. 1985. *Injury in America*.

National Fire Protection Association. *Risk Watch™*. Quincy, MA: National Fire Protection Association.

National Fire Protection Association. 1987. *The Learn Not to Burn Preschool Program*. Quincy, MA: National Fire Protection Association.

Nelson, Bob. 1994. *1001 Ways to Reward Employees*. New York: Workman Publishing.

New Jersey Dept. of Community Affairs, Division of Fire Safety. 1992. *New Jersey Fire Safety Skills: A Program of Fire Safety Instruction for Students in Grades K Through 11*. Trenton.

NFPA 1035, *Standard on Professional Qualifications for Public Fire and Life Safety Educator*, National Fire Protection Association, Quincy, MA.

Rausch, Erwin. 1985. *Win-Win Performance Management/Appraisal*. New York: John Wiley & Sons.

Schaenman, Phillip, et al. 1987. *Overcoming Barriers to Public Education*. Arlington, VA: TriData Corporation.

CHAPTER 5

ADDITIONAL READING

Rausch, Erwin, and Washbush, John B. 1998. *High Quality Management: Practical Guidelines to Becoming a More Effective Manager*. Milwaukee: ASQ Quality Press.

ENHANCING WORK SATISFACTION (PROVIDING RECOGNITION) AND PERFORMANCE EVALUATION

For lack of a fully meaningful term, we will use the word *recognition*, a popular but inappropriate word because it implies an audience and hints at an outstanding accomplishment. No single word conveys the expression of appreciation for all types of contributions, including the nonspectacular but consistent ones that are so important to staff member satisfaction.

Competent actions of appreciation for the contributions of company and department members provide frequent feedback on performance and thus create satisfying moments for the staff member. Positive messages balance, at least partially, the many negatives that occur in the workplace: the dissatisfying moments that come from monotonous or disliked tasks, errors, new restrictions, tight budgets, and conflicts.

Managers must take actions on many fronts to improve the work climate or maintain a highly satisfactory climate (see the climate guideline earlier in this chapter). Communications are crucial, keeping people abreast with what is happening, especially on matters that affect them.(See Additional Insight 2.1, Communications Techniques and Skills.) Managers should also try to reduce work-related stress to comfortable levels when it threatens to go well above them. They can show confidence by encouraging appropriate participation in decisions (see Additional Insight 1.1, Participation in Decision Making and Planning), providing honest feedback on performance, showing appreciation for contributions, and providing tangible rewards (compensation). A major result of positive actions is a high level of trust by department personnel in their officers and in the department.

This section will concentrate on two critical components of a high-quality climate: (1) competent performance evaluation, which can bring many opportunities for timely, honest feedback and (2) better-stated recognition and appreciation for individual contributions, even if they are not outstanding.

PERFORMANCE EVALUATION

Benefits

A performance evaluation provides many opportunities for expressing appreciation for a staff member's contribution to the company, bureau, or department. If an evaluation is conducted competently, its aftereffects can significantly enhance work satisfaction. An evaluation leads to these improvements:

1. Improved performance and higher quality of work life
2. Better control, though not necessarily tighter control, over the company's or department's activities
3. More competent fire fighters and officers because performance evaluation provides a sound basis for their development
4. More appropriate personnel actions
5. Compliance with regulations pertaining to equal employment opportunity

We will discuss each benefit in turn.

1. *Improved Performance and Higher Quality of Work Life.* Performance evaluations can improve the quality of work life by providing these advantages to department members:

- More recognition for their accomplishments
- More information about how their performance is seen by those to whom they report
- Specific suggestions for improving their performance
- Guidance on how they can reach greater competence in their current jobs and to further their careers
- Greater knowledge of the criteria by which they are evaluated

Performance evaluations result in improved performance for many of the same reasons and because they ensure that people understand clearly what needs to be done and how it is to their benefit to strive for superior performance. Performance evaluations also provide thorough documentation of performance deficiencies when needed to aid in developmental purposes and to prevent continued performance problems.

2. *Better Control.* Many managers and leaders, including officers in fire departments, equate tighter controls with better control. Still, they are aware that their ability to keep tabs on everything that happens is sharply limited by the situation and the available time. Managers who try to achieve tight personal control often fail to see that close supervision is resented, especially by the more capable people, and does very little to encourage the development of greater competence by others.

A soundly implemented performance evaluation achieves *better* control by relying on a system of performance reviews that provides regular feedback. It gives staff members freedom while clearly identifying limits. Experienced and competent staff members are given wider limits than those who are new or less competent. (See Additional Insight 1.1, Participation in Decision Making and Planning.)

By establishing and revising limits for appropriate behavior and by setting review periods, officers can maintain adequate practical control. At the same time they provide recognition and positive feedback, they can exercise discipline when necessary.

ADDITIONAL INSIGHT 5.1

Better control, if understood as a joint activity and appropriately implemented, leads to higher levels of competence and performance and greater work satisfaction for staff members.

3. *More Competent Fire Fighters and Officers and a Sound Basis for Competence Development.* A performance evaluation system provides the greatest benefits if it is clear that its primary emphasis is on competence development. A major step for managers should therefore be to develop a competence enhancement plan for each person whose performance is being evaluated. This plan should reinforce steps considered during performance and career counseling activities. The improvement plan may also include additional coaching. Performance evaluation thus serves to encourage department members to enhance their skills for their current positions and to prepare themselves for other desired positions.

4. *More Appropriate Personnel Actions.* Effective performance evaluation can be relied on to provide valid information about the performance and capabilities of individuals. It thus establishes a more solid foundation for personnel actions on compensation and promotion than exists when evaluations are not done or are not done competently. Competently administered performance evaluations also reduce the number of fire fighter complaints by uncovering dissatisfactions that can easily be corrected before they lead to grievances. At the same time, open communications are strengthened.

5. *Compliance with Regulations.* Many federal and state laws and regulations affect personnel actions. They are intended to ensure equal employment opportunities, prevent discrimination, and protect the employee against harsh and arbitrary personnel actions. Thorough performance evaluation practically ensures compliance with laws and regulations. Discrimination is nearly impossible if personnel actions are based primarily on factual performance data collected over time.

Obstacles

Despite the compelling case in favor of thorough performance evaluation, some experts recommend their complete abolishment (Deming 1986). They propose that some significant obstacles to competent and fair administration frequently lead to problems that more than offset the benefits (Meyer, Kay, and French 1965; Rausch 1985). By taking these obstacles into account, however, fire officers can minimize their effect.

Few organizations have well designed performance evaluation systems that satisfy all the requirements listed in the next section. Many impose counterproductive, complex procedures and documentation. They fail to rely primarily on relatively lean procedures and the thorough development of management/leadership skills for implementing fair, meaningful performance assessments. Fire officers can compensate for these by staying focused on competence development and recognition opportunities.

When ratings are necessary, it is difficult to avoid differences in ratings between different companies. Company officers can minimize this problem by communicating with the other company officers and their staff.

The evaluation skills required of officers are very complex and difficult to establish and maintain on an organizationwide basis. Officers must cope with the touchy situations and difficult decisions inherent in serious, compassionate, but tough-minded implementation of performance evaluation. To hold this problem to a minimum, fire officers can work on enhancing their own competence and use their influence to bring about departmentwide development programs.

It is easy to be generous when trying to be as fair as possible. However, generosity toward one staff member may be perceived as unfairness. It is difficult to maintain balance. Officers can, however, achieve a fine batting average if they devote thought and effort to ensure that their personnel actions are fully justified.

Because results are so important, it would, theoretically, be most desirable to base performance evaluations almost solely on results—on achievements of any goals or objectives and on the quality and quantity of effort devoted to them. However, rating solely on the basis of achieving goals and objectives may be unfair because staff members might be evaluated on matters that are not fully under their control. A sound evaluation system uses goals and objectives as guidelines, but places emphasis on the actions a staff member takes to achieve those goals as well as other important assignments. Thus, the process evaluates the quality and quantity of the effort that staff members devote to goal achievement and to performing all other functions of the job. (Understanding and following the approach suggested in Additional Insight 3.1, Goal Setting and Implementation, will minimize this problem or avoid it entirely.)

Two ways to ensure at least reasonable fairness of a performance evaluation program involve an appeals procedure and independent reviews of ratings by a second officer. There are, however, obstacles to their use:

- Few organizations have taken the tough step of setting up formal appeals procedures that are thoroughly understood by all staff members. Fire departments may be able to do that by using human resources personnel within the department or from other agencies. (See Additional Insight 4.1, Management of Potentially Damaging Conflict.)
- It is often very difficult to find an officer who has sufficient independent, detailed knowledge of the evaluated staff member's work to provide a fair, independent review.

Requirements

If a department is considering a new or revised performance evaluation system, it might benefit by looking at the following requirements of fair performance evaluations (adapted from Rausch 1985). The system is *fair*, overall, by being thorough, accurate and factual, and meaningful, and by satisfying needs.

ADDITIONAL INSIGHT 5.1

Thoroughness

- Performance of all staff members is measured similarly.
- Evaluations measure all responsibilities.
- Evaluations measure performance for the entire period.

Accuracy

- Performance standards cover all responsibilities.
- Performance standards are accurate and factual.
- There is little or no ambiguity.
- Evaluations are comparable between evaluators.
- Evaluations are reviewed by at least one officer who has detailed independent knowledge of the evaluated staff member's work or by a similarly independent source.
- An appeals procedure exists.

Meaningfulness

- Performance standards and evaluations consider the importance of each function.
- Performance standards and evaluations attempt to measure primarily matters under the control of the staff member.
- Evaluations occur at regular intervals or at appropriate moments.
- Officers continue to improve their skills in applying the requirements for effective evaluation.
- Evaluation results are used for important personnel decisions.

Satisfaction of Needs

- Performance standards are communicated in advance.
- Staff members are kept informed about their performance.
- Communications are factual, open, and honest.
- Performance standards are challenging but realistic.
- Staff members are directly or indirectly involved in setting performance standards and in the evaluations.
- Evaluations consider the adequacy of the officer's (or higher level officer's) support.
- Evaluation emphasizes competence development.

RECOGNITION

Expressing appreciation for the contributions of staff members is the primary way for most officers, as managers and leaders, to work toward achieving the climate guideline for an effective, successful organizational unit. Two skills and abilities are required: recognizing opportunities for showing appreciation and using appropriate techniques for showing appreciation (Didactic Systems 1977 and 1996; Nelson 1994).

Rewarding the members of the company or shift for both their visible and their less obvious accomplishments is one important way to foster greater mutual confidence. Officers are encouraged to show their appreciation for what their personnel do. When officers recognize the effort and thought their personnel devote to making the team successful, the personnel will have greater confidence in their officers. The officers' work life will be enriched by the pleasure they gain from showing appreciation, providing awards and possibly other rewards, and helping their staff grow.

A word of caution is in order. Providing evidence of appreciation for staff member contributions should not be used as a device to manipulate them to work harder. Many fire fighters and officers, like staff members in other organizations, are skeptical of "recognition" programs. Although workers are inevitably pleased when they receive recognition for one of their contributions, they can easily become cynical if they suspect that the purpose of showing appreciation is to manipulate them to greater effort.

Sincere, competently directed evidence of appreciation undoubtedly leads to a more efficient team. Appreciation can be specifically directed to foster improved teamwork, better quality, greater attention to detail, better methods (smarter ways of doing the work), and widespread willingness by staff members to overlook or downplay the many unavoidable annoyances in the daily work. If fire officers are perceived as being insincere or using superficial signs of appreciation to manipulate people, then showing appreciation brings little benefit. That is why competence is so important in identifying staff member contributions and in using the many possible ways to recognize company and department members.

Identifying Opportunities for Showing Appreciation

Noticing Achievements. Anybody can notice the outstanding, unusual contribution. Officers rarely have a tough time identifying the best performers in any activity and letting those staff members know (at least once in a while) that they consider them tops in that respect. What about the second best, or even the third best? What if all staff members are good, even though not as good as the best? How likely is it that officers will notice them and show some outward signs that their contributions are appreciated?

Becoming Alert to All Types of Achievements. Officers need to identify each staff member's unique contributions. The following list gives some functions, activities, and characteristics that may be considered individual achievements. Fire officers should be able to extend the list to make it appropriate for their own people. Moreover, fire officers may tailor the list to focus on one or more activities that they want to stress—

ADDITIONAL INSIGHT 5.1

for example, quality, safety, attendance, or relations with the public.

- Planning skills
- Ability to meet deadlines
- Use of safe practices
- Quantity of work
- Accuracy and quality of work
- Suggestions to improve the way work is done
- Skill in handling stressful situations
- Skill in handling routine tasks and nonroutine assignments
- Avoidance of waste and spoilage
- Activities involved in housekeeping and maintaining appropriate work-area appearance
- Care for equipment
- Accuracy in record keeping
- Accuracy in communicating information
- Competence in answering the telephone and using other communications equipment
- Contributions to prefire planning activities
- Contributions to fire and life safety education events and activities
- Quality of hose handling
- Adherence to SOPs or SOGs
- Data recording, arranging, and analyzing
- Activities related to budget, financial, and data-recording systems
- Modifications to a system
- Preparation of reports
- EMS activities
- Relations with the public
- Cooperation and teamwork
- Assistance to others
- Verbal expressions
- Effective use of questions
- Presentation before a group
- Functioning as a role model
- Listening skills
- Contributions at meetings
- Conduct in controversies
- Written communications (memos, letters, proposals)
- Attendance and punctuality
- Reliability in achieving results
- Availability

- Achievment of commitments
- Sound decision making
- High personal standards (quality, quantity, reliability, dependability)
- Creativity, innovation, and imagination
- Incisiveness
- Acceptance of and ability to cope with pressure
- Empathy
- Assistance in preventing and resolving conflict
- Assistance in ensuring open communications

Showing Appreciation for Staff Member Contributions and Accomplishments

Showing appreciation is hard for most managers and leaders, including fire department officers, and many spend too little time doing it. However, most are good at saying "thanks," paying personal compliments, talking about family matters, and joking about sports. They can say "That's great" when they notice or when a staff member reports some accomplishment that is nice to hear, that is needed for some project, or that they sense makes the staff member proud.

Some officers show their high regard for members of their teams by taking their suggestions seriously, by asking for their opinions, by making positive comments at team meetings about *specific* actions or behaviors, and by consulting with them on decisions. Whenever possible, these officers share the information they have about what is going on inside and outside the organization.

Officers should never take credit with higher-level officers for ideas or effort of fire fighters or other officers. Competent officers who are able to blur the distinctions between the job of the officer and the jobs of the other team members create a positive climate in which each member of the team feels secure and valued. This positive work climate adds many pleasant moments to the work day for team members and for the officer.

Reasons for Inadequate Recognition

Most officers could improve the work climate by regularly showing evidence of their appreciation for the positive things the members of their team contribute. However, there are at least five reasons so little attention is paid to this important need:

1. Officers are not aware of the many benefits such attention could bring.
2. The department does not provide the necessary support.
3. Officers are concerned that they might be perceived as showing favoritism if they then fail to show appreciation for someone else's contribution.
4. Officers may think that they don't have the time.

ADDITIONAL INSIGHT 5.1

5. Officers haven't developed the skills and habits needed to show appreciation for many of the valuable contributions by team members.

Let's look at these reasons more closely.

1. Inadequate Awareness of the Benefits. The benefits of recognition can be significant in both their immediate and delayed effects. The immediate effect is a pleasant moment for at least one staff member. Pleasant moments can have "contagious" effects, spreading throughout the organization and creating a constantly strengthening positive climate. Many positive results occur when staff members know that their officer appreciates their contributions.

- Staff members are likely to feel good about their work and challenge themselves to do at least as well in the future. They want to continue to deserve high regard, and they may hardly notice the extra attention their work demands because the sign of appreciation has made the work more pleasant.
- Staff members feel more confident in coming forward with suggestions. Action on these suggestions further strengthens the positive bond between staff member and officer, while bringing a direct benefit in effectiveness, quality, safety, or customer relations.
- Staff members develop a better understanding of the officer's views and are more comfortable and secure as a result of the increased trust.
- When an officer involves other companies or possibly other fire departments by extending the signs of appreciation to the members of the other organizations, interorganizational conflicts are less likely. Reciprocating actions by officers and members of other companies or departments add further positive moments.
- Sometimes officers can involve members of the public at fire and life safety education events, during inspections, and at other times with similar results in improved relations.

Even a healthy climate may gradually improve, and mutual understanding between officers and other staff members is strengthened. Some of these benefits may not come immediately when an officer begins to show more frequent signs of appreciation. The benefits will surely pile up, however, if the officer is sincere, develops skills for recognizing the wide range of performance and personality strengths for which appreciation can be shown, makes full use of the many ways to show appreciation, and perseveres.

2. Lack of Departmental Support. Strong, competent officers can accomplish much, even without higher-level support, by making effective use of the many psychological awards that are always available. They cannot do nearly as much alone, though, as they can do with department management backing. When the entire organization is actively involved in showing appreciation, commitment and full endorsement by top management intensify the value of the recognition. Equally or possibly more important, top management might be able to supply some funds for tangible awards, which officers can use in addition to the psychological ones. Organizational support also encourages uniform actions in all segments of the department.

3. *Concern About Showing Favoritism.* This concern, though possibly valid, is not a good reason for failing to show appreciation for contributions. If an officer regularly and sincerely gives recognition, staff members are likely to give appropiate credit for the effort and not worry about perceptions of favoritism. Benefits greatly outweigh any minor dissatisfaction by one individual. As with most officer actions, it's the batting average that counts.

4. *Not Enough Time for Additional Responsibility.* Even if providing signs of appreciation were to require additional time, the benefits would make such time highly productive. Thinking that this activity needs significant chunks of time, however, is a misconception. Award meetings, if any occur, are not the important issue; chances are they take place already. If they don't and time is really tight, meetings won't be missed much if an officer effectively uses other approaches. Saying something complimentary about work hardly takes any time. Even short handwritten notes take hardly any time.

5. *Inadequate Skills and Habits.* It may take a small investment of time for officers to get started, but the two skills involved in showing appreciation—identifying positive contributions and using many different ways to give recognition—will soon come naturally.

Ideas for Providing Signs of Appreciation and Awards

Signs of appreciation may be nontangible (psychological) or tangible. The nontangible ways for saying thanks vary in their level of seriousness and emphasis. An officer may simply say "thank you" for a task the staff member completed, possibly with different words like "That's great." A pleased expression, a friendly wave of the hand, a pat on the arm or shoulder, or a handshake, when added to the words, increases the emphasis.

The following list gives many basic ways to show appreciation in nontangible ways. The first items are relatively mild, and the last are the strongest. The list is not exhaustive, and the items are not mutually exclusive; most can be used in conjunction with one or more of the others. Fire officers have undoubtedly used every one of these techniques, either frequently or occasionally. The list may remind fire officers to use these simple means of showing appreciation more often.

ADDITIONAL INSIGHT 5.1

Verbal Expressions of Appreciation

- Pointing to the importance of a specific accomplishment
- Indicating awareness of an accomplishment that came to management's attention without the staff member's knowledge
- Offering help with an assignment
- Asking for an explanation of how a difficult task was accomplished
- Distinguishing the "thanks" from routine acknowledgments by setting them deliberately into a different place or time, such as the office, in front of other staff members or staff members from other department(s) informally, or even formally in front of some or all members of the team
- Asking for ideas and suggestions
- Asking for the staff member's opinion on ideas and suggestions from others (including officers)
- Providing prompt information on occurrences that affect the team
- Arranging for members or the officer of another company or of another function to provide any of the preceding items
- Arranging for members of the public with whom a staff member may be in regular contact to provide any of the preceding items
- Arranging for a higher-level officer to provide any of the preceding items
- Taking the staff member into your confidence on an event or a decision that is not yet public knowledge
- Implementing a suggestion or idea promptly
- Consulting regularly with a staff member who has the knowledge or experience to contribute effectively
- Delegating officer tasks that a team member can perform, such as leading or attending a meeting in the officer's absence
- Helping a staff member acquire skills that might be useful for positions up the career ladder

Fire officers have full control over nontangible signs of appreciation. They also have considerable control over those that come from higher-level officers since the latter are not likely to refuse a request from the supervising officer to say something positive to one or several members of the team.

Officers may be able to obtain signs of appreciation for their team members from other companies or even other fire departments, especially if fire officers are willing to initiate and/or reciprocate the action. All it might take is for fire officers and some team members to show appreciation for achievements of the other departments. To be even more certain of cooperation, officers can reach an informal agreement with the officers or possibly chiefs of the other organizations.

Gaining help from the public in showing appreciation to fire department members for their efforts and contributions is not so easy, of course, but it can sometimes be stimulated. When a staff member does something for a civilian, a request for a note to a specific fire fighter or officer, or even the chief, often brings a complimentary letter.

Tangible Signs of Appreciation

- Memos and letters
- Certificates and plaques
- Trophies

Memos and letters come from the same sources as nontangible signs of appreciation. Fire officers have full control over the memos and letters that they send. They also have considerable influence over those that come from higher-level officers; fire officers can make arrangements with other officers or fire departments to ensure their cooperation. Memos and letters from outsiders can also be solicited.

Certificates and plaques are most appropriate for achievements at emergency incidents or for the attainment of important individual, company, and department goals and objectives. A department should display these signs of appreciation where members of the public can see them. They can sometimes be mounted in a central, highly visible area, together with all other recent ones.

Frequency of Evidence of Appreciation

The frequency and number of both nontangible and tangible rewards are important issues. Too few is obviously not desirable. Too many of either, or both, can lessen their impact and value. No hard and fast rules determine the frequency of expressing appreciation for achievements. One instance, once a week on average, or even more is fine, but once every other month is undoubtedly too little.

Steps for Showing Appreciation of Accomplishments

1. Provide informal acknowledgments as soon as you become aware of a deserving accomplishment.
2. Set aside a few moments each week to write down what accomplishments of staff members deserve notice beyond the oral acknowledgment already provided. Consider all activities that enhance departmental or organizational performance:
   - General activities
   - Activities specific to the staff member's work
   - Personal characteristics

ADDITIONAL INSIGHT 5.1

Do not make the list overly long, and review the lists from previous months to avoid duplication.

3. Decide on the level of appreciation you might consider for each specific accomplishment.
4. Where applicable, enlist the aid of higher-level officers and officers in other departments for oral and written signs of appreciation.
5. Where applicable, review what additional steps you can take to gain cooperation from the public, and implement these steps.

ADDITIONAL INSIGHT 5.1

Prefire Planning and Related Loss Reduction Activities

INTRODUCTION

Prefire planning and related activities are critical to fire protection because they prevent some fires and limit the damage that fire causes. The purpose of a prefire plan is to enable fire department personnel to carry out attack preparations and fire-fighting operations at the scene of a fire as efficiently and effectively as possible. Based on data from prefire planning, fire departments often modify their operations and procedures so that the consequences of future incidents are less severe. At the same time, during prefire planning visits, fire officers may notice violations of fire safety codes or other potential hazards. The problems are communicated to the fire prevention staff. Follow-up visits by an inspector reduce the probability of a fire at that location. Prefire planning, in its full impact, prevents fires.

In this chapter we discuss prefire planning and related loss reduction functions, including prefire plans, water supply systems, fire investigation, and loss reduction information management systems. These functions come together to help a community plan its fire response and fire prevention activities based on actual conditions in the community, both past and present.

SCENARIO
THE PRODUCTIVITY DILEMMA

"Here's the situation." Captain Warner, the department's fire prevention officer, began the meeting with the four inspectors. "The repercussions from the Shmieg dry-cleaning plant fire over in Territown are beginning to hit us, too. People are upset by reports about the confusion at a fire that might have been prevented if the department had done a more thorough inspection. Their town council is up in arms because the fire not only demolished the entire plant but also destroyed part of the building next door. They want to know why there wasn't a better prefire plan."

"Well, you know most of that, so, to make the story short, let me tell you more about what has happened here. The mayor and the safety commissioner held a meeting with the chief late yesterday. They discussed our inspection record and came to the conclusion that we can and really should do more. The chief did all he could to show them how we're already putting a fair amount of personal time into inspections and certainly into public education. Still, they want more inspections, more careful and thorough inspections, with better records and better coordination with the prefire plans. They also want even more public education in the schools, along with events for adults. They'll help with publicity, they say, and maybe they'll get around to doing it, but they want us to lead this effort."

"With the decrease in the number of fires—and they gave us credit for that—the department should have the time. They want us to get organized to put greater emphasis on fire prevention and loss reduction. So, the pressure is on the five of us, first of all, and also on the fire companies. They've got to get their prefire plans into better shape and get them into the computer, which still doesn't have much in it. I asked you all to come here this morning so we could talk about this and see what we can do."

George, the most senior of the four part-time inspectors, spoke up. He was clearly upset. "Why us? Why not the whole department? Shouldn't they carry most of the load? We've got over 400 inspections to do, not counting the new construction inspections, which you do. And all the stuff we are doing with the kids. It's ridiculous to ask us to come up with still more work without an offer of help from the others. The companies can take over much more of the inspection load. They're barely doing all the inspections they're supposed to do—you know, the mom-and-pop stores and the small offices." All the inspectors were on regular fire-fighting duty, including responses and training, and were exempt from only apparatus maintenance and housekeeping duties.

The captain looked at the others: Mary, their newest member; Frank, another old-timer; and Sam. Their heads nodded in agreement with George. "You know that I agree with you, George, but that doesn't help us. I am sure when I go to the companies, they'll tell me their sad tales about how the fallout from Territown is hitting them, too. Maybe I'll get something from them. But first I've got to get prepared so I know what I want from them and so I can tell them what we're going to do. I don't want them to think that we are trying to put the whole load on them. That'll just get their backs up, and I'll accomplish nothing."

Mary commented, "Who's going to train them if they'll do more of the inspections? Once they get into new types of occupancies, other than the small stores and offices they are doing now, they've got to know more about codes. They'll have to be sharper on retrofits. You know what I mean. Who'll come up with a training program, and who'll get them up to speed?" She had a tendency to look at the whole picture.

"Why don't we get them to do the multifamily apartment houses—we've got a whole bunch of them. That way they won't have too much new to learn, and it'll give us a fair amount of relief. We can then upgrade the inspection procedure, catch up with the reinspections, and maybe do a little more with the education effort," suggested Frank.

"Sounds like a good idea to me," added Sam, who felt he had to say something.

"OK. They've probably got the message about a tighter schedule on mom-and-pops without my bringing it up. I'll ask them to take on the multifamilies. If they don't get too upset about that, I'll talk to them about doing some of the fire safety education work in the schools. Now, what about us? What are we going to do to fill the time that their help will free up?" The captain tried to keep the discussion focused on what his own group would do.

"Are they going to get us the little handheld computers, so we can enter the information from the inspections and it can be transferred directly into the database back here at headquarters? That would save some time." Mary continued her positive thinking.

But not George. "Why don't you first see whether we'll get some help from the companies, and then we can talk about what we can do so the total package will be acceptable to the chief?"

"Because I don't think that I'll cut much ice if I go there empty-handed," said the captain. "I feel I have to be able to say that we are going to do this and that, and they have to contribute because they are the main beneficiaries of the reduced fire incidence. As far as your question is concerned, Mary, they're still considering where they'll get $15,000 so they can buy the little computers and the programming. Don't forget that they need almost as much for the laptops and phone connections, which they want to install on each engine and in the command car."

"You know what they're going to say—they'll tell you that they've been cut back and that's taken up any slack there might have been—but there wasn't any because they were understaffed to begin with. I'm just being the devil's advocate." Sam thought it was useful for the captain to be prepared for that response.

"OK. Let's get back to what we are going to do. Let me make some suggestions here and see where you may have a problem," the captain continued. "First, I think we can reorganize our schedules so we don't have to run back and forth so much. By making the routes more logical than they are now, we can do all the inspections and reinspections in one area, street by street. That'll take a little juggling, but I think it'll help. Second, even without the little computers, we can revise our inspection form so it is more in line with the computer screen, and we can get a little more disciplined about the way we enter the information on the form. You know what I mean—clearer entries on the form for us to transfer into the computer. How do you feel about that? Would you like to make some other suggestions?"

"Reorganizing the schedules is easier said than done, but we can try," was George's response. No one else had a comment.

"OK then," the captain said. "Would you, George and Frank, get together and come up with a plan on how we can best reorganize the schedules. Then give your suggestions to Mary and Sam for their comments. Discuss the comments and finalize the proposed plan. If you can't agree, let me know. Otherwise, let me see the plan you would recommend. Mary and Sam, would you start on the other project—the form revisions? Come up with a draft, and then get George and Frank's suggestions. Let me see what the four of you agree on. Once I see what we can do, I'll go to the company commanders. If I need help, I'll get the chief involved. I'll probably prepare him in advance, though, so I'll have his views and I hope his support. When do you think you can have all that finished?"

"It'll probably take at least two or three months, and possibly longer." George was cautious about committing himself. "It all depends on when we can get together and what interruptions come along. We still have to keep up with our inspections and the school commitments we've made. There's a lot to do—we can't drop everything just because there was a problem in Territown."

"Two or three months is a long time. I don't think we should take that long to come up with something concrete. How about shooting to get it done in one month?" The captain was not happy with George's suggestion. "What do you all think?"

"We'll try" was Frank's noncommittal response, and the others nodded, George with a doubting look on his face.

"OK, thanks. Call on me if you need me." With that, the captain closed the meeting.

It so happened that the next day the chief scheduled a meeting with the officers for Thursday of the following week. He had been requested to provide information about the department's plans in response to the request for greater emphasis on fire prevention and loss reduction.

At the meeting, Captain Warner had a chance to bring up the request for assistance on inspections. The companies had no problem with the mom-and-pop places; the commanders knew that they had to commit to that. Getting cooperation on the multifamilies was not so easy. Still, because the chief gave him some support on that, the captain got agreement, but only after extensive discussions about help to the companies on prefire plan updates.

The prefire plans needed extensive updating. They would be converted to computerized versions so they could be transmitted to the responding apparatus after laptops had been purchased. The companies wanted the fire prevention staff to help them by reviewing prefire plans during inspections and providing input for revisions. The captain had to agree to set up some way to do that. He did, however, get a promise from each of the companies to take over a portion of the fire safety education work in the schools.

At the end of the meeting, the chief asked for all the plans to be finalized with specifics so he could inform the safety commissioner.

Scenario Analysis: Fire Service Function Perspective

As in earlier chapters, this analysis is based on the goal and reasonable guidelines for the chapter's topic of prefire planning and related loss reduction activities.

Please read the objective and the guidelines. Then, before reading the analysis, you may want to give some thought to the strengths and weaknesses of the way the fire department discharged its responsibility for prefire planning and related loss reduction activities. These are the relevant goals:

The primary goals of prefire planning and related loss reduction functions are to ensure that department members have thorough plans for attacking fires most effectively, with the available built-on protection systems, water supply, and other extinguishing agents, and that the members are knowledgeable, skilled, and equipped to implement the plans.

The guideline to support this objective follows:

What else needs to be done to ensure that prefire planning for fire suppression is as thorough and useful as possible? Specifically:

1. What else needs to be done to provide responding companies with adequate information, appropriately analyzed and formulated into plans?
2. What else needs to be done so that prefire plans are used most effectively in staff development (that is, training and drills), in fire investigations, and in water supply review and testing?
3. What else needs to be done so that the related information management systems are as effective as possible?

Keep in mind that you do not need to remember the detailed wording of the guideline when making decisions or even during occasional reviews of the guideline. Only in the most important decisions might that be worthwhile. However, memorizing a personalized, abbreviated (summarized) version and applying it with every relevant decision are critical for developing the necessary habit that will ensure the highest quality performance possible under the existing circumstances. We will discuss these three segments of the guideline in relation to the scenario.

1. *What else needs to be done to provide responding companies with adequate information, appropriately analyzed and formulated into plans?* The planned extensive updating and conversion of prefire plans to computerized versions and their transmittal to responding apparatus should go a long way to satisfying this portion of the guideline. Still, while they update and convert the system and develop standard operating guidelines, the department personnel who will be involved in these activities should keep this guideline in mind.

Specifically, because fire prevention personnel will assume a more active role in prefire planning, if they use guidelines when making decisions, they would ask themselves questions such as these from time to time:

- Will the information we generate be reviewed by the respective companies and immediately incorporated in the prefire plans?

- Will we get feedback on the usefulness of the information we have provided and on what other information is needed?

2. *What else needs to be done so that prefire plans are used most effectively in staff development (that is, training and drills), in fire investigations, and in water supply review and testing?* The scenario was essentially silent on these issues, so we cannot draw any conclusions about the respective activities in the department. However, the chief and other involved officers should ask themselves questions such as the following:

- How could prefire plans be used more effectively in staff development?
- To what extent are prefire plans used during fire investigations? How effectively is the information developed during the investigations used to improve prefire plans?

3. *What else needs to be done so that the related information management systems are as effective as possible?* Here too the scenario did not provide any information on which to base specific comments. However, the chief and other involved officers should continue to ask themselves questions such as the following:

- What are we doing to analyze complaints about inadequacies in the information management system and to consider possible improvements?
- What can we do to compare the department's information management system with those of other departments about which there is information in the literature or that are known to have unusually fine systems?

Scenario Analysis: Management/Leadership Perspective

We will base our analysis of the scenario in this chapter and in the remaining chapters on the guidelines for the 3Cs. We begin this section by reviewing all three guidelines and their expansion with specific subguideline questions. Then in Chapters 7–11, we will present only the specific questions.

Before reading the analysis, you may want to give some thought again to the strengths and weaknesses of the way Captain Warner handled the situation from the management/leadership perspective.

We first recall the control guideline:

> What else needs to be done to ensure effective control and coordination, so that the decision will lead to the outcome we seek and so that we will know when we have to modify our implementation or plan because we are not getting the results we want?

The specific questions follow:

1. Are the goals and objectives appropriate and effectively communicated?
2. Is the participation in decision making appropriate?
3. Are coordination, cooperation, and inter- and intra-unit communications being stimulated?
4. Is full advantage being taken of positive discipline and performance counseling?

The competence guideline is next:

> What else needs to be done so that all those who will be involved in implementing the decision and those who will otherwise be affected will have the necessary competencies to ensure effective progress toward excellence in fire department operations and service to the community.

These are the specific questions:

1. Are changes needed in the recruiting and selecting for vacancies?
2. Are management of learning concepts applied effectively?
3. Are coaching and counseling on self-development being used to their best advantage?

As with the functional guideline, we should keep in mind, that recalling the detailed wording of the guideline when making decisions is not necessary except in the most important decisions. However, memorizing a personalized, abbreviated (summarized) version and applying it with every relevant decision is critical for developing the necessary habit that will help to ensure the highest quality performance possible under the existing circumstances.

The climate guideline follows:

> What else needs to be done so that all those who will be involved in implementing the decision and those who will otherwise be affected will be in favor of it or at least have as positive a view as possible so there will be a favorable climate?

The specific questions related to the climate guideline are as follows:

1. Are appropriate psychological and tangible rewards offered and provided effectively and efficiently to bring the highest possible level of satisfaction from the creation and use of the product/service?
2. Are policies in place to help reduce work-related stress?

Discussion of the Control Guideline Questions.

1. *Are the goals and objectives appropriate and effectively communicated?* We know little about the chief from the scenario. We do know, however, that he had failed to set goals for fire prevention activities long before the Territown fire. A competent manager would have realized the need for such goals, certainly when submitting the request for the small computers and the programming and possibly even earlier.

Captain Warner did try to have his staff suggest specific objectives for completing the revised schedule and the recommended changes in the inspection form, but he gave up a little too soon. If he had tried harder, he may have obtained a tentative timetable. The reluctance on the part of the team members indicated that they were apprehensive about committing to goals or even to objectives. Perhaps negative consequences followed when goals or objectives were not met in the past. The captain should have developed an understanding with his staff about his responsibility in meeting goals and objectives. An officer should really expect only two things from staff members who are working on a goal or objective: their best effort and early notification if it appears that a goal or objective may not be achieved as planned. Then the officer can decide whether to devote additional resources (in this case, possibly giving personal help) or to accept a change in the objective (completion date, in this case). (See Additional Insight 3.1, Goal Setting and Implementation.)

If Captain Warner had applied this guideline segment, he might have asked himself these questions:

- How important is it that I set objectives on these matters?
- What do I have to do so that the inspectors will not be apprehensive about setting challenging, yet realistic, objectives?
- To what extent might it be useful for me to meet with each of the two teams once to help them get started?

2. *Is the participation in decision making appropriate?* On the surface, Captain Warner's participation practices were sound. He seemed to involve the inspectors appropriately in decisions that affected them. However, participation is more than skin deep. We would need to know more about the relationship between the captain and the inspectors to make a judgement.

If the captain had applied this guideline segment, he would have asked himself questions:

- Do the inspectors feel free to express their views with respect to any decisions that have to be made when I request their input?

- Do the inspectors feel that what they say has an impact on the decision, or at least is given serious consideration?

3. *Are coordination, cooperation, and inter- and intra-unit communications being stimulated?*   Although there seemed to be fairly good coordination and cooperation on the planned changes, if the captain used guidelines, he might have asked a few questions such as:

- Do new plans contain thorough inter- and intra-unit communications procedures?
- What else can the fire prevention staff do to achieve even better coordination with the companies through improved procedures, possibly jointly developed?
- What else can the fire prevention staff do to achieve even better cooperation with and from the companies without having to do more than their fair share of the work?
- Does the fire prevention staff do all it can to prevent conflict with the companies? (See Additional Insight 4.1, Management of Potentially Damaging Conflict.)
- When should I review progress with each of the two teams?

4. *Is full advantage being taken for the potential of positive discipline and performance counseling?*   More can always be done to gain the *full* benefit of truly positive discipline and performance counseling. Positive discipline rests on the five foundations listed in Exhibit 6.1 The captain could ask himself one or more specific questions pertaining to practices in his department and the fire prevention bureau with respect to all the items in Exhibit 6.1 except possibly the second one.

Exhibit 6.1 The Five Foundations of Positive Discipline

1. The culture encourages open two-way communications.
2. The same standards and norms apply to all.
3. Department members who deserve commendation and privileges receive them. Most privileges in the fire service are likely to involve some preferential treatment in work assignments; however, there may also be some leeway in trading timeoff for duty time or other such privileges.
4. Those who violate accepted rules or fail to adhere to reasonable norms and standards, cooperatively determined, receive help at first and then gradually more stringent warnings until their behavior conforms or disciplinary steps are invoked.
5. Counseling is used competently to reduce, to a minimum, the use of the organization's disciplinary procedure.

Discussion of the Competence Guideline Questions

1. *Are changes needed in recruiting and selecting for vacancies?* This segment of the guideline does not have any direct application to the scenario because all the inspectors appeared to be competent and positively oriented.

2. *Are management of learning concepts applied effectively?** The captain did not show any concern about the ability of the inspectors to work effectively on the two assignments they had accepted. To use this guideline segment, he may ask himself these questions:

- To what extent should I meet with each of the two teams once to help them get started, so I can see what different skills they need to complete the task competently?
- To what extent do I need to provide more support for learning, such as determining what each inspector needs to know or be able to do?

3. *Are coaching and counseling on self-development being used to their best advantage?* Questions such as the following apply here:

- Where would coaching be appropriate on matters related to the new assignments?
- What would working with the inspectors on these assignments reveal about their self-development? What counseling on self-development might be advisable?

Discussion of the Climate Guideline Questions.

1. *Are appropriate psychological and tangible rewards offered and provided effectively and efficiently to bring the highest possible level of satisfaction from the creation and use of the product/service?* The scenario was silent on rewards in the department and specifically in the fire prevention bureau. In implementing this guideline segment, the captain could ask himself these questions:

- For what contributions by the inspectors should I show appreciation? [See Additional Insight 5.1, Enhancing Work Satisfaction (Providing Recognition) and Performance Evaluation.]
- What other ways can I use to give recognition (show appreciation) for contributions?

---

*Management of learning strives to shift the responsibility for acquiring competence to the learners. We will discuss it in detail in Chapter 11 and in Additional Insight 11.1, Management of Learning and Coaching.

- What more can I do to ensure that the inspectors are aware that I appreciate their efforts?

2. *Are policies in place to help reduce work-related stress?*   The captain appeared to consider the effects of stress on the inspectors, and he attempted to relieve the stess with his efforts to involve the companies in inspections. The scenario gave no information about any steps to speed up the computer purchases, which would, over the long run, help relieve the pressure. We know nothing about what else the chief planned to do, such as requesting additional staff or even making it clear to the inspectors and companies that he fully understood their situation and would continue to do whatever he could to bring substantive relief.

In conclusion, from this analysis we see that the need for and benefits of fire prevention and loss reduction are so compelling that fire departments are well advised to increase their emphasis on activities that will bring results in these two areas. The department in the scenario should probably not have waited for the disaster in Territown to initiate action on enhancing its fire prevention efforts, but at least something will be done now to bring benefits to the community. If the officers and members of the department had used functional guidelines with relevant decisions and reviewed them from time to time, it would have taken action long ago. If the captain had developed the habit of considering the 3Cs guidelines or similar ones with every decision, in addition to the relevant functional guidelines, his actions would have been more effective, as was shown by the questions the 3Cs analysis raised.

The steps that Captain Warner and his group are taking and the additional workload that the companies assume will unquestionably result in higher productivity. Depending on the severity of the previous staff reductions, they may, however, place too much stress on the inspectors and fire fighters and lead to negative reactions that might ultimately offset any gains.

It might have been worthwhile for the town leadership to check the staffing adequacy of the department, in light of the demands for more service. At the least, they should move quickly on the planned computer purchases and programming services. Such action will help productivity gains and support the fire fighters and inspectors.

PREFIRE PLANNING

A prefire plan is a survey of a location and a plan for fighting a fire that might strike that particular occupancy. The plan includes the key matters that influence a fire attack.

At the fireground, fire fighters may begin attack preparations more quickly if they know details about the fire site before they arrive and if they have pre-determined advantageous positions for apparatus and hose layouts, at least for single-building fires. For fires in multibuilding compounds, specifying apparatus placement may not be advantageous because that depends on the specific location and extent of the fire. The detail in a prefire plan is limited by the extent to which it is advisable to specify features of the plan without knowing the specific location of the fire on the property involved. With effective prefire plans, officers spend much less time making decisions concerning the fire site during and after the sizeup process. In fact, sizeup can and should begin with the facts already known from the prefire planning visits.

The Prefire Planning Process

The extent and detail of prefire planning for a site depend, of course, on the site's size, complexity, and special hazards (if any). Prefire planning should involve all first-due companies and special hazard teams (such as hazardous materials or confined space) on an annual basis. It spells out a course of action against a potential fire that is based on the collective experiences of those involved in the planning process, on known existing conditions, and on reasonable expectations of what suppression activities will occur. NFPA 1620, *Recommended Practice for Pre-Incident Planning*, provides guidance on not only prefire planning but also other related aspects such as planning for hazardous material involvement and for casualties.

Prefire plans are necessary for all target hazards and special risks. For single- and two-family dwellings and other small occupancies, however, prefire plans are limited to specifying the locations of the nearest hydrants and possibly route information. For these risks, a standard operating procedure (or guideline) should be sufficient.

Prefire planning is a necessary adjunct to tactical operations and, if used, should increase operational efficiency, reduce fire losses, and provide an optimal level of fire protection. The prefire planning process involves four steps: (1) information gathering; (2) information analysis; (3) information dissemination; and (4) review and drill.

1. *Information gathering.* Pertinent information is collected at the selected site, including building construction features, occupancy, exposures, utility disconnects, fire hydrant locations, water-main sizes, and anything else that affects fire-fighting operations if a fire occurs.
2. *Information analysis.* The information must be analyzed in terms of what is pertinent and vital to fire suppression operations. An operable pre-

fire plan is then formulated and organized into a format that is usable on the fireground.

3. *Information dissemination.* All companies that might respond to a pre-planned location should receive copies of, or otherwise have access to, the plan, so that they can become familiar with both the plan and the pertinent factors relating to it.

4. *Review and drill.* Each company that might be involved at the preplanned location should review the plan from time to time, preferably on a pre-determined schedule. For major hazards, periodic drills with all companies involved should be scheduled at the property.

Description of the Prefire Plan

Data in the Prefire Plan.   A prefire plan usually consists of a data sheet or building survey form (Exhibits 6.2 and 6.3) and usually also a building layout. Data sheets contain the necessary information for fire department use during emergencies. A wide range of information, such as building layout, access, utilities, and process hazards, are listed in data form. Other data that might be needed are details about the type of occupancy, special volatile materials, hours that the building is normally occupied, and construction of any adjacent buildings. A building layout is provided to portray the information, to the greatest extent possible, on the form. This pictorial information allows fire fighters to review the occupancy while on location.

Together, data sheets and layouts should contain information on the exact location of a property and the best response routes at various times of the day and in different weather conditions (see Figure 6.1). Such information helps eliminate delays on steep hills or delays caused by heavy traffic or flooding. It also provides fire fighters with the knowledge of the best places to station apparatus in relation to water supply, space, grade, and traffic conditions.

To be most useful, a prefire plan should contain only the most valuable information that is not likely to change. Secondary information or items that might become outdated should be noted in pencil. The inclusion of too much information makes a prefire plan too complicated and less effective for the user.

The building layout for a school shown in Figure 6.2 details access routes; dimensions; construction materials; locations of stairs, windows, exits, sprinklers and their valves, standpipes, hydrants, main power switches, utility shutoffs, and all other fire protection equipment; and other pertinent information about building construction. Layouts may also indicate supplementary

Exhibit 6.2 Prefire Plan Data Sheet

1. General Information

   Address:_____

   Type of occupancy:_____

   Business name:_____

   Name and address of owner:_____

   _____

   Name and address of occupants:_____

   _____

   Number of stores:_____Type of construction:_____

   Width:_____Height:_____Length:_____

   Structural conditions, parapets, fire walls, etc.:_____

   _____

   Heating: Type:_____Size:_____

             Location:_____

   Cooling: Type:_____Size:_____

   Remarks about owner:_____

   _____

2. Outside Information

   Plot Plan (Draw on a separate piece of graph paper. Show all detail on street location, hydrants, sprinkler, and standpipe siameses, etc.)

   Indicate private protection available:_____

   _____

   Indicate fire department connections:_____

   _____

   Location of public utilities, valves, and switches

   Water:_____

   Gas:_____

   Electric:_____

   Obstructions:_____

3. Floor Plan for Each Floor

   (Draw a floor plan for each floor (on graph paper). Show the rooms, halls, doors, windows, etc.)

   Rescue matters:_____

   _____

   Forcible entry problems:_____

   _____

   (continued)

Exhibit 6.2 Prefire Plan Data Sheet (continued)

Possible points for ventilation:_____

Determine possible routes of fire travel and how to stop it:_____

Locate man-traps (things that can surprise a fire fighter and cause injury, such as holes, drop-offs, open vats, etc.):_____

Locate hazardous materials:_____

4. Housekeeping and Fire Protection Issues

   Fire protection equipment:
   a. Extinguishers:
      In place: Yes:_____ No:_____
      Tested annually: Yes:_____ No:_____
   b. Automatic sprinklers:
      In place: Yes:_____ No:_____
      Tested annually: Yes:_____ No:_____
   c. General housekeeping comments:_____

5. Roof Plan (Draw on graph paper. Show vents, hatches, skylights etc.)
   Type of roof construction:_____

   Are there heavy objects on the roof? Yes:_____ No:_____
   If yes, list them here and where located:_____

   Are there parapets? Yes:_____ No:_____
   If yes, where are they located and what is their condition?_____

6. Special Elevation Sketch (Use a cross section or cutaway view if there are special features of building elevation or terrain concern. Draw on graph paper.)

7. Remarks:_____

Source: Adelphi Fire Company

Exhibit 6.3 Prefire Plan Building Survey Form for a Commercial Property

Date_____

Name of business_____
Address of business_____
Type of occupancy_____
Telephone number_____
Person to notify in case of emergency
Name_____Telephone number_____

Regular Business Hours

Mon.___to___Tues.___to___Wed.___to___Thur.___to___Fri.___to_____

Sat.____to___Sun.____to___Are there night security people? Yes ( ) No ( )

If yes, number of people:____

Potential number of customers_____

Protective Systems

Is there a fire detection system? Yes ( ) No ( )
Agency supervising system_____Phone_____
Is the building protected by an automatic sprinkler system? Yes ( ) No ( )
If yes, type: Dry ( ) Wet ( ) Chemical ( ) Combination ( ) Other ( )
Agency supervising system_____Phone_____
Is there a standpipe system? Yes ( ) No ( )

Other Necessary Information

Gas company_____Shut off location_____
Electric company_____Shut off location_____
Number of stories_____Type of construction_____
Are there any attached buildings? Yes ( ) No ( ) If so, describe in drawing on rear of this form.
Type of heating system_____ Is there on site fuel storage? Yes ( ) No ( )
If so, describe on drawing on rear of this form.
Are there any special hazards? Yes ( ) No ( ) If so, describe and show on drawing on rear of this form.
_____

Source: Newark, NJ, Fire Department.

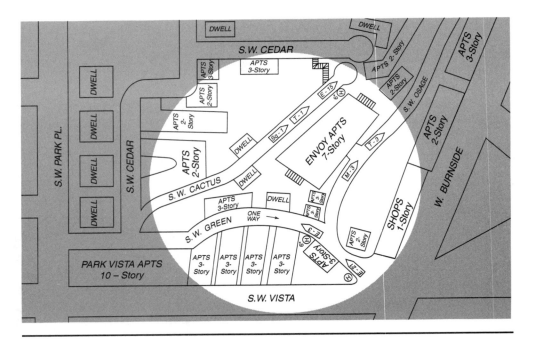

Figure 6.1 Prefire plan layout for an apartment complex shows access routes and water supplies to be used by apparatus. (*Source:* Portland Fire Bureau, Portland, Oregon)

water supplies from additional mains for companies other than those making the first response.

The sketch in Figure 6.2 relies on symbols to convey much of the necessary information for the prefire plan. The symbols that are usually used on prefire plans are an important shorthand method for providing information as quickly and simply as possible. Because they are used to save time and space, symbols should be understood more easily and more rapidly than words. Complicated symbols are not useful. Different fire departments and organizations use different symbols. In cases where a symbol does not describe the situation sufficiently, it may also be necessary to provide ancillary information. For guidance, see NFPA 170, *Standard for Fire Safety Symbols*.

The symbols and plans presented in this book allow any inspector or fire fighter involved in prefire planning to develop an acceptable plan in accordance with convention. Many organizations use slightly different symbols and layouts, however. When taking information from a plan developed by another source, department personnel should refer to the legend of symbols used by that organization to accurately interpret the information.

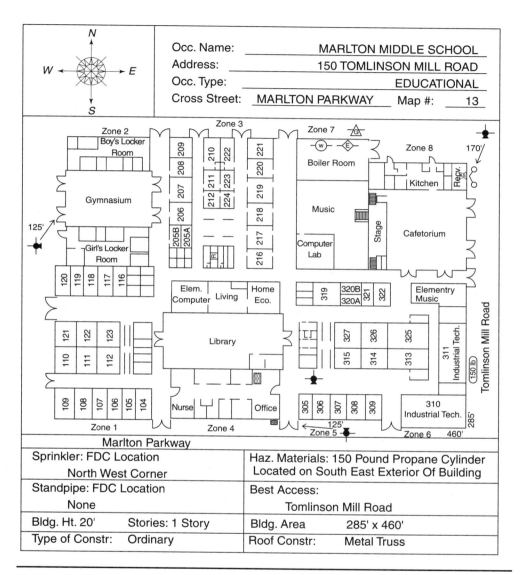

Figure 6.2 This diagram illustrates a typical building layout for a prefire plan. (Courtesy of Evesham Township, NJ, Fire Department)

The features of any property can be placed into one of four categories, nicknamed COPE: construction (C), occupancy (O), protection (P), exposure (E). The standard plan symbols shown in Figure 6.3 can be used to represent the general location and type of COPE features found in a facility. The standard abbreviations are listed in Exhibit 6.4. Figure 6.4 shows a typical site sketch containing the symbols and abbreviations.

The Survey Process.   The entire fire company actively participates in the survey process to gather data for the prefire plan. The survey for prefire planning is very similar to an inspection. The department usually supplies checklists and reporting forms that make the information-collecting process more efficient (see Exhibit 6.3).

Fire fighters may gather data in teams or separately. Sometimes the company officer may need to collect or verify certain pieces of information. When the building or property has been surveyed thoroughly (the amount of time varies depending on the size and complexity of the property) and the entire company is satisfied that the survey information is correct, a final map of the building and a data sheet showing the specific hazard(s) in the building can be prepared. The map or data sheet may indicate the best response route and initial attack positions at this stage.

A fire department or company that is formulating a program of prefire planning might use a procedure similar to the following:

1. Determine the order of priority for the properties to be surveyed.
2. Plan a schedule for surveys.
3. Notify owners or managers of properties about the prefire plan program and request permission to make a survey. As with inspections, this request is not legally necessary, but it maintains good community relations.
4. Decide which unit or fire fighters will carry out the survey on the specific property.
5. After the survey data are gathered, incorporate them into the prefire plan.
6. Modify prefire plans whenever there is information about changes in occupancies or buildings, and incorporate all changes into the plan as soon as possible.

Changing traffic patterns, improvements in apparatus and equipment, and new concepts in fire-fighting operations also may require the updating of plans. Information about changes can come from many sources, including inspections by the company or fire prevention inspectors, newspaper reports, scheduled periodic updates of prefire plans, or incidental sources, such as conversations with friends, acquaintances, or other fire fighter contacts.

Relationship Between Prefire Planning and Inspections.   Prefire planning and inspections have some overlap. Like inspections, prefire plan surveys

SYMBOLS FOR PRE-FIRE PLANNING

Triangle symbols can point at a specific location or direction.

Diamond symbols identify a specific location by touching a wall.

Circle symbols are used for all piping system appendatures, such as valves, since most pipes are round.

Square symbols are used for all room designations, as they represent most rooms having four sides.

Access features, assessment features, ventilation features, and utility shutoffs.

Access features

Fire department access point — FD

Fire department key box — K

Roof access — RA

Assessment features

Fire Alarm annunciator panel — AP

Fire alarm reset panel — RP

Fire alarm voice communication panel — CP

Smoke control and pressurization panel — SP

Sprinkler system water flow bell — WB

Utility shutoffs

Electric shutoff — E

Domestic water shutoff — W

Gas shutoff — G

Specific variations

LP-gas shutoff — LPG

Natural gas shutoff — NG

Ventilation features

Sky light — SL

Smoke vent — SV

Detector/extinguishing equipment.

Duct detector — DD

Heat detector — HD

Smoke detector — SD

Flow switch (water) — FS

Manual pull station — PS

Tamper switch — TS

Halon system — HL

Dry chemical system — DC

CO₂ system — CO2

Wet chemical system — WC

Foam system — FO

Clean agent system — CA

Beam smoke detector — BSD

Water flow control valves and water savers.

Post-indicator valve — PIV

Riser valve — RV

Sprinkler zone valve — ZV

Hose cabinet or connection — HC

Wall hydrant — WH

Test header (fire pump) — TH

Inspector's test connection — TC

Fire hydrant — FH

Fire department connection — FDC

Drafting site — DS

Water tank — WT

Equipment rooms.

Air-conditioning equipment room Air-handling units (AHUs) — AC

Elevator equipment room — EE

Emergency generator room — EG

Fire pump room — FP

Telephone equipment room — TE

Boiler room — BR

Electrical/transformer room — ET

Figure 6.3 Standard plan symbols are used to represent the general location and type of features. (*Source*: Cote, *Fire Protection Handbook*, 1997, Table 11-14B)

SYMBOLS FOR FIRE FIGHTING EQUIPMENT, INCLUDING STANDPIPE AND HOSE SYSTEMS

Hose station, charged standpipe

Hose station, dry standpipe

Monitor nozzle, dry

Monitor nozzle, charged

CO_2 reel station

Dry chemical reel station

Foam reel station

SYMBOLS FOR SPECIAL HAZARD SYSTEMS

Agent storage container.
Specify type of agent and mounting.

Special spray nozzle.
Specify type, orifice, size, other
required data (shown here on pipe).

SYMBOLS FOR FIRE EXTINGUISHERS

Water extinguisher

Foam extinguisher

Dry chemical
extinguishers
for fires of liquid,
gas, electrical types

Dry chemical extinguisher
for fires of all types,
except metals

CO_2 extinguishers

Halon extinguishers

Extinguisher for metal
fires

SYMBOLS FOR FIRE ALARMS, DETECTION, AND RELATED EQUIPMENT

Control panels.

Fire alarm control panel [FCP]

Fire system annunciator [FSA]

Fire alarm transponder [FTR]
or transmitter

Elevator status/recall [ESR]

Fire alarm communicator [FAC]

Halon control panel [HCP]

Control panel for heating, [HVA]
ventilation, air conditioning,
exhaust stairwell pressurization,
or similar equipment

Manual stations.

Foam $\square_F$

Wet chemical $\square_W$

Pull station $\square_P$

Halon $\square_H$

Carbon dioxide $\square_C$

Dry chemical $\square_D$

Fire service or emergency telephone station.

Accessible [C]$_A$ Jack [C]$_J$ Hand-set [C]$_H$

Indicating appliances.

Bell (gong)

Speaker/horn
(electric horn)

Horn with light
as separate
assembly

Mini-horn [M]

Horn with light
as one assembly.

Light (lamp, signal light,
indicator lamp. strobe)

Related equipment.
Door holder

**Heat detector
(thermal detector).**

Combination — rate of
rise and fixed temperature R/F

Rate compensation R/C

Fixed temperature F

Rate of rise only R

Gas detector

Flame detector
(flicker detector)
Indicate UV, IR, or
visible radiation type

Smoke detector.

Photoelectric products P
of combustion
detector

Ionization products of I
combustion detector

Beam transmitter BT

Beam receiver BR

Smoke
detector in duct

SYMBOLS FOR SMOKE/PRESSURIZATION CONTROL

Purge controls.

Manual control

Pressurized stairwell

Ventilation openings —↑—

Fans

General

Duct

Roof

Wall

(Orient as required for
base or head injection.)

(Orient as required for intake
or exhaust.)

Dampers

Fire

Smoke

Fire/smoke

Barometric

Figure 6.3 (continued)

SYMBOLS RELATED TO MEANS OF EGRESS

Exit signs. Indicate direction of flow for each face.

Illuminated exit sign, single face

Illuminated exit sign, double face

Emergency lights. Indicate if light head (lamp) is remote from battery.

Emergency light, battery powered, one lamp

Combined battery powered emergency lights and illuminated exit signs, two lamps

Emergency light, battery powered, three lamps

SYMBOLS FOR WATER SUPPLY AND DISTRIBUTION

Mains, pipe. Indicate pipe size and material.

Public water main

Private water main

Water main under building

Suction pipe

Hydrants. Indicate size, type of thread, or connection. Symbol element may be utilized in any combination to fit the type of hydrant.

Private hydrant, one hose outlet

Public hydrant, two hose outlets

Public hydrant, two hose outlets and pumper connection

Thrust block

Wall hydrant, two hose outlets

Private housed hydrant, two hose outlets

Others:

Riser

Meter

Valves. Indicate valve size.

Valves (general)

Valve in pit

Post-indicator valve

Key-operated valve

OS&Y valve (outside screw and yoke, rising stem)

Indicating butterfly valve

Nonindicating valve (nonrising-stem valve)

Backflow preventer—double-check type

Backflow preventer—reduced pressure zone (RPZ) type

Fire department connections. Indicate size and type.

Siamese fire department connection

Free-standing siamese fire department connection

Single fire department connection

Fire pump. For test headers. Indicate number and size of outlets.

Fire pump with drives

Freestanding test header

Wall test header

SYMBOLS FOR SPRINKLER SYSTEMS

Piping, valves, control devices. Indicate size.

Sprinkler riser

Check valve, general

Alarm check valve

Dry-pipe valve

Dry-pipe valve with quick opening device (accelerator or exhauster)

Deluge valve

(Arrow indicates direction of flow)

Preaction valve

Alarm/supervisory devices.

Flow detector/switch (flow alarm)

Pressure detector/switch (Specify type —water, low air, high air, etc.)

Water motor alarm/water motor gong (shield optional)

Bell (gong)

Level detector/switch

Tamper detector or tamper switch

Valve with tamper detector/switch

SYMBOLS FOR EXTINGUISHING SYSTEMS†

Water-based systems:

Wet (charged) system.

Automatically actuated

Manually actuated

Dry system.

Automatically actuated

Manually actuated

Foam system.

Automatically actuated

Manually actuated

Dry chemical systems for liquid, gas, and electrical-type fires.

Automatically actuated

Manually actuated

For fires of all types, except metals.

Automatically actuated

Manually actuated

Systems utilizing a gaseous medium.

Carbon dioxide system.

Automatically actuated

Manually actuated

Halon or clean agent extinguishing system.

Automatically actuated

Manually actuated

Supplementary symbols

Nonsprinklered space

Fully sprinklered space

Partially sprinklered space

† These symbols are intended for use in identifying the type of installed system protecting an area within a building.

Figure 6.3 (continued)

SYMBOLS FOR SITE FEATURES

Buildings.

1. The exterior walls of buildings are outlined in single thickness lines if other than fire-rated, and double thickness lines if fire-rated.

2. The perimeter of canopies, loading docks, and other open-walled structures are shown by broken lines.

Railroad tracks. Railroad tracks are shown by a single line with crossed dashes.

Streets. Streets are shown, usually at property lines.

10   12-14
Downing Street

Bodies of water. Rivers, lakes, etc., are outlined.

POND

CREEK

Fences.

Fences are shown by lines with "x's" every in. (25.4 mm).

Gates are shown.

Property lines.

Fire department access.

F.D.

SYMBOLS FOR BUILDING CONSTRUCTION

Types of building construction. Types of construction are shown narratively.

| FIRE RESISTIVE CONST. (TYPE I) | WOOD-FRAME CONST. (TYPE V) |

Height. Indicate number of stories above ground, number of stories below ground, and height from grade to eaves.

A	B	C 1=2 24 ft	F
3 40 ft	1B 20 ft	1 D	13B
	1 E		UNDER-GROUND

G

A Three stories, no basement, 40 ft to eaves.
B One story with basement, 20 ft to eaves.
C One equals two stories, no basement, 24 ft to eaves.
D One-story open porch or shed.
E One-story addition.
F Thirteen stories with basement.
G Underground structure.

(Includes copyrighted material of Insurance Services Office with its permission. Copyright, Insurance Services Office 1975.)

Walls. Indicate construction.

Wall

——S——	Smoke barrier
——▶——	1/2-hr fire-rated
——▶S——	1/2-hr fire-rated/ smoke barrier
——◆——	3/4-hr fire-rated
——◆S——	3/4-hr fire-rated/ smoke barrier
——◆——	1-hr fire-rated
——◆S——	1-hr fire-rated/ smoke barrier

◆◆◆	2-hr fire-rated
◆◆◆S	2-hr fire-rated/ smoke barrier
◆◆◆◆	3-hr fire-rated
◆◆◆◆S	3-hr fire-rated/ smoke barrier
◆◆◆◆	4-hr fire-rated
◆◆◆◆◆S	4-hr fire-rated/ smoke barrier

Parapets. The symbol for parapets utilizes one cross for each 6 in. (152 mm) that the parapet extends above the roof. The cross is drawn through an extension of the wall line for the parapeted wall (in plan view).

Symbol used to note wall ratings and parapets on life safety plans and risk analysis plans/cross sections.

Floor openings, wall openings, roof openings, and their protection.

Opening in wall. Indicate floors.	— —	Opening hoistway	E
Rated fire door in wall (less than 3 hr). Indicate floors.	/ —	Escalator	
Fire door in wall (3-hr rated). Indicate floors.	/ C —	Stairs in combustible shaft	
		Stairs in fire-rated shaft	
Elevator in combustible shaft.	E	Stairs in open shaft	
Elevator in non-combustible shaft.	E	Skylight	SL

Roof, floor assemblies. These symbols indicate features in cross sections. Descriptive notes are often required.

Fire-resistive floor or roof	———	Floor/ceiling or roof/ceiling assembly. Details indicated as necessary.	═══
Wood joisted floor or roof		Floor on ground	
Other floors or roofs. Note construction.	(Steel deck on steel joists)	Truss roof. Note construction.	

Miscellaneous features. For tanks, indicate type, dimensions, construction, capacity, pressurization, and contents.

Boiler		Horizontal tank, above ground.	
Chimney. Describe height and construction.		Vertical tank, above ground.	
		Horizontal tank, below ground.	
Fire escape			

Figure 6.3 (continued)

Exhibit 6.4 Legend of Common Abbreviations Used with Standard Plan Symbols

Above	ABV	Joist, Joisted	J
Accelerator	ACC	Liquid	LIQ
Acetylene	ACET	Liquid Oxygen	LOX
Aluminum	AL	Manufacture	MFR
Asbestos	ASB	Manufacturing	MFG
Asphalt-Protected Metal	APM	Maximum Capacity	MAX CAP
Attic	A	Mean Sea Level	MSL
Automatic	AUTO	Metal	MT
Automatic Fire Alarm	AFA	Mezzanine	MEZ
Automatic, Sprinklers	AS	Mill Use	MU
Avenue	AVE	Normally Closed	NC
Basement	B	Normally Open	NO
Beam	BM	North	N
Board on Joist	BDOJ	Number	No
Brick	BR	Open Sprinklers	OS
Building	BLDG	Outside Screw & Yoke Valve	OS&Y
Cast Iron	CI	Partition (label composition)	PTN (e.g., WD PTN)
Cement	CEM	Plaster	PLAS
Centrifugal Fire Pump	CFP	Plaster Board	PLAS BD
Cinder Block	CB	Plafform	PLATF
Composition Roof	COMPR	Pound (unit of force)	LB
Concrete	CONC	Pressure	PRESS
Construction	CONST	Unit of Pressure (psi)	PSI
Corrugated Iron	COR IR	Protected Steel	PROT ST
Corrugated Steel	COR ST	Private	PRIVATE
Diameter	DIA	Public	PUB
Diesel Engine	D ENG	Railroad	RR
Domestic	DOM	Reinforced Concrete	RC
Double Hydrant	DH	Reinforcing Steel	RST
Dry-Pipe Valve	DPV	Reservoir	RES
East	E	Revolutions per Minute	RPM
Electric Motor Driven	EMD	Roof	RF
Elevator	ELEV	Room	RM
Engine	ENG	Slate Shingle Roof	SSR
Exhauster	EXH	Space	SP
Feet	FT	South	S
Fibre Board	FBR BD	Stainless Steel	SST
Fire Escape	FE	Steam Fire Pump	SFP
Fire Department Pumper Connection	FDPC	Steel	ST
Fire Detection Units	FDU	Steel Deck	ST DK
Products of Combustion	POC	Stone	STONE
Rate of Heat Rise	RHR	Story	STO
Fixed Temperature	FTEP	Street	STREET
Fire Pump	FP	Stucco	STUC
Floor	FL	Suspended Acoustical Plaster Ceiling	SAPL
Frame	FR	Suspended Acoustical Tile Ceiling	SATL
Fuel Oil (label with grade number)	FO #——	Suspended Plaster Ceiling	SPC
Gallon	GAL	Suspended Sprayed Acoustical Ceiling	SSAL
Gallons per Day	GPD	Tank (label capacity in gallons)	TK
Gallons per Minute	GPM	Tenant	TEN
Galvanized Iron	GALVI	Tile Block	TB
Galvanized Steel	GALVS	Timber	TMBR
Gas, Natural	GAS	Tin Clad	TIN CL
Gasoline	GASOLINE	Triple Hydrant	TH
Gasoline-Engine Driven	GED	Truss	TR
Generator	GEN	Under	UND
Glass	GL	Vault	VLT
Glass Block	GLB	Veneer	VEN
Gypsum	GYM	Volts (indicate number of)	450V
Gypsum Board	GYM BD	Wall Board	WLBD
High Voltage	HV	Wall Hydrant	WLH
Hollow Tile	HT	Water Pipe	WP
Hose Connection	HC	West	W
Hydrant	HYD	Wire Glass	WGL
Inch, Inches	IN	Wire Net	WN
Iron	IR	Wood	WD
Iron Clad	IR CL	Wood Frame	WD FR
Iron Pipe	IP	Yard	YD

Notes: Some words that have a common abbreviation, e.g., "ST" for "street," are spelled out fully to avoid confusion with similar abbreviations used herein for other terms. For SI units: 1 gal = 3.785 L; 1 psi = 6.895 kPa/1lb (force) = 4.448N.
Source: Cote, Fire Protection Handbook, 1997, Table 11–14A.

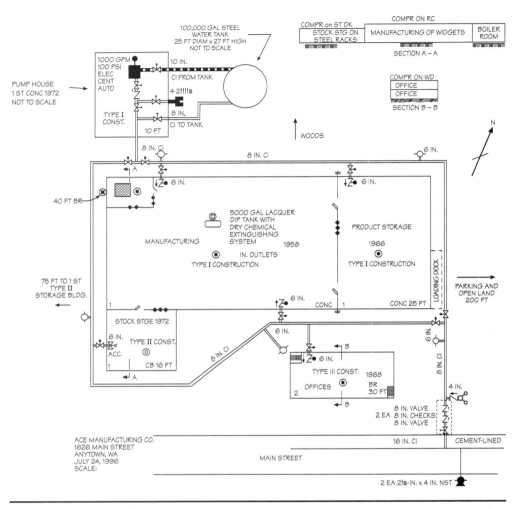

Figure 6.4 This diagram illustrates a typical site sketch. (*Source*: Cote, *Fire Protection Handbook*, 1997, Figure 11-14A)

often uncover code violations. Reports of violations should be sent to the fire prevention bureau or code official. If inspectors review prefire plans during inspections, they can provide the company that protects the inspected property with information for updating the plans. Prefire plan surveys and inspection reports help fire fighters become familiar with the buildings in their districts.

Before a company performs an inspection, the company officer should review the prefire plan with the fire fighters who will participate in the inspection. Then, if an inspection indicates that certain features of the plan have changed (for example, if an opening has been filled in or the location of hazardous material storage has been changed), the information can be incorporated into the prefire plan.

The public recognizes the advantages of advance planning, especially for emergencies, and they are usually impressed by prefire plans. Although a department's direct contact with the public is limited to the initial visit and follow-ups, newspaper articles about prefire plans can help make the general public more aware of the services of a fire department.

Generally, a fire department cannot efficiently cover an entire district as frequently as it should in the limited time available. Therefore, priority is given to certain areas or occupancies of high value or to those with a great potential for loss of life or property.

Use and Application of Prefire Plans

The time spent on prefire planning brings full benefits only if the plans are integrated with daily operations. Here are some suggestions:

- Planners should maintain good coordination with fire prevention activities so there is adequate follow-up on any unsafe conditions uncovered during site visits.
- Inspectors in fire prevention and in the companies should report any relevant changes in locations that might affect the prefire plans.
- All operating companies should have easy access to the latest prefire plan when they respond to an emergency.

Many fire departments that do not have on-board computers carry prefire plan books in all of their vehicles. Responding officers and fire fighters can then refresh their memories about the details of an occupancy while traveling to the site of an emergency. When departments have on-board computers in their fire command vehicles, incident commanders and subordinate officers are able to conduct more informed fire-fighting operations through ongoing reference to prefire plans in the system. Other departments use on-board fax machines to send prefire plans from central locations, to update information, or to forward the latest specialized data to the emergency scene. Computers and fax machines can pay off in enhanced operations, especially fire fighter safety.

Prefire planning achieves the maximum benefit when the plans serve as the basis for periodic training sessions. An increased awareness of sites, combined with on-site practical drills, translate into improved operational techniques.

Using prefire plans for training purposes makes drill simulation more stimulating. Fire fighters take a great interest in things that they can see, feel, and experience. Prefire plans bring life and relevance to their job performance. It is critical to rotate the facilities used for these drills. Personnel quickly tire of traveling to the same tank storage facility every few months. They must see the relevance of their drill work. We will discuss the use of prefire plans for training purposes further in Chapter 11.

Role of the Company Officer and the 3Cs

The company officer has the responsibility to manage the process of developing and maintaining up-to-date prefire plans. The officer can also use the plans to increase interest in fire prevention and in the details of the hazards in the district.

Fire fighters are generally interested in the initial survey of a building. However, repeat surveys for updating and subsequent inspections are likely to be treated less enthusiastically. This is especially true if the same personnel have inspected the same building for violations on a number of occasions. Officers have to apply considerable management and leadership competence to get the greatest community service and work satisfaction out of fire fighters during the time spent on prefire planning. The 3Cs or similar guidelines can be of help. Consider the following related issues:

• Every fire department needs to have goals and objectives, competently set with appropriate participation. The desire to work toward achieving challenging goals and objectives, freely or even enthusiastically set, motivates participants to action.

• Gaining full cooperation with setting and achieving goals and objectives requires that the members of the company have the competence to perform each required task. Work assignments need to place equal demands on all fire fighters, considering their respective skills and talents. Assigning to individual fire fighters those tasks that provide the greatest satisfaction to them and rotating the tasks that are widely desired or that present opportunities for increasing competence can also help with assuring high-quality plans and a high-quality work life.

• Continuing and effective communications ensure that everyone is aware of the importance of the service they perform and of any developments that add to this awareness. In effect, the officer who communicates well ensures that the norms that bring full cooperation are reinforced whenever an opportunity presents itself. (See Additional Insight 9.1, Positive Discipline and Counseling.)

• Coordination needs to be ensured with sound procedures for inspections and surveys, which have also been communicated thoroughly.

• Continuing competence development (and self-development by the officer) may be needed to ensure that fire fighters gain ever greater competence in understanding the procedures, laws, regulations, and codes and become comfortable with new data-collecting and data-recording equipment and programs. Fire fighters should receive the necessary training to correct any knowledge and skill deficiencies in drawing maps that show building dimensions, understanding the use of symbols, and deciding which details about a building are important enough to record. A competent fire fighter is less likely to feel apprehensive or inadequate. (See Chapter 11 and especially Additional Insight 11.1, Management of Learning and Coaching.) The frequent use of prefire plans to create interesting simulations during training can also be helpful in enhancing competence. The simulations add in-terest to prefire plans and the surveys and inspections that provide the foundation for them.

• The company's or the bureau's climate should be supportive and satisfying. To ensure such a climate, officers should prevent disruptive conflict (see Additional Insight 4.1, Management of Potentially Damaging Conflict) and assure fire fighters that their contributions are appreciated [see Additional Insight 5.1, Enhancing Work Satisfaction (Providing Recognition) and Performance Evaluation].

Use of Electronic Data Processing Equipment in Prefire Plans

In today's environment of rapid technological developments, the use of computers in prefire planning brings continual changes in the way a department organizes itself to prepare, maintain, update, and provide access to prefire plans. Increasingly sophisticated computer programs allow more and more departments to generate layouts and maps of properties in addition to the core information that is in pure data form. Commercial software enables departments to computerize prefire planning functions. Information can be found in trade publications such as *Firehouse*, *Fire Chief*, and *Fire Engineering*.

A mix of central and satellite electronic data processing equipment allows planners to gather information on location and enter it into notebook computers or other input devices. The information on the portable unit can then be transferred to the department's main database. When several steps of paperwork are eliminated, the inspection activity becomes more productive. Graphics programs are now available that will allow for the creation of diagrams on location. In this way, the drawing can be compared with the actual conditions when the plans are completed. Multiple visits are no longer necessary, which saves time for the fire department and reduces the disruption to those being visited.

Without computers, the information and graphics in prefire plans are available only on paper, possibly in binders at headquarters (for transmission via radio), or in the response or command vehicles. With electronic data processing equipment, responding forces can access copy and graphics via a portable computer on the command vehicle or on apparatus—which contains all the information for the entire response area—or via fax or modem on the vehicle from a central dispatch location. Electronic devices can help the incident commander size up and manage a fire emergency by providing the latest available information and instant clarification if, or when, needed.

The use of computers for accessing prefire plans can improve the effectiveness of responses to fires by both career and volunteer departments. Computers also assist in other ways and with other types of emergencies by making critical information rapidly available in such areas as hazardous materials, medical emergencies that are encountered only infrequently or rarely, topography for flood and forest fire emergencies, emergency management information for evacuation routes, and regional police department response. Developing prefire planning strategies for the most important or likely hazards can be a complex and time-consuming task, especially for those fire departments that have many industrial complexes, institutions, and commercial occupancies in their district. In the past, one barrier to extensive prefire planning was the inability to gather, sort, and store the data generated by a comprehensive prefire planning program. The barrier was compounded by the difficulties in quickly recovering the information for use during an emergency.

Rapid advances in computer hardware and software technology enable the fire service to maintain the data generated by a communitywide prefire planning system. Departments that have data-receiving equipment on board responding vehicles or that have complete mobile computers can now review prefire plans prior to personnel's arrival at the scene of a fire (see Figure 6.5). Gathering the data is still highly labor intensive, however, and many fire departments are unwilling or unable to devote the necessary time and energy to it, despite the fact that the dividends in operating efficiency, fire fighter safety, and protection for the community are extensive.

The success of fire department suppression operations in the future will depend even more heavily than today on the quality of prefire planning because the number of sites covered by a company is increasing and the availability of accurate information will make a critical difference in saving lives and property.

WATER SUPPLIES AND SYSTEMS

Water is still the most common fire extinguishing agent. Planning for its availability is therefore a critical aspect of the prefire planning process. Fire fighters must have an adequate and reliable water supply to prevent the spread of fires and extinguish them. Traditionally, the Insurance Services Office (ISO) grading schedule has placed emphasis on a municipality's ability to deliver sufficient water for the control of fires.

A thorough knowledge of water supplies and systems is necessary for fire service officers as detailed in NFPA 1021, *Standard for Fire Officer Professional Qualifications*. Therefore, a review of the importance of water supplies and systems to the fire service is appropriate here to provide perspective on the topic. For more detailed information on water supplies for fire protection, consult the *Fire Protection Handbook* (Cote 1997) or NFPA 1142, *Standard on Water Supplies for Suburban and Rural Fire Fighting* (formerly NFPA 1231).

Figure 6.5 A fire fighter operates an on-board computer system. (Courtesy of Newark, New Jersey, Fire Department)

Elements of the Water Supply

The modern water supply system, whether publicly or privately owned, should be adequate and reliable. Water systems designed today for municipal use have two functions: They supply potable water for domestic consumption, and they supply water for fire protection. Domestic water consumption includes water used for human consumption, sanitation, industrial processes, gardening (such as irrigating and lawn sprinkling), agricultural irrigation, and air conditioning and similar water-consuming processes.

Industrial sites often have separate systems for supplying process water and water for fire protection. Any dual-purpose system should be able to supply enough water for fire protection and at the same time meet the maximum anticipated consumption for other purposes. A good working relationship between the fire department management and the water company management will ensure that the available supply of water is used most efficiently.

Adequacy and Reliability of the Water Supply

The adequacy of any given water supply system can be determined by engineering estimates.* The source, including storage facilities in the distribution system, must be sufficient to furnish all the water that combined fire and domestic uses might require at any one time. Poor arrangement of the supply works and deficiencies such as undersized piping, excessive numbers of bends in the system, and outmoded pumping facilities may limit the adequacy of the supply or affect its reliability. In a pumping system, a common arrangement is to have one set of pumps that takes suction from wells or from a river, lake, or other body of water. If the water does not have to be filtered, the pumps can discharge directly into the distribution system. Where filtration or other treatment is necessary, pumps take suction from the primary or raw water source and discharge to sedimentation basins or other facilities from which the water then flows to filter beds. After processing, the water flows to clear-water reservoirs from which a second set of pumps discharges the water directly into the supply system. Unfortunately, failure of any part of the equipment can affect the entire system. Good prefire plans should reflect any situations that are uncommon for the department or that might present difficulties during fire suppression. For instance, prefire plans should indicate, for major potential risks, how the water supply could be augmented, if it is marginal. They should reflect water supply

---

*The information in this section is adapted from the NFPA's *Fire Protection Handbook* (Cote 1997, p. 6-72).

disturbances that may require supplemental water sources. In these situations, flexibility and the rapid transmission of electronic information can be of great help.

When department personnel assess the reliability of the supply works, they should evaluate the following features:

- Minimum yield
- Frequency and duration of droughts
- Condition of intakes
- Possibility of earthquakes, floods, and forest fires
- Ice formations and silting up or shifting of river channels
- Absence of guards or lookouts where needed to protect the facility from physical injury

Reservoirs out of service for cleaning and the interdependence of parts of waterworks also affect reliability. The condition, arrangement, and dependability of individual units of plant equipment, such as pumps, engines, generators, electric motors, fuel supply, electric transmission facilities, and similar items, are additional factors. Pumping stations that are combustible are subject to destruction by fire unless they are protected by automatic sprinkler systems.

Duplicating pumping units and storage facilities and arranging mains and distributors so that water can be supplied to them from more than one direction are measures that can ensure continuous operation. The importance of duplicate facilities is shown by the frequency of their use.

Standpipes and Automatic Sprinkler Systems

Standpipes and automatic sprinkler systems are sometimes considered the first line of defense against the spread of fire. Standpipe and hose systems provide a means for the manual application of water to fires in buildings. Although standpipes are required in buildings of large area and height, they do not take the place of automatic sprinkler systems. For further information, consult the latest editions of NFPA 13, *Standard for the Installation of Sprinkler Systems*, and NFPA 14, *Standard for the Installation of Standpipe and Hose Systems*. They provide guidance for questions in these areas.

In his book *Automatic Sprinkler and Standpipe Systems*, John L. Bryan comments on the efficiency of automatic sprinkler systems. Sprinklers cut the chances of dying in a fire by one-third to two-thirds and cut property loss per fire by one-half to two-thirds (1997, p. 37). Sometimes the savings in annual insurance premiums are sufficient to pay for the installation of a sprinkler system in a few years.

When sprinklers produce unsatisfactory results, which occurs rarely, one of these reasons is usually involved:

- Partial, antiquated, poorly maintained, or inappropriate systems
- Explosions or flash fires that overpower the system before it reacts
- Fires so close to people or to sensitive, valuable properties that fatal injury or expensive damage, respectively, occurs before a system reacts (Bryan 1997, p. 37)

The overall effectiveness of a sprinkler system usually depends on the adequacy and reliability of the public water supply system, even when a water tower is on top of or adjacent to the building. Supplies often are limited in quantity and alone might not be adequate for a large fire. When a sprinkler system exists, the probability is very small that any fire can spread. For additional guidance, we recommend that you consult NFPA 13, *Standard for Sprinkler Maintenance*; NFPA 14, *Standard for Inspection and Maintenance of Standpipe and Hose Systems*.

Areas of Overlapping Responsibility

The water company and the fire department cooperate in ensuring an adequate and reliable water supply. In the following situations, their responsibilities somewhat overlap:

- When extensive new construction is considered, both the water company and the fire department are involved in determining whether the water supply system will provide the required fire flow.
- The fire department often has maps of the complete water system. Hydrant checks of the entire district are usually carried out on a rotating basis so that every hydrant is checked at least once a year (more frequently in high-value areas) to determine whether the hydrant is in operating condition and whether it can deliver the required fire flow. You may wish to consult the standards of the American Waterworks Association* regarding the design, placement, and maintenance of hydrants and water supply systems. The water company usually is responsible for promptly correcting any problems that are found.

- When changes of occupancy occur before permits are issued, inspectors should consider the adequacy of the water supplies relative to the haz-

*The address is 666 W. Quincy Avenue, Denver, CO 80235.

ards of the new occupancy. For guidance in determining water supply requirements, see NFPA 1, *Fire Prevention Code*, which gives water supply requirements for sprinklers, airport terminals, and buildings under construction.

- Where the relationship between the fire department and the water utility is good, the utility immediately informs the fire department about problems that affect the water supply. If these communications do not occur reliably (and even if they do), all affected segments of the fire department need to be vigilant in noticing water supply problems that the appropriate fire department liaison officials can address with the water utility. Conversely, when there is a major fire, the department should inform the water utility so that it can increase pressure, if necessary, or open emergency valves.

Auxiliary Water Supplies: A Fire Department Responsibility

The fire department has the responsibility to arrange for auxiliary water supplies. When the supplies are located on private property, the department must get the cooperation of the property owner before the need arises. When the water supplies are under the jurisdiction of a public authority, agreements for their use must be established before the need arises. The fire department must be thoroughly conversant with the provisions of NFPA 1142 (formerly NFPA 1231).

Some rural and suburban fire departments maintain large-capacity tank trucks and tanker shuttles for the transportation of water when there are no hydrants. Other departments have apparatus specially designed to draw water by suction from static water supplies. (Pumpers—especially new ones—should meet the requirements of NFPA 1901, *Standard for Automotive Fire Apparatus*.) A department must, of course, arrange for access to the water supply for every possible fire in the district. Sometimes the arrangements take considerable planning and negotiations with those who control the supplies.

In the past, lack of hydrants adversely affected insurance rates because the Insurance Services Office (ISO) considered hydrants to be the preferred source of water. Changes in the grading schedule have eliminated this severe bias against alternative water supply methods.

Department Organization to Ensure Adequate Water Supplies

Of major concern to fire department management is departmental organization to ensure adequate water supplies. For example, a large fire department might appoint a special committee to work with the water company to

ensure adequate water supplies for fire protection. This committee could include a water liaison officer from the fire department, a water supply company representative, and representatives from government agencies affected
by the decisions. Smaller departments usually do not need to organize committees for such functions. They generally delegate these responsibilities to
specific personnel within the department.

Regardless of the way the water supply planning is handled, fire companies must know how to obtain the supply they need at every point in the
district. In most departments, at least one person in each first-due company
knows the water supply system well. These people should know the general
range of pressures in the water mains and the size of the mains available for
fire fighting. If the locations of hydrants and mains are noted on the prefire
plans, as they should be, then pumpers may be located without unnecessar-
ily long hose lays. Prefire planning for all large buildings or plants should
emphasize the water supply.

Role of the Company Officer in Ensuring Adequate Water Supply

The company officer has the responsibility to become familiar with the water supply system maps and the locations, flow capabilities, and operation
of hydrants in the district. The company officer should make sure that
scheduled inspections of hydrants in the district are done and that reasonable priority is given to inspecting all water supply system components in
the department's district.

A fire department connection is mandatory on all standpipe systems
and on sprinkler systems. The connection provides the only means of supplying water to the dry standpipe system. The fire department connection
should always be inspected by fire companies during in-service inspections.
Regular inspection and testing of fire department connections for both
standpipe and sprinkler systems and the complete testing of both systems
are requirements for every fire department. It is the company officer's responsibility to see that tests, inspections, and evaluations for all standpipes
and sprinkler systems in the department's district are carried out in accordance with accepted procedures (see Figure 6.6).

Automatic sprinkler systems, one of the greatest aids to fire departments, can function effectively only with sufficient water pressure. The
major cause of unsatisfactory performance of automatic sprinkler systems
is human error (Cote 1997, p. 1-30). Water supply control valves may be
closed when the fire starts or before the fire is completely extinguished.
Therefore, upon arrival at a sprinklered building, the officer in command

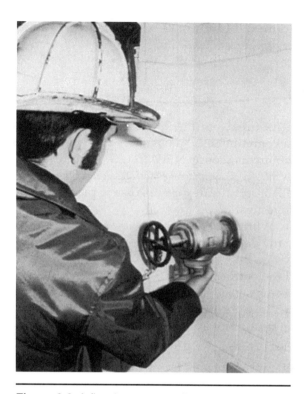

Figure 6.6 A fire department officer checks the condition of the discharge valve from a standpipe system. (Courtesy of Stephen C. Leahy, College Park Volunteer Fire Department, College Park, Maryland)

should supplement the sprinkler system's water supply with hose lines from adequate public water mains or natural sources to the fire department connection on the sprinkler system, so that adequate pressure is ensured. Another responsibility for the officer upon arrival is either to station someone at the sprinkler system supply valve to prevent the water from being turned off or to otherwise ensure that it remains open until the fire is effectively controlled. These essential and simultaneous procedures are determined, in part, by information obtained during fire department prefire planning.

Details on the testing, maintenance, and periodic inspection of standpipe systems are given in NFPA 14. Details on acceptance tests and water supplies for sprinkler systems can be found in NFPA 13. Information on inspections and prefire planning for both automatic sprinkler and standpipe systems is covered in NFPA 13E, *Guide for Fire Department Operations in Properties Protected by Sprinkler and Standpipe Systems.*

FIRE INVESTIGATION

An important job for fire departments is to identify the causes of fires that occur within their communities as a basis for determining what needs to be done to reduce the fire incidence. NFPA 921, *Guide for Fire and Explosion Investigations,* and NFPA 1033, *Standard for Professional Qualifications for Fire Investigator,* are available for fire department use in planning, conducting, and managing a fire investigation program.

Investigations of fires that are not considered to have suspicious origin are conducted by fire department officers or fire prevention inspectors. The investigations provide information about the causes of fires, the reasons for their spread, and the performance of whatever automatic fire suppression equipment had been installed in the building. Much can be learned from the causes of these fires, and their effects on the buildings involved.

The four basic reasons for investigating fires are

- To develop new fire prevention methodology for addressing special causes identified by the investigation and for fine-tuning strategies for similar situations in the future
- To determine whether fires are caused by arson so that appropriate authorities can be notified
- To determine compliance with codes and permit compliance enforcement steps
- To provide documentation for insurance company inquiries

For fires of unknown or suspicious origin, a higher degree of sophistication is required from the fire investigators. For suspicious fires, investigators must gather additional information from investigations and interrogations to strengthen a criminal case.

Arson Investigations

The authority for investigating fires of a suspicious nature rests primarily with the state fire marshal's office or, in the absence of this position, with the state police. In practice, however, many states delegate this authority to local fire department officials who cooperate fully with the police. Some large departments have arson squads with the special task of investigating suspicious fires (see Figure 6.7). Some of these squads have officers, trained by the police department, who have the power of arrest and the responsibility for preparing cases for prosecution. Depending on local ordinances, fire prevention officers may also have such police powers. Other squads are composed of officers from both fire and police departments.

The special training needed for fire investigation might not be available in smaller fire departments. In such instances, when arson is suspected, the task is often turned over to the local police department or to the police department in the nearest large community.

The investigation of a suspicious fire usually begins with a search for evidence of the way the fire started. The investigator searches the fire scene for clues and obtains statements from witnesses and others who can contribute ideas and information. Sometimes the officer in charge of the investigation calls on the fire fighters for their assistance. Care should be taken to ensure that all possible evidence is collected and that pictures are taken of all

Figure 6.7 Fire investigators examine debris in a nursing home fire. (*Source*: Desert News, Salt Lake City, Utah)

significant places. As part of the investigation, department records are searched for relevant information. The insurance companies are contacted for additional information.

Because arson is a felony in every state, investigators attempt to determine the ignition sequence after arson has been determined as the cause. During the arson investigation, police authorities normally have primary responsibility, much the same as for other felonies that occur within their jurisdiction. In many areas of the United States, however, particularly at the state level or in larger cities, state and municipal laws give fire authorities the power to perform the police function of an arson investigation. If a suspect is apprehended and charged, the fire service has the primary responsibility to prove that the fire was willfully and maliciously caused. This may involve activities that are not usually considered fire department responsibilities, such as interviewing witnesses and uncovering leads on possible motive, opportunity, and intent.

When the fire service is developing an arson case, it must establish three facts:

- The fire caused actual charring or destruction.
- The ignition sequence of the fire was the result of willful and malicious design or intent.
- The suspect, or an agent of the suspect, had the opportunity to cause the fire.

When prosecutors present an arson case, each of these elements must be established to the satisfaction of the court in the order listed. An established motive is of great value in an arson case, although motive is not, from a legal standpoint, considered an essential element of arson. Many incendiary fires are set with no known motive.

Fire Loss Reporting

A fire protection duty that is as important as comprehensive investigation is the accurate reporting of such investigations. This duty is seldom recognized as an essential part of fire protection. In the past, fire departments and fire protection interests have often had to depend on unreliable or incomplete projections and reports to justify corrective action for fire problems.

NFPA 901, *Standard Classifications for Incident Reporting and Fire Protection Data*, establishes uniform language, methods, and procedures for fire loss reporting and coding. This system is designed so that all information can be coded for electronic or manual data processing. The use of computerized data processing is essential for the collection and rapid retrieval of meaningful fire loss data.

Computers have become essential. Even small departments now use personal computers to maintain their records and reports. They provide an almost unlimited capability for storing, organizing, and retrieving the great volume of reports generated by large, and even smaller, jurisdictions, from ignition sequence investigation. These data include not only point-of-origin information, utility status, and involvement of flammable materials and accelerants, but also the detailed description of specific evidence supporting the conclusions.

As discussed early in this chapter and elsewhere, through the use of database programs, fire departments are able to gather, analyze, and utilize great quantities of information and statistics. They can use spreadsheet analyses to identify fire hazard trends, which help authorities pinpoint problems, suggest answers, and avoid misdirecting fire protection dollars. Computer analyses have a major effect on fire suppression procedures and the enforcement of fire prevention codes and fire investigations.

Role of the Company Officer in Fire Investigations

The company officer has the responsibilty to try to determine the origin of any fire immediately upon arrival at the scene and to continue the investigation, if necessary, after the fire has been brought under control. If an inspector from the fire prevention bureau is available, that inspector should assist with or, where department policy calls for it, take charge of the investigation.

Fire investigators may suspect, but not prove, arson from conditions such as an absence of evidence that the fire was due to accident, a smell of gasoline or kerosene, simultaneous start of the fire in several different parts of the building, and other clues. In such instances, the officer should call the arson squad, the fire prevention bureau, or the chief immediately, so that a more detailed investigation can be made.

As long as fire fighters are at the scene, the fire department retains control of the property and can prevent members of the public, including the owner, from entering and possibly disturbing any evidence. Thus, fire fighters and other officials who are conducting investigations can search the premises and collect necessary evidence without obtaining legal sanction.

Company officers should schedule training in fire investigation from time to time so that fire fighters can learn to find clues to the causes of fires. Such training also ensures that fire department personnel do not disturb pieces of evidence in the clean-up process after a fire.

LOSS REDUCTION INFORMATION MANAGEMENT SYSTEMS

In loss reduction, as in other fire department operations, the development of an effective information management system is crucial. Although the system should be as thorough as possible, it should also provide a simple and easy-to-use procedure for gathering and retrieving information necessary to manage the loss reduction program. To assist in future efforts, justify past actions, and assist outside agencies, the system must be able to record, retrieve, and rapidly utilize data that are gathered on a continual basis.

To manage the large volume of data in a loss reduction information system, most fire departments use computers. When a fire occurs, information about that incident is recorded along with all pertinent data on the structure or occupancy in question. This information has many and varied uses:

- It provides updates for existing prefire plans and helps with the development of new plans. A department can reduce the amount of time needed to complete a survey and produce the document.
- It helps in the development of a history of code compliance for specific properties. The data assist with scheduling inspections and designating problem occupancy classes (types) for special inspection efforts.
- It assists insurance companies and ISO grading personnel in determining rates and ratings.
- It serves as the basis for statistical analyses to show what changes in practices and regulations might strengthen fire prevention activities or reduce the probability of fires through structural changes.

Rather than using a series of standard reports, a fire department should determine what it needs and set up a system for gathering and using the information. The best systems are designed to meet a specific department's needs. If a department already has an information system, it should be screened for extraneous data. If setting up a new one, the department should organize a committee to study the needs and design a system to meet the precise needs identified for the community served.

CONCLUDING REMARKS

The probability is low that a specific prefire plan will actually be used. That fact has a strong demotivating impact on work that is related to prefire planning. It is reinforced by the belief that a competent sizeup does not require much in the way of prefire planning for most fires. Furthermore, unless prefire planning is carefully integrated with inspections, updating of the plans is a time-consuming activity that brings few rewards. And then there is the impact of the steady flow of new communications and electronic data processing equipment that is continually changing the ways in which prefire plans are developed and what they contain. Rapid change further strengthens the feeling that work on prefire plans is of little value, because the plans will likely have to be revised again in the near future.

Many functional and management/leadership decisions are involved in ensuring that useful prefire plans will be available when they are needed. Company officers are fully aware that thorough prefire planning is well worth the effort, even if it saves only one life, prevents one serious injury, or significantly reduces property losses. Still, they have a tough assignment to translate that awareness into action. Reviewing the fire service function guidelines from time to time will help to spark ideas.

The small likelihood of benefits from prefire planning, which may not even be reaped by their company but by a company in another station or on another shift, are not enough to overcome the typical lethargic attitudes toward prefire planning.

To achieve enthusiasm, or even willing cooperation, requires thorough attention to all 3Cs.

- Goals and objectives have to be set cooperatively on scheduling plan development and updating it for individual properties.
- Obstacles to cooperation with other shifts have to be eliminated.
- Obstacles to thorough coordination with inspections have to be identified and eliminated.
- Competence in preparing plans and entering them into whatever system exists has to be sharpened.
- Competence in upgrading the system to the best that can be done with available equipment must be ensured.
- Rewards, at least nontangible ones, for effort on prefire planning have to be considered.

The goals and guidelines given at the beginning of this chapter are valuable to company command officers and chief officers alike. They provide perspective for those who want to strike the best balance between the costs of developing, maintaining, and updating prefire plans and the benefits that the company, department, and community will gain from them.

STUDY QUESTIONS AND ACTIVITIES

If you are working alone, prepare your own written responses to these questions. If you are studying with a team or working in class, discuss the questions with the group and write a consensus answer.

1. Review the prefire planning process in your department and the prefire plans for a few high-risk occupancies. Write a short evaluation of prefire planning in your department, including what you consider to be the strengths and weaknesses.
2. Your department's existing prefire plan must be expanded to include a recently constructed chemical storage building. This building is located at the top of a steep hill in the outskirts of your community. What special information do you need to prepare a fully useful prefire plan for this facility?
3. Review the ways in which inspections are used in your department to provide information for updating prefire plans, how the information is communicated, and what is actually done to upgrade the plans. Suggest how these procedures could be improved to achieve more useful prefire plans.
4. Describe the role of the company officer in ensuring that all fire fighters are familiar with the prefire plans for all important hazards.
5. Discuss why it is important to a conduct a fire investigation after every fire. How can a department use the information obtained during such an investigation?
6. What should a company officer know about the water supply in the district? What alternatives are available to ensure adequate water to fight fires where the primary supply is marginal or inadequate?
7. Prepare specific contingency plans for one or two important hazards in your district where the primary water supply is marginal or inadequate.
8. What are your department's standard operating procedures or guidelines for actions to be taken with respect to sprinklers and standpipes during inspections and for the first company arriving at a fire emergency? Are these procedures adequate? Where should changes be considered?
9. Review NFPA 901 and compare your department's standard and actual reporting procedures with NFPA 901. Where, if anywhere, are there differences, and why?
10. What are the specific responsibilities of a company officer with respect to prefire planning and related loss reduction activities, considering the objectives and guidelines?

11. In what ways can a company officer apply the 3Cs guidelines to prefire planning and related loss reduction activities and thereby improve the performance of the company?

CHAPTER 6
REFERENCES

Bryan, J. L. 1997. *Automatic Sprinkler and Standpipe Systems*, 3rd ed. Quincy, MA: National Fire Protection Association.

Cote, A. E., ed. 1997. *Fire Protection Handbook*, 18th ed. Quincy, MA: National Fire Protection Association.

Lakein, Alan. 1973. *How to Get Control of Your Time and Your Life*. New York: P. H. Wyden.

Lieberman, H. R., and Rausch, Erwin. 1976. *Managing and Allocating Time*. Cranford, NJ: Didactic Systems, Inc.

NFPA 1, *Fire Prevention Code*, National Fire Protection Association, Quincy, MA.

NFPA 13, *Standard for the Installation of Sprinkler Systems*, National Fire Protection Association, Quincy, MA.

NFPA 13E, *Guide for Fire Department Operations in Properties Protected by Sprinkler and Standpipe Systems*, National Fire Protection Association, Quincy, MA.

NFPA 14, *Standard for the Installation of Standpipe and Hose Systems*, National Fire Protection Association, Quincy, MA.

NFPA 170, *Standard for Fire Safety Symbols*, National Fire Protection Association, Quincy, MA.

NFPA 901, *Standard Classifications for Incident Reporting and Fire Protection Data*, National Fire Protection Association, Quincy, MA.

NFPA 921, *Guide for Fire and Explosion Investigations*, National Fire Protection Association, Quincy, MA.

NFPA 1021, *Standard for Fire Officer Professional Qualifications*, National Fire Protection Association, Quincy, MA.

NFPA 1033, *Standard for Professional Qualifications for Fire Investigator*, National Fire Protection Association, Quincy, MA.

NFPA 1142 [formerly NFPA 1231], *Standard on Water Supplies for Suburban and Rural Fire Fighting*, National Fire Protection Association, Quincy, MA.

NFPA 1620, *Recommended Practice for Pre-Incident Planning*, National Fire Protection Association, Quincy, MA.

NFPA 1901, *Standard for Automotive Fire Apparatus*, National Fire Protection Association, Quincy, MA.

Rausch, Erwin. 1984. *Balancing Needs of People and Organizations*. Washington, DC: Bureau of National Affairs, 1978; Cranford, NJ: Didactic Systems, Inc.

CHAPTER 6
ADDITIONAL READINGS

Didactic Systems, Inc. 1971. *Effective Delegation: A Didactic Simulation/Game.* Cranford, NJ: Didactic Systems, Inc.

Didactic Systems, Inc. 1974. *Managing Time Effectively: A Self-Study Unit.* Cranford, NJ: Didactic Systems, Inc.

Didactic Systems, Inc. 1975. *Time Management: A Didactic Exercise.* Cranford, NJ: Didactic Systems, Inc.

Didactic Systems, Inc. 1976. *Managing and Allocating Time: A Didactic Simulation Exercise.* Cranford, NJ: Didactic Systems, Inc.

TIME MANAGEMENT AND DELEGATION

This section briefly covers major principles for making effective use of time. One of the most important, of course, is delegation of work that others can do, whenever there just is not enough time. That is why we combine delegation with time management in this section. Time management is a skill that was on the top of the lists for management development programs during the 1970s and has since almost disappeared. Still, we should consider its importance.

Alan Lakein, the top guru on the subject, wrote a comprehensive book on effective time management (Lakein 1973). Two important suggestions are to set priorities and avoid time wasters. Little needs to be said about time wasters. We all know what they are—unimportant phone calls, interruptions, uninvited visitors, unnecessary or poorly conducted meetings, and so forth.

More important, however, are priorities. They determine what work managers (and officers) do themselves and what they delegate or assign. Lakein suggested three groupings: Priority 1 items are done by the manager or officer, those in priority 2 are delegated, and (tongue in cheek) those in priority 3 are placed in the bottom drawer of the desk: If they don't rear their "ugly" heads within a month, they are thrown away.

Though interesting, these priorities do not provide much guidance. A diagram of the importance/urgency separation first published in a simulation exercise can shed some light (Lieberman and Rausch 1976). The diagram in Figure 6.8 illustrates all four possible combinations of importance and urgency.

- C: Matters that have low importance/low urgency can be postponed until there is some spare time available, or they can be assigned to anyone who has or will have time and the necessary competence; sometimes they can even be ignored.
- D: Matters that have high importance/low urgency should be delegated (often with goals—see Additional Insight 5.1, "Making a Goals Program Work") to someone who has the necessary competence and preferably is not busy with tasks that are more important and at least equally urgent.
- A: Matters that have low importance/high urgency should be done if time is available; otherwise, they should be assigned or delegated.
- B: Matters that have high urgency/high importance should be done personally, with portions assigned or delegated.

With these four categories in mind, officers, as managers/leaders, can concentrate on those issues that most deserve their personal attention. At the same time, the added impetus to delegate encourages developmental assignments that will strengthen the organizational unit's competence.

Setting Priorties
Importance

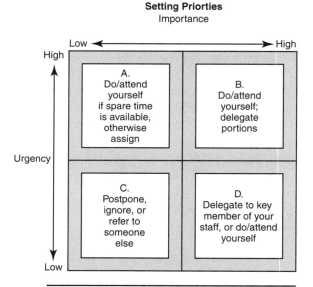

Figure 6.8 The diagram illustrates four possible combinations of importance and urgency. (*Source:* Lieberman and Rausch 1976; Rausch 1984)

It is worthwhile to think of delegation and task assignment in a similar vein as participation in decision making and even, to some extent, as goals or objectves and action steps. (See both Additional Insight 4.1, Participation in Decision Making and Planning, and Additional Insight 5.1, Making a Goals Program Work.) In delegation, as in setting goals or objectives, the officer and the delegatee jointly set them on what is to be achieved, and the delegatee has wide authority to make the necessary decisions, with the officer providing all needed and requested support. Both parties understand that it is the delegatee's responsibility to achieve the delegated results or, if obstacles develop that may prevent such achievement, to give the officer adequate notice and information so that timely intervention with additional resources is possible. In effect, this defines the responsibility that remains with the officer: to intervene and accept responsibility for the outcome when the results are threatened.

Task assignment, in contrast, involves action steps. The officer and the assignee agree on the definition of the task and on the resources that may be required. Then the assignee is expected to complete the task.

Clear definition and communication of the limits of authority and the way responsibilities are shared are of great importance in delegation and even in task assignment. When a delegated assignment requires the efforts of several staff members at the same organizational level, and especially when staff members from different companies or bureaus are involved, the officer or officers must clearly define the limits of authority and reach thorough mutual understanding on when the officer should be asked for support and involvement.

Management of Physical Resources

INTRODUCTION

Physical resources are the facilities, apparatus, equipment, and expendable supplies needed to conduct operations—both emergency and nonemergency. Because a fire department's physical resources are so important to its ability to protect life and property, the fire insurance rates for a community are affected by the quality of these resources.

Throughout this chapter we refer to NFPA standards for fire department managers to use to ensure that the department is appropriately equipped and that the physical resources are adequate for to the needs of the community. We also refer to ratings that use these standards and others. These ratings are developed by an insurance rating organization, the Insurance Services Office (ISO), and used by insurance companies as a basis for premiums.

Our review of the fire department's physical resources in this chapter takes into account the needs of the community and the resources that are required for the fire department to serve the community effectively. We also look at the functional considerations in managing a fire department's physical resources. The standards and ratings are discussed toward the end of this chapter.

SCENARIO

THE TURNAROUND CHALLENGE

The doctors told Captain Murd that he would be fine. Still, after almost 35 years of service, he felt it was time to hang up his cap. When he informed Battalion Chief Leonta, the commander of the first shift, of his decision to retire, the chief again expressed his relief that the captain would fully recover but did nothing to try to change the captain's mind. In fact, Chief Leonta may have been pleased to have the opportunity to appoint someone else to head up the first shift in Station 3. He had received a number of complaints, like low air in breathing apparatus when the next shift checked them, a hose line not connected when the second shift unloaded it, and even an almost empty booster tank after the shift had extinguished a minor brush fire. Problems like these along with less serious ones had occurred even before Captain Murd became ill. Chief Leonta had himself found that minor procedures were not always followed, especially in the way the shift performed station maintenance.

In their frequent discussions about these matters, Captain Murd had always found some excuse that seemed to satisfy him, though not the chief. Chief Leonta had finally warned the captain that his position was in jeopardy if his company did not improve its performance. Soon after that, the captain found out that he had both a duodenal ulcer and prostate cancer. The captain's illness did not make it easy for the chief. Chief Leonta pitched in to help the acting company commander, a fire fighter who was up for promotion and who was filling in as the fourth member of the company. That was when the chief became aware of how ineffective Captain Murd had really been as a leader and how he had contributed to the low morale of the three fire fighters on the shift.

The department was heading into a difficult time. The latest contract with the union stipulated that fire department personnel would paint the three fire stations and refinish the floors. That work had been done previously by outside contractors. Fire suppression personnel would also assume wider responsibility for conducting field inspections in their respective districts and for completing the expanded inspection schedules within a specified time.

Nobody in the department was happy with the situation. However, being residents of the town themselves, the fire fighters knew how strong the pressures were to reduce taxes. The members of the town's new administration had been voted in with large majorities just before negotiations for the new contract started. They had campaigned vigorously on a platform that promised to hold the line on taxes. The fire fighters understood that they could not have reached agreement on the pay increase, which the union won, without the concessions on fire station maintenance.

Two new challenges now faced the department:

- What steps should be taken to meet the inspection commitments? In the past, inspections had received low priority. The least important locations were often not inspected, or reinspections were not done.
- How should the maintenance work be done, and how should it be apportioned among the staff members in each station house without cutting into training time?

The department chief moved quickly to appoint a new captain and to hire a volunteer fire fighter as a career member to fill the vacancy, in line with the department's promotional policy. The chief also set up meetings with the battalion chiefs and the fire prevention bureau to prepare realistic inspection schedules for all shifts.

Battalion Chief Leonta recommended that an experienced, strong company commander from another station be assigned to his shift in Station 3 rather than the new officer, at least temporarily until things settled down. He felt it would be difficult for a newly appointed officer, or even one of the captains from the same shift, to overcome so many challenges at once. The department chief agreed and Captain Sharpe, who had headed up the second-shift company at the headquarters station house, was brought in to take charge. A new officer replaced him there, under the watchful eye of the second shift battalion chief.

When Captain Sharpe reported to his new post, he considered his top priorities:

- Reviewing the equipment and apparatus routines to ensure that they were thorough, complete, and followed
- Reviewing the station maintenance routines assigned to his shift to ensure that they too were appropriate and adhered to
- Revising the inspection schedules as soon as he received the new ones from the battalion chief
- Arranging for the painting and floor refinishing share of the work to be assigned to the company

The captain spent most of the first day touring the station house and getting to know the fire fighters. He deliberately did not check apparatus or equipment, and he did not seem to notice where cleanliness left something to be desired.

Toward the end of the day, he called his team together in the day room to bring them up to date and to start a discussion on what and when things would be done and who would do them. First, he expressed his pleasure that Captain Murd apparently had gained control over his illnesses. Then he briefly told the team what he believed they already knew: the contract concessions on inspections, painting, and floor refinishing, as well as a few of the complaints that had been voiced by the second shift. The fire fighters listened silently. Captain Sharpe continued, "So that's what we have to do. There's a lot of work involved and I count on your cooperation, just as you can be sure that you can count on me to help out wherever necessary. There'll have to be some changes, of course, so we will not get complaints. In fact, I hope that you will work with me to make this tour, in Station 3, a fine example of the way things should be done. But, before we get down to specifics, let me hear your thoughts on all this."

With that he stopped talking and looked around. There was Frank, the old-timer who had been with the department almost 30 years; John, who was a steward in the union; and Peter, who had joined the department full-time two years ago, after four years as a volunteer.

John was the first to speak up. "We got a bum rap with some of those complaints. Yes, we did leave the pumper half empty, but we got in late after the fire. Captain Murd was supposed to tell the next shift and ask them to fill the tank. He forgot, and we got the black eye. And when it comes to the hose that was not connected, maybe it was us, but maybe it was another shift. It may just be that, when it was our turn to lay out the hoses, we had to cut it short and never got to the section that was not connected. Don't take everything you hear at face value."

Frank and Peter just nodded their heads, possibly waiting to see how the captain would respond.

"You know, that may be what happened, but it really doesn't matter. What's in the past is in the past. I'm only concerned with the future," the captain said.

John replied again. "Well, if it's the future you're interested in, then let's get to work on having headquarters supply us with a compressor like the one they gave to Winster Street. It's less likely that we'll have trouble with the SCBAs once we have that kind of compressor."

"That's something that's already in the works," Captain Sharpe answered. "The headquarters station is going to get the next compressor and, unfortunately, we are last. You know that we can spend just so much money each quarter, so not all equipment can be bought at the same time. But then, as you may remember, Station 3 was first when it came to upgrading the face pieces."

"When I say the future, though," the captain continued, "I'm talking more about what we are going to do here, how we are going to handle the new work, together with the old work, without getting bent out of shape. Let me be specific. Let's take a look at what we've got to do when we start and end our shift; what's involved in apparatus, equipment, and station maintenance. Then we can lay out how the new work will fit into that and agree on who'll do what. How do you feel about that?"

"Looks like a good way to start," Frank spoke up. "We've got to figure out how to handle these new assignments, and still get our regular work done."

"I guess that's right," Peter added. John remained silent.

"All right, we don't have an SOP or even a suggested operating guideline, but we can make up a rough one. You tell me what we will do each time we come on duty and before we close." Captain Sharpe opened the folder he had in front of him, pulled out the pencil, and looked around.

This time Peter was the first to speak. "We start by checking the personal equipment, like the SCBA, the hood, the face piece, coat, safety glasses, PASS-alarm, the pressure gauge, the coat, and the pants and boots. I don't want to go on a run and have trouble getting a boot because someone put a small rock into it."

Frank laughed. "That was to teach you a lesson about checking your boots every time. I saw you weren't doing it, and I thought you'd get the message that way for sure."

"I figured it was you. Just wait."

The implied intent of more horseplay did not sit well with Captain Sharpe. "All right, guys, I'm all for having as much fun as we can possibly squeeze into a day, but please, let's make it clean fun and not something that can cause us trouble if we have to pull out in a hurry. Will you promise that?"

He looked straight at Peter, who answered quietly, "You're right, Cap'n."

"OK, what else in the way of personal equipment checkout? When we are finished, I'll write all this up, and we'll each have a copy so we can refer to it to make sure we don't forget something. In effect it'll be an operating guideline."

Frank offered an item: "There is one other item that I check frequently, and that is my personal tools. It's easy to forget one or the other item, and over the years I've lost a few by not checking every time, when I can still remember where they might be."

"OK, would we check all these items again, before we sign off from the shift?" Here the captain didn't really expect a positive answer.

Sure enough, Pete was honest. "To be frank, we don't really do it all the time. It depends on whether we had a run and how busy we are at the end of the shift."

"Should we?"

The alarm went off. The dispatcher's voice came over the loudspeaker: "Engine 3, mutual aid, staging, Rivertown High School."

"OK. That does it for today; if there is time when we come back, let's clear up our thinking on what to do at the end of the shift. Then we'll make a list for the apparatus checks. If we don't have time, we'll do it as soon as we can on the next shift. Let's get going." With that the captain walked to his equipment and prepared to board the engine.

Scenario Analysis: Fire Service Function Perspective

Before reading the guidelines and analysis, you may want to give some thought to the strengths and weaknesses of the way Captain Sharpe handled matters related to the management of physical resources. Consider the goal for the management of physical resources:

> The primary goal for management of physical resources is to ensure the avoidance of wasteful practices and the acquisition of the best possible equipment and facilities, obtainable with available budget funds and maintained and deployed effectively, to support the department's functions and mission statement so the department can respond effectively to all fire service requirements.

A guideline supports this goal:

> What else needs to be done to avoid wasteful practices and to acquire the best possible equipment and facilities, obtainable with available budget funds and maintained and deployed effectively, to support the department's functions and mission statement so the department can respond effectively to all fire service requirements?

The following specific questions are related to this guideline:

1. What else should be done to identify and eliminate wasteful practices so that the department's resources are used most effectively?
2. What else should be done to ensure that the department has the optimum number of fire stations, considering available standards (such as NFPA and ISO), that these fire stations are located to meet changes in

the community's needs, and that they are in condition to meet apparatus and personnel needs?*

3. What else should be done to ensure that the department has the optimum number and types of apparatus for fire and other emergencies, considering available standards (such as NFPA and ISO), that the equipment meets the needs all the situations that arise and are likely to arise in the community, and that each piece of apparatus is maintained in top operating condition and appearance?

4. What else should be done to ensure that the department has optimum administrative and logistic facilities, repair facilities, alarm receipt and dispatch facilities and other communications equipment, personal protective equipment, and supplies for fire and other emergencies, considering available standards [such as NFPA, ISO, and FEMA (Fire Equipment Manufacturers Association)]; that the equipment meets the needs of all situations that arise and are likely to arise in the community; and that all are maintained in top operating condition?

Keep in mind that it is not necessary to remember the detailed wording of the guideline when you are making decisions. Only in the most important decisions might that be worthwhile. However, memorizing a personalized, abbreviated (summarized) version and applying it with every relevant decision are critical for developing the necessary habit that will help to ensure the highest quality performance possible under the existing circumstances.

This guideline is a tall order and requires constant attention to the physical resources by the chief and the other officers. In the turnaround challenge scenario, department top management seemed to be aware of its role in managing physical resources. The are purchasing compressors as rapidly as financial resources permit and paying attention to the physical upkeep of the stations. We can reasonably assume that they are paying similar, if not greater, attention to the station size and location issues and to issues that pertain to apparatus and equipment.

We will now look at the specific questions related to the guideline.

1. *What else should be done to identify and eliminate wasteful practices so that the department's resources are used most effectively?* The scenario gave little information on this question, which applies mostly to the use of supplies, utilities, and other expendable commodities that are part of the budget. Still, the

---

*Guidance is available from the NFPA's *Fire Protection Handbook* (Cote 1997, Section 10) and from such standards as NFPA 1201, *Standard for Developing Fire Protection Services for the Public,* and the Insurance Services Office publication *Grading Schedule for Municipal Fire Protection* (ISO 1980b).

department chief, the battalion chief, and especially the captain might have asked themselves questions such as these:

- Do we know whether we are using the supplies and utilities as effectively as possible?
- How can we identify potential wasteful practices, even the small ones that add up in the course of the year?
- What can we do to ensure that all department members are alert to wasteful practices and that they assist with identifying and eliminating them?

2. *What else should be done to ensure that the department has the optimum number of fire stations, considering available standards (such as NFPA and ISO), that these fire stations are located to meet changes in the community's needs, and that they are in condition to meet apparatus and personnel needs?*   Again, the scenario gave little information for this question, except that painting and floor refinishing had been done by outside contractors in the past. Still, the department chief, the battalion chief, and the captain might have asked questions such as these:

- To what extent is the layout of each station most appropriate for the current needs of apparatus and equipment?
- Have there been developments since the station was built or modified that might make a different layout more effective?
- When was the location of the station in relation to the district it serves evaluated the last time? If there have been significant changes in the district, should the appropriateness of the location be reviewed again?
- Are changes in the apparatus likely in the next few years? If so, will the station be able to accommodate the apparatus being considered, or will revisions be required?
- If changes in the station layout are indicated, what preparatory steps should be taken?
- Are station maintenance procedures fully appropriate for the station(s) in light of the different alarm loads? Are they being implemented properly? If not, what should be done to ensure that the station maintenance procedures are fully appropriate and followed? Should there be written, departmentwide standard operating guidelines?
- What maintenance inspection/monitoring steps should be taken by the shift and company commanders?

3. *What else should be done to ensure that the department has the optimum number and types of apparatus for fire and other emergencies, considering available stan-*

dards (such as NFPA and ISO), that the equipment meets the needs of all situations that arise and are likely to arise in the community, and that each piece of apparatus is maintained in top operating condition and appearance? Little information in the scenario related to this question. The lack of response to the captain's request to make a list for apparatus checks, however, indicated that apparatus maintenance was not completely satisfactory. This assumption was supported by the lack of standard operating guidelines for apparatus maintenance. The department chief, the battalion chief, and the captain might have asked themselves questions such as the following:

- Has the adequacy of the apparatus been evaluated in the last few years? If there have been significant changes in the district, should the appropriateness of the apparatus be reviewed again and if so, how?
- Will the apparatus meet the district's needs effectively in the next few years? Is it likely that changes will be indicated? If yes, what preparatory steps should be taken?
- Are the maintenance and inspection procedures adequate? How thoroughly are these procedures followed? Are changes indicated on the basis of operational difficulties that have been encountered? Should there be written, departmentwide standard operating guidelines?
- What else needs to be done to ensure that the procedures are effectively communicated and that the fire fighters are fully competent to carry them out?
- What inspection/monitoring steps should be taken by the shift and company commanders?

4. *What else should be done to ensure that the department has optimum administrative and logistic facilities, repair facilities, alarm receipt and dispatch facilities and other communications equipment, personal protective equipment, and supplies for fire and other emergencies? Is there adequate consideration of available standards [such as NFPA, ISO, and FEMA (Fire Equipment Manufacturers Association)]; that the equipment meets the needs of all the situations that arise and are likely to arise in the community; and that all are maintained in top operating condition?* In the scenario, the routine aspects of managing equipment were handled in a variety of ways. The department and at least some shift commanders failed to provide standard operating guidelines and they did not insist that thorough procedures be adhered to. At least one company commander, Captain Murd, was quite lax, and the shift commander, though aware of the problem, did not seem to do much to correct it. The department chief, the battalion chief, and the captain might have asked these questions:

- What should be done to ensure that adequate equipment procedures exist and that they are followed? What departmentwide standard operating guidelines are needed?
- What inspection/monitoring steps should be taken by the shift and company commanders?

Scenario Analysis: Management/Leadership Perspective

The analysis of management/leadership of this scenario will be based on the guidelines for the 3Cs (see Appendix A, Decision Guidelines). We will repeat only the specific questions in this chapter, however. After reading the questions for each guideline, you may want to give some thought to the strengths and weaknesses of the way Captain Sharpe managed the physical resources from the management/leadership perspective.

The questions that expand on the basic control guideline are

1. Are the goals and objectives appropriate and effectively communicated?
2. Is the participation in decision making appropriate?
3. Are coordination, cooperation, and inter- and intra-unit communications being stimulated?
4. Is full advantage being taken of positive discipline and performance counseling? (See Additional Insight 9.1, Positive Discipline and Counseling.)

The questions from the competence guideline are

1. Are changes needed in recruiting and selecting for vacancies?
2. Are management of learning concepts applied effectively?
3. Are coaching and counseling on self-development being used to their best advantage?

The questions from the climate guideline are

1. Are appropriate psychological and tangible rewards offered and provided effectively and efficiently to bring the highest possible level of satisfaction from the creation and use of the product/service?
2. Are policies in place to help reduce work-related stress?

We repeat that it is not necessary to remember the detailed wording of the guideline when making decisions. Memorizing a personalized, abbreviated (summarized) version and applying it with every relevant decision are the critical steps to ensure the highest quality performance.

Discussion of the Control Guideline.   In this section we will look at each of the specific questions related to this guideline.

1. *Are the goals and objectives appropriate and effectively communicated?* The scenario did not directly discuss goals and objectives, though it mentioned Captain Sharpe's priorities, most of which were objectives. It is quite likely that the captain had other objectives, and possibly some goals in mind. In addition, there appeared to have been department objectives which were not explicit, but still operative, such as the objective to install a new compressor in each station. No evidence indicated that Captain Sharpe communicated goals or objectives—even those with his highest priority—to his crew or took steps to ensure their full acceptance.

Still, the captain was aware that he had to provide adequate support for the goals and objectives. He assured the fire fighters: "you can be sure that you can count on me to help out wherever necessary."

Questions that the captain might have asked himself include the following:

- Should I set goals and objectives formally with the fire fighters—with appropriate participation, of course—or at least communicate them clearly to the fire fighters? If so, which ones should I set?
- What else can I do to gain the team's full and unreserved cooperation in striving for the goals?

2. *Is participation in decision making appropriate?* A new commander who has to set direction would appropriately suggest plans with specific steps, although that may seem to allow little participation in decision making for fire fighters. The captain laid the foundation for appropriate participation by clearly indicating direction and then sharing the steps for specific implementation with the group. He offered opportunities for comments and developed the equipment check procedure with extensive input from the team. That is what appropriate participation is all about in this situation.

Questions here could include the following:

- To what extent could or should I give greater authority for decisions to the fire fighters? For instance, should one fire fighter lead the discussion on equipment checks and another on apparatus checks? How can I ensure that the results will be high-quality lists?
- What other matters pertaining to participation should I consider when the more difficult issues pertaining to inspection schedules and sharing of maintenance tasks come up for discussion?

3. *Are coordination, cooperation, and inter- and intra-unit communications being stimulated?* Procedures are a critical element for effective coordination. The captain was taking reasonable steps to assure coordination on the equipment checks, and he would undoubtedly do more when the team reviews what to do on the apparatus because that is a shared (and probably rotated)

activity. Captain Sharpe also took an appropriate step in appealing for cooperation. Still, we know too little about what happened after the scenario ended to be sure that he continued effectively on course. In decisively showing disapproval of pranks, he did take a step to prevent conflict, an essential ingredient of good cooperation.

The questions that the captain and the chiefs might ask themselves concern the potential for conflict within the company and with the other shifts.

- How can I reduce the potential for intra- and inter-company conflict that is likely to arise as the additional work in inspections and recurrent major station maintenance is distributed?
- What mechanism can I use to best resolve conflicts that might occur, to have the least amount of interference with high-quality performance on these tasks?

4. *Is full advantage being taken of positive discipline and performance counseling?* Again, the scenario contained little about positive discipline and nothing at all on performance counseling. The only hint on how the captain would approach positive discipline was the firm but friendly way in which he reminded Peter of the potential consequences of horseplay. (Also see Chapter 9, Fire Personnel Management.)

Following are the types of questions that officers might ask and that could contribute to high-level performance:

- What other steps can I take to ensure the best possible positive discipline environment? (See Additional Insight 9.1, Positive Discipline and Counseling.)
- When is performance counseling indicated? Only when there are serious deficiencies, or also in less significant instances? Should I also use it when performance is acceptable but could be improved?
- To what extent can I improve the performance of the entire team through effective joint exploration of performance standards and expectations?

Discussion of the Competence Guideline.   In this section we consider the questions that expand the basic competence guideline.

1. *Are changes needed in recruiting and selecting for vacancies?*   The department appeared to take an appropriate approach to assigning new officers, and presumably it is just as thoughtful in recruiting and selecting, within the limits of applicable regulations and union contract stipulations.

2. *Are management of learning concepts applied effectively?*   (See Additional Insight 11.1, Management of Learning and Coaching.) In light of the newness of the relationship, the captain was wise not to probe individual fire fighters about proper equipment checking procedure. When working on apparatus checks, he will gain insight into the knowledge and skills of the fire fighters on more complex issues. Since there was no evidence of inadequate performance in emergencies, we can assume that training in fire fighting used sound procedures. Still, the following questions might also be applicable for fire-fighting skills:

- How can I adapt training better to individual learning needs? Should I develop KSA (knowledge/skill/abilities) profiles for the various JPRs (job performance requirements) with and for fire fighters and then adapt training accordingly?
- What else can I do to make training for routine activities more stimulating?
- To what extent can and should fire fighters be involved in instruction?
- What else should I do to monitor KSA levels and relevant JPRs to ensure that fire fighters retain even rarely used knowledge and skills?

3. *Are coaching and counseling on self-development being used to their best advantage?*   It was too early for the captain to work on these managerial functions, so the scenario was silent on matters related to them. Still, he might ask himself questions:

- When and where should I use one-on-one coaching to help a fire fighter acquire knowledge or a skill?
- To what extent and for what purposes should fire fighters work on sharpening knowledge or skills by themselves on their own initiative?
- How can I use counseling to help them achieve career goals and overcome KSA deficiencies in the relevant JPR?
- Where do I have to strengthen KSAs for my own JPRs?

Discussion of the Climate Guideline.   The two questions that expand on the climate guideline are dicussed here.

1. *Are appropriate psychological and tangible rewards offered and provided effectively and efficiently to bring the highest possible level of satisfaction from the creation and use of the product/service?*   On this topic the scenario was completely silent. The captain had no opportunities to provide evidence of his appreciation for accomplishments or even contributions of effort. As time goes on, he might, however, ask questions such as the following:

- What type of effort by individual fire fighters should I recognize with comments and possibly with other intangible signs of appreciation?
- In what ways can I show such appreciation without seeming to be superficial, insincere, or repetitive? [See Additional Insight 5.1, Enhancing Work Satisfaction (Providing Recognition) and Performance Evaluation.]

2. *Are policies in place to help reduce work-related stress?*   Here, too, the scenario was almost silent on policies or practices. The planned changes in work loads and work schedules clearly will add to any existing work pressures. The change in company leadership adds a certain amount of temporary stress, and the new inspection and major maintenance work increase it. One positive aspect discussed in the scenario relates to this question. The captain appeared to be sensitive to how his words or actions can influence stress. Before mentioning the complaints he had heard, he spoke about other things. Then, when the discussion did turn to the complaints, he quickly redirected it toward the future, dismissing the past as unimportant. He thus concentrated on the positive, on the opportunities ahead, and thereby helped to defuse the stressful aspects inherent in past problems. The captain might still ask what he could do to ensure as little additional work-related stress as possible under the circumstances, and what steps he could take to provide as much relief as possible.

In conclusion, this case, like those in the previous chapters, highlights the benefits of thinking with guidelines. Such is the nature of the fire service that, in all likelihood, not much will go wrong even if there are no SOGs for equipment and apparatus checks or for the appearance of the station house's interior. Still, fire fighters find much satisfaction in being part of a department that takes pride in its image as well as in its ability to deal effectively with emergencies.

There is a connection between conscientious attention to the maintenance of physical resources (in addition to ensuring that they are adequate) and service to the community. The connection holds even for responses to emergencies. Well-maintained apparatus and equipment are more reliable, which in turn leads to greater confidence and respect.

Guidelines are one way to ensure that all bases are covered—that the department meets the requirements of the goals of the specific function and performs the tasks efficiently and competently. At the same time, guidelines are likely to bring and maintain a higher level of staff satisfaction than would exist if officers did not keep them in mind.

PHYSICAL RESOURCES OF THE FIRE DEPARTMENT

The most valuable resources in any fire department are the people who staff its agencies and equipment. Yet no matter how well staffed the fire department or how competent the people, members cannot do their jobs without the necessary physical resources. A fire department uses three types of physical resources: facilities (the real estate, including land, building, and other improvements), apparatus, and equipment and supplies. These resources make it possible for a fire department to work toward its goals.

Facilities

A fire department's facilities include buildings or areas for personnel, apparatus, equipment, and supplies; administrative offices; communications functions; training facilities; and maintenance. In smaller organizations, these functions often are contained in one building; larger organizations, however, might use several facilities in different locations.

The Fire Station.   The fire station is the single, most vital unifying element in a fire department. As the center of a community's fire-fighting operations, the station is a symbol of the protection of lives and property.

A fire station's upkeep involves routine activities and management decisions by either the chief or another officer or bureau whose specific duties include handling matters related to the station, such as major maintenance and renovation. In volunteer departments, a committee representing the membership might manage the building and might even be able to recommend building a new station.

If the site and capabilities of an existing fire station have no inherent problems, management has the responsibility to properly maintain the building so that it will remain functional in the future. If the present building is not adequate to meet fire fighter and/or community needs, and if funds have not been budgeted to build a new station in the near future, then the department must "make do" with the facilities and initiate steps to alleviate some of the problems that make it important to rebuild or relocate.

For example, the department in a growing community may need another pumper for effective suppression of fires and may be able to purchase it. The existing station, however, might not have enough space to house the new pumper. If the existing station is near a municipal maintenance facility that houses large trucks, the department might be able to park the new pumper at that location until a new station can be built or the existing station be enlarged. Storing the pumper at a municipal facility would be a simple, though temporary, solution.

If the protection capabilities of the present fire station have become insufficient because of the community's growth, the construction of a new fire station may be considered a necessity. Major emphasis is placed on obtaining approval to build a new fire station. A fire department relies on recommendations based on the type of hazards it serves to determine whether another fire station is needed and to support a request to authorities. Once a department obtains approval to build a new station, many related decisions follow, primarily concerning the location and design specifications of the station.

Considerations for New Stations.   The number and location of fire stations must be reevaluated periodically, but at least annually, as a community's structures and population change. The number of stations a department should have depends, like everything else, on a balance between the costs of the stations and their maintenance, on the one hand, and the need for more stations, on the other. If a station is located near the high-response section of a community (such as a heavily populated area of multiple-occupancy or wood-frame structures), that location is probably appropriate. Station relocation is necessary over time if the types of hazards and the locations of most fires move to a significant distance from the station.

If a department finds that relocation or construction of a new fire station is necessary, the three issues to consider are location, station design, and funding.

Location.   The location of a station in a community directly affects the total response time needed to combat fires effectively. For example, although a fire station is centrally located in a community, the majority of the responses might be at substantial distances from the station. Therefore, an evaluation of the time from receipt of an alarm to the arrival at a fire plays an important part in determining the need for relocating a fire station. The total time is the sum of the time it takes to complete each of the following five fire-fighting processes.

1. *Detection* is the time it takes to detect a fire. Automatic fire detection systems, such as smoke and heat detectors, give early warnings of fire and save considerable response time. Some detectors are connected directly to a fire station through a central station signaling system, whereas others sound only in the building in which there is a danger. In the latter case, detection time depends on human response and then on the number of people who are in the vicinity of the fire, how rapidly they respond, and the time of day.

2. *Alarm* is the time that elapses between detection of the fire and transmission of the alarm to the fire station. It depends on the availability of

alarm boxes, directly connected alarms, telephones, the extent of automation, reliability, and the speed of transmission.

3. *Dispatch* is the time required to alert responding companies. If information is recorded automatically and if dispatchers have the most modern communication equipment, the time needed for dispatch is minimal.

4. *Turnout* is the speed with which personnel—paid, off-duty, and volunteers—can report for duty. Turnout depends on the location of the personnel at the time of the alarm, whether at the station, at work, or in their homes.

5. *Response time* is the travel time for the apparatus and on-duty personnel from the station to the fire. It depends on the distance from the station to the emergency and on the topographic, traffic, and weather conditions. When traffic is particularly heavy, the police department might be needed to aid in traveling to the fire and in beginning evacuation.

Fire department officers should also be aware of the "minor" changes in a community, changes that occur so gradually that most people are barely aware of them. For example, vacant lots are filled in, industrial interests are relocated, a small farm is sold to a real estate developer, and zoning ordinances are changed to attract more business and people. Such changes directly affect fire spread and fire-fighting abilities. They should be taken into consideration by fire department officers when they look at recommending possible relocation.

Station Design.   Design considerations involve the apparatus, equipment, and personnel to make the station appropriate for the purposes it will serve in the near and distant future. Relevant issues are the space and height of apparatus areas considering the needs of specialized equipment that might be purchased in the future, training area, living quarters, communications center, office space, repair shop, parking areas, and potential expansion for laundry, infection control, and storage.

Whether the department's staff is currently paid or volunteer, station design should consider the needs of paid staff because a volunteer department may switch to paid fire fighters in the future. Although questions about future changes should be resolved in discussion with municipal officials and other interested parties, the ultimate responsibility for location, design, and staffing lies with the department officials and their advisors.

Other important areas of consideration when planning station specifications are the heating and ventilating needs of the station and the setup of the watch-desk area. The design of the station should incorporate a heating system that is able to recover rapidly after apparatus responds, especially during the winter when the doors to the station are not closed immediately. In addition, adequate ventilation should be provided in the apparatus room before the doors are opened to avoid a buildup of carbon monoxide during the usual engine warmup, drilling, or servicing.

The proper setup of the watch-desk area is a major concern because that is where alarms are received. The watch desk should be a desk–console arrangement with wall space for maps, schedules, and instructions and with ample room for the necessary radio equipment, alarm control devices, floor controls, and traffic signal controls. The watch-desk area should be as sound-proof as possible and should allow for clear visibility of the entire apparatus room. A desirable location for the watch desk is near the front entrance of the station, where visitors enter and seek information (Hickey 1985).

Before final decisions can be made, department personnel must thoroughly research and understand traffic flow, terrain features, area characteristics, weather peculiarities, and other special considerations, in addition to the space needs, so they can communicate them to the architects (see Figures 7.1–7.4).

Funding. In the never-changing environment of tight restrictions on spending, a department might consider proposing a public relations campaign

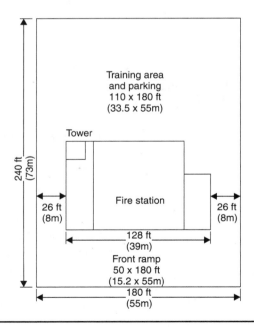

Figure 7.1 A plot plan for a typical district fire station for urban and suburban services has a minimum recommended plot size of 43,200 square feet.

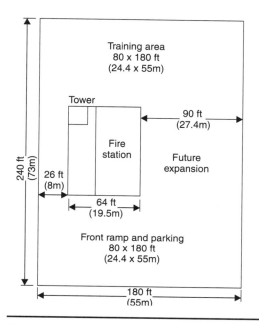

Figure 7.2 A plot plan for a typical rural fire station has a minimum recommended plot size of 43,200 square feet.

and launching it, with approval from the appropriate authorities, before going public with a formal proposal for major changes such as extensive renovations or a new station. The purpose is to communicate the reasons why more facilities and possibly additional staff are needed before submitting the request for the necessary budget.

The community relations effort should address all the factors that influence the reaction of the community to the request: (1) the professionalism exhibited by the fire department, (2) the ISO rating of the present fire station (see the end of this chapter for a discussion of ISO ratings), (3) the proposed location of a new station, (4) the procedure for presenting proposals, and (5) the amount of money involved.

Funding and other aspects of financial management will be discussed in greater detail in Chapter 8, Management of Financial Resources.

Apparatus

Whether or not fire stations are adequate, a fire department needs to continually analyze and evaluate the adequacy and effectiveness of its apparatus for combating the diverse emergencies it is called on to meet now or in

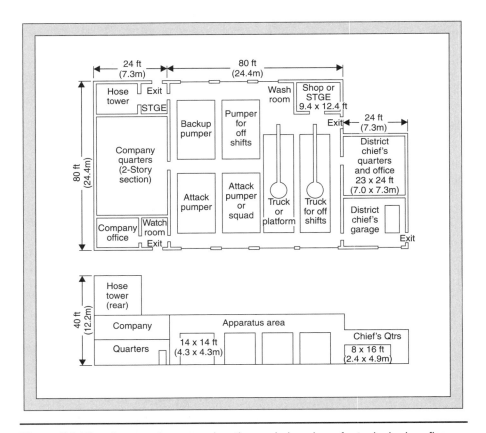

Figure 7.3 The diagram shows an elevation and plan view of a typical urban fire station.

the future. The Insurance Services Office (ISO) grading schedule and survey are useful for determining whether to maintain present apparatus, modify, or purchase additional or newer equipment.* Remember that the ISO requirements are used to determine insurance rates. Any decisions to acquire new apparatus and equipment for a fire department should not be based solely on ISO requirements. They are one part of an evaluation system that also includes locally conducted cost–benefit analyses by department leaders.

---

*For information on the interpretation of the ISO grading schedule, you should consult the *Fire Suppression Rating Schedule Handbook* (Hickey 1993). Also, for more information on the ISO, see the end of this chapter and Chapter 6 on prefire planning or contact Insurance Services Office, Inc., at 7 World Trade Center, New York, NY 10048-1199.

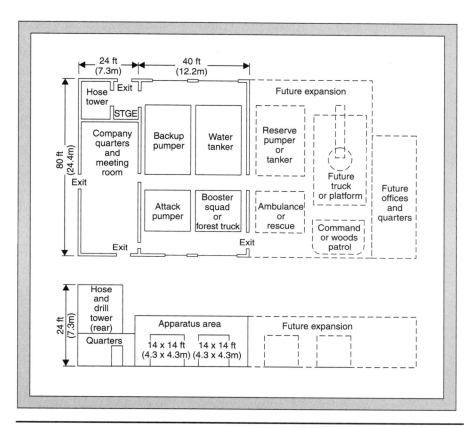

Figure 7.4 The diagram shows an elevation and plan view of a typical rural fire station.

Number and Effectiveness of Needed Apparatus. With respect to the number of apparatus needed, either in the department or readily available through mutual aid, the ISO schedule provides guidance. It is based on flow rates that have been determined as the amount of water needed to confine a large fire to the city block where the fire originated or to a large building standing alone. Because the flow rates vary widely based on different structural and occupancy variables, several tables are available.

The tables shown in Exhibit 7.1 pertain to initial and sustained fire attack. All built-up areas should have an engine company within 1½ miles and a ladder/service company within 2½ miles. Response areas with five buildings that are three stories tall or 35 feet or higher, with five buildings that have a required fire flow greater than 3,500 gallons per minute, or with any combination of these criteria should have an aerial ladder company.

Exhibit 7.1 ISO Rating Schedule for Water Flow Rates for Initial and Sustained Fire Attack

NEEDED FIRE FLOW

300. GENERAL:

This item develops Needed Fire Flows for selected locations throughout the city which are used in the review of subsequent items of this Schedule. The calculation of a Needed Fire Flow (NFF$_i$) for a subject building in gallons per minute (gpm) considers the Construction (C$_i$), Occupancy (O$_i$), Exposure (X$_i$) and Communication (P$_i$) of each selected building, or fire division, as outlined below.

310. CONSTRUCTION FACTOR (C$_i$):

That portion of the Needed Fire Flow attributed to the construction and area of the selected building is determined by the following formula:

$C_i = 18F (A_i)^{0.5}$

F = Coefficient related to the class of construction:

F = 1.5 for Construction Class 1* (Frame)
= 1.0 for Construction Class 2* (Joisted Masonry)
= 0.8 for Construction Class 3* (Non-Combustible) and
    Construction Class 4* (Masonry Non-Combustible)
= 0.6 for Construction Class 5* (Modified Fire Resistive)
    and Construction Class 6* (Fire Resistive)

A$_i$ = Effective* area

In buildings with mixed construction a value, C$_{im}$, shall be calculated for each class of construction using the effective area of the building. These C$_{im}$ values are multiplied by their individual percentage of the total area. The C$_i$ applicable to the entire building is the sum of these values. However, the value of the C$_i$ shall not be less than the value for any part of the building based upon its own construction and area.

The maximum value of C$_i$ is limited by the following:
8,000 gpm for Construction Classes 1 and 2
6,000 gpm for Construction Classes 3, 4, 5 and 6
6,000 gpm for a 1-story building of any class of construction.

The minimum value of C$_i$ is 500 gpm. The calculated value of C$_i$ shall be rounded to the nearest 250 gpm.

320. OCCUPANCY FACTOR (O$_i$):

The factors below reflect the influence of the occupancy in the selected building on the Needed Fire Flow.

Occupancy Combustibility Class*	Occupancy Factor (O$_i$)
C-1*(Non-Combustible)	0.75
C-2*(Limited Combustible)	0.85
C-3*(Combustible)	1.00
C-4*(Free Burning)	1.15
C-5*(Rapid Burning)	1.25

330. EXPOSURES (X$_i$) AND COMMUNICATION (P$_i$) FACTORS:

The factors developed in this item reflect the influence of exposed and communicating buildings on the Needed Fire Flow. A value for (X$_i$ + P$_i$) shall be developed for each side of the subject building:

$(X + P)_i = 1.0 + \sum_{i=1}^{n} (X_i + P_i)$, maximum 1.75, where n = number of sides of subject building.

A. **Factor for Exposure (X$_i$):**

The factor for X$_i$ depends upon the construction and length-height value* (length of wall in feet, times height in stories) of the exposed building and the distance between facing walls of the subject building and the exposed building, and shall be selected from Table 330.A.

---

*When an asterisk is shown next to a term in this item, the term is defined in greater detail in the Commercial Fire Rating Schedule.

Exhibit 7.1 (continued)

NEEDED FIRE FLOW

TABLE 330.A
FACTOR FOR EXPOSURE (X_I)

Construction of Facing Wall of Subject Bldg.	Distance Feet to the Exposed Building	Length - Height of Facing Wall of Exposed Building	1,3	Construction of Facing Wall of Exposed Building Classes		
					2, 4, 5, & 6	
				Unprotected Openings	Semi-Protected Openings (wired glass or outside open sprinklers)	Blank Wall
Frame, Metal or Masonry with Openings	0-10	1-100	0.22	0.21	0.16	0
		101-200	0.23	0.22	0.17	0
		201-300	0.24	0.23	0.18	0
		301-400	0.25	0.24	0.19	0
		Over 400	0.25	0.25	0.20	0
	11-30	1-100	0.17	0.15	0.11	0
		101-200	0.18	0.16	0.12	0
		201-300	0.19	0.18	0.14	0
		301-400	0.20	0.19	0.15	0
		Over 400	0.20	0.19	0.15	0
	31-60	1-100	0.12	0.10	0.07	0
		101-200	0.13	0.11	0.08	0
		201-300	0.14	0.13	0.10	0
		301-400	0.15	0.14	0.11	0
		Over 400	0.15	0.15	0.12	0
	61-100	1-100	0.08	0.06	0.04	0
		101-200	0.08	0.07	0.05	0
		201-300	0.09	0.08	0.06	0
		301-400	0.10	0.09	0.07	0
		Over 400	0.10	0.10	0.08	0
Blank Masonry Wall	Facing Wall of the Exposed Building Is Higher Than Subject Building: Use the above table EXCEPT use only the Length-Height of Facing Wall of the Exposed Building ABOVE the height of the Facing Wall of the Subject Building. Buildings five stories or over in height, consider as five stories.					
	When the Height of the Facing Wall of the Exposed Building is the Same or Lower than the Height of the Facing Wall of the Subject Building. $X_I = 0$.					

330. EXPOSURE (X_I) AND COMMUNICATION (P_I) FACTORS: (Continued)

   B. **Factor for Communications (P_I):**

      The factor for P_i depends upon the protection for communicating party wall* openings and the length and construction of communications between fire divisions* and shall be selected from Table 330.B. When more than one communication type exists in any one side wall, apply only the largest factor P_i for that side. When there is no communication on a side. $P_i = 0$.

      *When an asterisk is shown next to a term in this item, the term is defined in greater detail in the Commercial Fire Rating Schedule.

5

Exhibit 7.1 (continued)

NEEDED FIRE FLOW

TABLE 330.B
FACTOR FOR COMMUNICATIONS (P$_I$)

Description of Protection of Passageway Openings	Fire Resistive, Non-Combustible or Slow-Burning Communications				Communications With Combustible Construction					
	Open	Enclosed			Open			Enclosed		
	Any Length	10 Ft. or Less	11 Ft. to 20 Ft.	21 Ft. to 50 Ft. +	10 Ft. or Less	11 Ft. to 20 Ft.	21 Ft. to 50 Ft. +	10 Ft. or Less	11 Ft. to 20 Ft.	21 Ft. to 50 Ft. +
Unprotected	0	+ +	0.30	0.20	0.30	0.20	0.10	+ +	+ +	0.30
Single Class A Fire Door at One End of Passageway	0	0.20	0.10	0	0.20	0.15	0	0.30	0.20	0.10
Single Class B Fire Door at One End of Passageway	0	0.30	0.20	0.10	0.25	0.20	0.10	0.35	0.25	0.15
Single Class A Fire Door at Each End or Double Class A Fire Doors at One End of Passageway	0	0	0	0	0	0	0	0	0	0
Single Class B Fire Door at Each End or Double Class B Fire Doors at One End of Passageway	0	0.10	0.05	0	0	0	0	0.15	0.10	0

+   For over 50 feet, P$_i$ = 0.

+ + For unprotected passageways of this length, consider the 2 buildings as a single Fire Division.

Note: When a party wall has communicating openings protected by a single automatic or self-closing Class B fire door, it qualifies as a division wall* for reduction of area.

Note: Where communications are protected by a recognized water curtain, the value of P$_i$ is O.

*When an asterisk is shown next to a term in this item, the term is defined in greater detail in the Commercial Fire Rating Schedule.

Exhibit 7.1 (continued)

NEEDED FIRE FLOW

340. CALCULATION OF NEEDED FIRE FLOW (NFF$_i$):

$$NFF_i = (C_i)(O_i)(X + P)_i$$

When a wood shingle roof covering on the building being considered, or on exposed buildings, can contribute to spreading fires add 500 gpm to the Needed Fire Flow.

The Needed Fire Flow shall not exceed 12,000 gpm nor be less than 500 gpm.

The Needed Fire Flow shall be rounded off to the nearest 250 gpm if less than 2500 gpm and to the nearest 500 gpm if greater than 2500 gpm.

Note 1: For 1- and 2-family dwellings not exceeding 2 stories in height, the following Needed Fire Flows shall be used.

Distance between buildings	Needed Fire Flow
Over 100'	500 gpm
31-100'	750
11-30'	1000
10' or less	1500

Note 2: Other habitational buildings, up to 3500 gpm maximum.

In the ongoing evaluation of needed apparatus, all possible emergencies—and especially the cost-effective containment and extinguishment of fires—have to be considered. Fire officers must compare the benefits of using existing apparatus, of exchanging equipment, and of purchasing new apparatus. Although it might seem that the latest piece of equipment is most desirable and the oldest piece should be sold, closer examination of apparatus efficiency and future needs might reveal otherwise. (For further discussion, see Chapter 8, Management of Financial Resources.)

For example, in a small community, the average age of apparatus might be as high as eight to ten years. The newest engine might be at least three years old. When considering the ages and capabilities of these pumping apparatus, the department's analysis should go beyond looking at the capabilities of new apparatus, such as greater acceleration provided by a new pumper in extinguishing a fire. After all, the oldest pumper is not likely to be the first-in engine. It is used as a backup unit when newer engines are not sufficient to cover a specific fire. The focus then shifts from how much better a new engine can fight a fire, when compared with the oldest one, to how often the old engine will be used in a fire attack.

If the old engine is used regularly in a fire attack, how much damage could be prevented if a more modern engine were used? Although this question has no easy answer, an analysis of difficulties that were encountered in the past and estimates of damage that might have been prevented might lead to a rough figure. If that figure, when projected to the future, provides sound justification for a new engine, then it could support a proposal for one.

Determining the benefits that a new engine will bring is difficult. The following questions have to be considered in making the decision:

- How much higher is the annual maintenance cost of the old engine?
- When will the old engine be so outdated that it has to be scrapped, and how much less will its resale value be at that time compared with to-day's value?
- How long will it be before the newest engine will be outdated, making fire attack less effective than it would be if a newer engine could assume the lead position?
- Will the new equipment require fewer personnel?
- What is the fuel efficiency?
- How effective is the equipment layout on the engine?

In the private sector, equipment replacement decisions involve many complex questions, and decisions are usually made by considering long- and short-term profitability. The basis for the purchase of a new piece of equipment usually is whether or not it will "pay for itself" in a specified period of time. Cost and resale value factors are considered to determine whether a new piece of equipment is worth buying. In addition, however, questions pertaining to productivity receive high priority, such as how much more productive it will be than existing equipment and how long it will be before still more productive equipment is available.

As difficult as it is to answer these questions in the private sector, it is often more difficult in the fire service. For example, in view of the possibility that a life might be lost without a new piece of apparatus, how should one evaluate the probability that the person might instead suffer a minor or a serious injury? Most people agree that an aerial ladder is justifiable equipment if it can be credited with saving a life; however, is it justifiable equipment if it prevents third-degree burns over most of an arm? These questions are vastly more difficult than those faced in industry because they concern issues that go beyond simple dollars and cents. Thus, traditional objective business analysis has to be augmented by subjective considerations in making fire service apparatus replacement decisions.

Rebuilt and Specialized Apparatus. Although standard pumpers and ladder trucks carry their own specialized equipment for dealing with specific hazards and situations (such as vehicle rescue and salvage operations), many larger fire departments use vehicles that are outfitted for only one or several types of special operations or purposes. Special-equipment vehicles can be trucks that carry extra hose and nozzles to lend greater flexibility to fire-fighting operations, trucks with large capacities to generate power and lighting to supplement portable generators on standard equipment, or trucks that are specifically designed for particular hazards. They might also be apparatus with special fire pump capabilities, with an elevating platform, or with a special capacity water tank. Many fire departments use the NFPA 1900 series of standards as general guidelines for making decisions concerning design specifications.

A department that has the capability to design and modify a piece of apparatus from used equipment might consider doing so. Refurbishing presents an opportunity to upgrade the department's fire-fighting capability at lower cost than alternative approaches when the used equipment runs well, its body is not damaged, and complete, satisfactory maintenance records exist. For example, a locality might need a specialized piece of apparatus, such as one with special fire pump capabilities, with an elevating platform, or with a special capacity water tank, but would use it so infrequently that it cannot justify large expenditures on new apparatus. Or, a department might find that it needs an extra backup piece of standard equipment to replace one that it can no longer maintain economically. In these situations, purchasing used apparatus or rebuilding the chassis of an older piece of apparatus might be a satisfactory solution to a serious problem.

Many fire departments use the following NFPA standards as a general guideline for making decisions about design specifications:

- NFPA 1901, *Standard for Automotive Fire Apparatus*
- NFPA 1906, *Standard for Wildland Fire Apparatus*

Selection of Apparatus Based on Fire District Influences.   The type of apparatus that is most appropriate for a district is determined by two considerations: size and type of district and the hazards in the district.

Size and Type of District.   Even though two fire departments might have the same number of people to protect, their individual apparatus needs could vary greatly due to the particular area and type of district served by each. For example, districts that have areas in which the population is widely dispersed might have to pay particular attention to the driving characteristics of the vehicles they purchase, with emphasis on a faster response time. In contrast, a district with steep grades, narrow winding roads, or difficult terrain might be especially interested in a vehicle's shifting characteristics. In this case, a four-wheel-drive vehicle with extra large tires or a tanklike base can reduce response times. A vehicle with smaller overall dimensions might be easier to maneuver around curves.

Fire departments in rural areas need large water tank trucks, portable suction basins (which allow tank trucks to unload their water next to the pumper and go back to the nearest hydrant or static source for more), portable pumps, and suction hose because most of the water has to come from a static source, such as a pond, or from water carried to the fire scene. Brush fire trucks or ladder trucks in rural areas often have auxiliary booster pumps with capacities of less than 500 gallons per minute. Auxiliary pumps are useful on small, nonstructural fires.

For suburban or semirural areas with a high incidence of brush fires or fires in rubbish dumped in vacant lots, purchasing a grass or brush fire truck instead of a regular pumper might be appropriate, or such a vehicle might be considered as an addition to the standard pumper fleet. Equipped with booster pumps, these trucks are better adapted to fighting small fires than regular pumpers. They are less expensive than regular pumpers because they can be built from regular commercial truck chassis. Grass or brush fire trucks are easier to drive and faster to place in position than are regular pumpers, and they eliminate the need to take a standard pumper away from fighting a structural fire.

Fire departments in urban or suburban areas have to consider the advantages of different aerial devices to protect buildings more than three stories high. In addition, they must take into account constraints to the use of mechanized aerial devices, such as obstacles above roadways, width of streets, and parking spaces available near buildings. The number of people who can be rescued, potential use of such equipment, and frequency of use should be major considerations before any purchase. A district where there are many fire hazards or where the fire department is frequently called to help in other emergencies might need additional rescue vehicles.

Hazards in the District.   Hazards affect ratings, fire station locations, personnel, apparatus, and equipment. The following hazardous conditions may bring about the need for additional companies and fire department personnel:

- Numerous wood-shingle roofs
- Large concentrations of closely spaced, wood-frame buildings
- Large blocks that have structures weak in fire resistance and lack adequate private fire protection
- Large, individual structures with inadequate private fire protection
- Narrow streets and traffic congestion, including parked cars
- Severe weather conditions, such as frequent droughts or heavy snows

All of these conditions are usually considered in the ISO survey and are taken into account in establishing the grade classifications. However, a fire department's management may have a valid perspective that differs from the conclusions of the ISO survey. In such instances, the department can appeal to the ISO for a review, but it is likely to achieve a more favorable rating only if it can show objectively why the need is more or less serious than ISO engineers found it to be. For example, a series of flow tests by the local water utility system may provide different results from the ISO, thus requiring a review and reconciliation. Discrepancies do not happen often, but they are possible. Records that are dated or inadequate may also create a need for a review of the municipal data.

Although districts with flammable liquid fire hazards are likely to have specially equipped apparatus for delivering foam or other chemical extinguishing agents, most districts are confronted with explosives or flammable liquids fires at some time. Gasoline station tanks must be filled from commercial gasoline tankers, which often catch fire when they overturn on a slick surface or at an angular bend in the road. A fire department should be equipped to fight such fires, and its apparatus must carry an adequate number of extinguishers that can deliver foam.

Some community airports and industries with unusually high risks maintain their own fire-fighting units. These occupancies require adequate pumpers primarily equipped for fighting specific types of fires. Whether or not to obtain and use specialized types of vehicles and equipment is a management decision.

A municipality that has a large waterfront area with commercial docks or marinas might purchase one or more fireboats that are specially equipped with turrets that can throw heavy streams or include an elevating platform (see Figure 7.5). Because fireboats represent an enormous financial investment in relation to their limited use, districts with small areas of waterfront might purchase a used fireboat or make arrangements with neighbors for joint purchase and use.

In dry climates with heavily wooded areas, forest and brush fires are a serious hazard. Departments confronted with this problem usually either have their own special equipment or make arrangements for leasing forest fire protection devices. Aircraft and helicopters are used extensively in the control of forest fires, particularly by the U.S. Forest Service. Experiments are also being conducted in the use of large-capacity foam trucks on such fires.

Consideration of Apparatus Features.   In addition to ISO grading schedules, NFPA 1901 and NFPA 1906 have standard specifications for the most useful features of various types of apparatus. These standards list the

Figure 7.5 The large waterfront area is protected by a fireboat.
(Courtesy of Newark Fire Department)

features of the apparatus and the equipment that must be carried on each different type. Each of the standards discusses vendor and purchaser responsibilities for the design of engines, carrying capacity, and the cooling, lubrication, fuel exhaust, and electrical systems. The specifications in these standards provide the essential requirements that must be adhered to by the department to provide better protection, promote fire fighter health and safety, and avoid potential legal problems.

Specific recommendations for new apparatus features may come from the battalion or company level of a particular department or from the fire fighters who use the equipment. A well-managed department can rely on the experience of the fire officers and fire fighters because they have specific knowledge of features needed on existing apparatus. Fire fighters might request extra capabilities for preconnected hose, additional intakes, or larger compartments for a newer piece of apparatus. They might suggest relocating various appliances and controls, new facilities for storing respiratory equipment, new locations for lights, or greater availability of lights and generators. If the climate in a department supports effective two-way communications, firsthand recommendations for improving fire-fighting capability and efficiency are encouraged and better decisions result than in an environment where communications are less open.

Equipment and Supplies

In addition to a functional fire station and an efficient range of apparatus, fire departments need the right equipment. Fire officers must make decisions pertaining to all types of new and used equipment, specifically personal equipment used by fire fighters, support equipment carried on apparatus and maintained in the station, central and distributed communications equipment, and auxiliary equipment.

Personal Equipment. Most personal equipment is intended to protect fire fighters as they perform their work. Breathing apparatus protects against the danger of inhaling the products of combustion. Protective headgear, footwear, and clothing insulate the body from the most adverse environmental conditions, such as intense heat and cold, and from exposure to radiation contamination, chemicals, and water. Helmets safeguard against injury from falling objects. To assist fire departments in meeting their personal equipment needs, NFPA has developed standards for selecting and using personal safety equipment:

- NFPA 1971, *Standard on Protective Ensemble for Structural Fire Fighting* (incorporates NFPA 1972, NFPA 1973, and NFPA 1974)
- NFPA 1981, *Standard on Open-Circuit Self-Contained Breathing Apparatus for the Fire Service*
- NFPA 1982, *Standard on Personal Alert Safety Systems (PASS)*
- NFPA 1983, *Standard on Fire Service Life Safety Rope and System Components*

Fire departments once believed that uniformity of the personal equipment was of great significance. This view has gradually changed, so that today, at least in more enlightened departments, fire fighters are allowed considerable freedom in selecting their personal equipment.

Support Equipment Carried on Apparatus and Maintained in the Station. (Equipment carried on the apparatus includes ladders, nozzles, hoses, various tools, etc.). The NFPA has issues specific standards for the various types of equipment:

- NFPA 10, *Standard for Portable Fire Extinguishers*
- NFPA 1931, *Standard on Design of and Design Verification Tests for Fire Department Ground Ladders*
- NFPA 1961, *Standard on Fire Hose*
- NFPA 1963, *Standard for Fire Hose Connections*

Equipment for special uses, such as materials and clothing for hazardous material response and various rescue operations, have no standards at this time. Support equipment also includes confined-space and high-angle rescue equipment and life guns, which are used for shooting rope to people in distress and also have some use in water rescues and rescues from cliffs or canyons. Finally some equipment is used solely for support functions, such as repair and training equipment.

Central and Distributed Communications Equipment.   Communications equipment falls into two groups: the equipment used in central communications operations and the equipment that is distributed for use in the field.

Central communications equipment.   The equipment required by the communications center must operate reliably so that emergency and urgent messages are transmitted and received on a priority basis. A communications center should meet the following needs:

- The public must have 24-hour access to headquarters to initiate an alarm or to obtain information.
- Headquarters must be able to contact its fire fighters rapidly, both in the fire station and in their homes.
- Headquarters must be able to communicate with personnel on the apparatus.
- The chief officer must be able to communicate with remote units and sometimes with individual fire fighters.
- Communication must be possible between pieces of apparatus and between apparatus and individual fire fighters.
- Officers must be able to contact headquarters to request aid or further instructions.
- A regional communications system must exist for mutual aid calls.
- All wireless elements of the systems must operate with minimum interference so that emergency and urgent messages are easily understood.

Central communications equipment focuses on the operations room where all calls and alarms are received and from which all alarms are transmitted to companies or stations. There might be a desk or console for the dispatcher or dispatchers on duty if dispatching is not handled by an external dispatch site.

An operations room should contain at least the following:

- Equipment to receive calls from the public from stationary and wireless (mobile) telephones

- Equipment to receive alarms from street boxes and private alarm systems
- Radio transmitting and receiving
- Tone alerting equipment (pagers, Plectrons)
- Computer terminals
- Video display units (for computerized dispatch when the alarm assignments are made by computer)

Equipment requirements are dictated by the size of the operation and the type of services that are provided. Centers in communities that have communitywide alarm systems require space for alarm equipment and standby batteries. Space is also required for radio and telephone equipment. If computer-assisted dispatching is used or anticipated, additional space is required (see Figure 7.6).

Large communities might require one large center and possibly satellite centers to handle message loads, whereas smaller communities might require only a watch desk located in a fire station. In addition to receiving calls

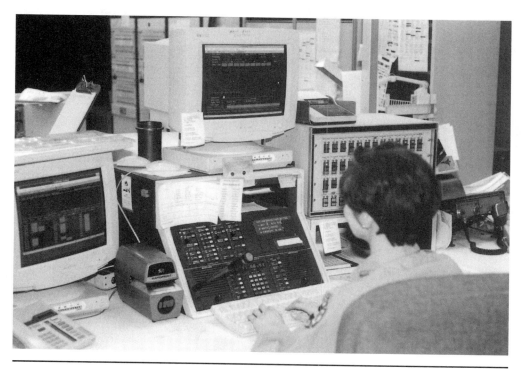

Figure 7.6 A computer terminal is used to dispatch apparatus. (Courtesy Newark Fire Department)

for assistance, the alarm center must be capable of handling all radio communications for the department, keeping a continuous record on the status of all companies, handling communications to each station, and maintaining current files and maps of streets.

A central station might have a status board that presents a visual indication of the location of the various stations and the equipment housed in each station (see Figure 7.7). Indicating lights at each location show the current disposition of apparatus—for example, in service by radio, out of service, and so on.

Distributed communications equipment.   Because the noise level is high when apparatus is in motion or pumping, radio equipment must be loud enough to be clearly audible above the noise. Multiple speakers and earphones may be necessary. For guidance on hearing standards, refer to the latest edition of NFPA 1500, *Standard on Fire Department Occupational Safety and Health Program.*

Many fire officers now carry cellular telephones when responding to an emergency. These phones can reduce the need for radio communications.

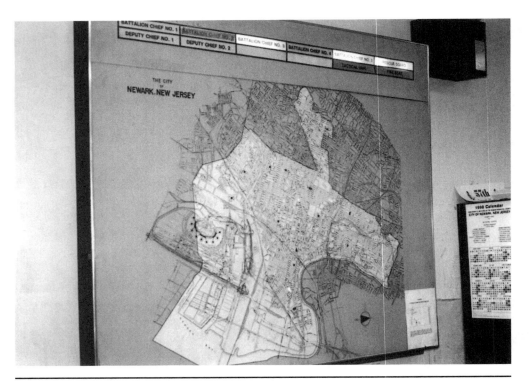

Figure 7.7 A status board is found in a large communications center. (Courtesy Newark Fire Department)

Together with mobile computers on apparatus, they can keep the radio frequencies clear for fireground communications and speed the flow of data to and from the fire scene.

Computers are also available on some apparatus. They are especially useful in the field to make prefire plans more detailed and accessible while a company is traveling to the emergency scene. The cost of personal computers has decreased enough so that most fire departments are now able to afford them.

Auxiliary Equipment.   The catchphrase "auxiliary equipment" refers to additional equipment that is used in the station and carried on the apparatus. Equipment used in the station includes compressors for recharging the breathing apparatus, generators, and maintenance equipment. Auxiliary equipment that is carried on the apparatus includes, besides portable lights, those tools, attachments, and small pieces of apparatus that are not an integral part of the vehicle body or hose. NFPA 1901 and NFPA 1906 list the standard equipment items that should be included on each specific piece of apparatus. The NFPA listings are similar to the ISO lists that are used for the purpose of grading apparatus. Older apparatus might not have all of these items, and departments are sometimes downgraded for not having them. In such cases, there may not be a convenient way of adapting old apparatus to fit modern requirements.

The list of equipment specified in NFPA 1901 and NFPA 1906 is only the minimum required. A department that frequently needs certain pieces of equipment that are not listed should not feel constrained by the list. For instance, departments that are often involved in vehicle rescue work usually carry medical equipment and tools to extricate people from wrecked automobiles and trucks. Rural and suburban departments that are not specially equipped with trucks to fight brush fires but that are frequently called to grass and brush fires might include such items as hay forks, buckets, metal rakes, back tanks, fire brooms, and portable pumps on their standard pumpers.

Hose in standpipe-protected properties sometimes rots because of age (or might be cut or stolen by vandals). Many fire departments therefore carry portable kits containing 1½-inch hose that can be connected to standpipes.

MANAGING THE PHYSICAL RESOURCES

A department's overall effectiveness and the degree to which it can control, extinguish, and help prevent fires are greatly influenced by the management and coordination of its resources. The two major segments of the management of physical resources are ensuring adequacy and maintenance.

Ensuring adequacy is primarily the responsibility of top management in the department. It concerns the following issues:

- Monitoring the operational effectiveness of existing resources
- Staying attuned to new developments in the community such as housing development, approved, pending, or possibly even already in construction, which may require new types of resources or more of them
- Keeping abreast of technology improvements such as self-lubrication devices for motors
- Obtaining funds for new or updated resources that may be required or advantageous
- Supporting companies in their maintenance and readiness requirements

Maintenance is primarily the responsibility of officers at the company level. Company-level officers serve as the representatives of top-level management in the conservation of station-level resources. They may be called upon to manage a number of station-level programs, including the ones listed here:

- Maintaining the physical plant at each station, including repairs and housekeeping
- Ensuring the readiness of apparatus through periodic maintenance, repairs, and daily and weekly routine maintenance (such as cleaning and protecting against rust and abrasion)
- Ensuring the readiness of equipment, including requisitioning needed new equipment, staying abreast of technological improvements in equipment, and doing routine maintenance

Other issues related to the management of the physical resources are procurement and use of the resources.

Procurement of Physical Resources

Expenditures related to procurement of physical resources for a fire department are a large part of the annual operating budget. Fire officers should know how their agency's procurement system functions. At the department level, one individual is usually assigned the responsibility for the acquisition of resources. Supplies and equipment, for nonemergency as well as for emergency needs, must be maintained at adequate levels for each division of the fire department. In many departments, a central control location is supplied by using expenditures from the appropriate lines of the agency budget.

Some form of inventory control records are needed. They usually involve sending out requisitions and/or purchase orders, checking in materials when delivered against these, and keeping physical or electronic "books" in which receipts and disbursements of materials are entered and balances maintained. Fire officers need to be aware of their role in maintaining adequate supply levels in their respective areas. Periodic physical counting of inventories of all items is needed to assure that control is maintained and resource availability is appropriate.

Use of Physical Resources

Resources should be used in ways that are cost effective. Important for getting the most value from each physical resource involves preventive maintenance. Officers must be aware of these procedures for all equipment, not only apparatus, and the appropriate methods for taking care of property placed in their care. If departmental SOGs are not in place, individual stations or even officers can prepare them for use with their companies, stations, or bureaus.

Problems that can increase a fire department's operating costs include inadvertent loss of physical resources, waste, theft of resources, abuse of resources, and improper, inadequate, or nonexistent preventive maintenance.

Training and self-development by officers are the keys to success in the proper management of physical resources. If officers and other personnel know how to acquire, maintain, and properly use various physical resources, they can maximize their service life and use fire department financial resources more efficiently.

Challenges in the Management of Physical Resources

Fire-fighting apparatus and facility needs change as technological advances lead to new construction materials and more hazardous substances. Combined with budget constraints, this constantly changing nature of the job of fire fighting requires that facilities, apparatus, and equipment be updated and modernized. Consider these examples:

- The increased use of plastics and their many derivatives in construction requires different strategies, tactics, and materials for containing and transporting them.
- The risks involved in the manufacture, storage, and transit of hazardous materials have led to local, state, and federal regulations that complicate the challenges of being prepared for and fighting many fires.
- The advent and popularity of high-rise living and working have complicated the task of fire fighting and called for expensive aerial apparatus.

- Loss of life and severe economic losses associated with arson fires have created the need for new communication and detection equipment, such as air- and gas-sampling devices and cellular phones, in addition to more effective inspection and investigative procedures.

Changes in the physical resources required to do the job are therefore necessary. Changes might involve purchasing aerial ladders to fight fires in tall buildings, or they could involve decisions about the location and construction of a new station or the modification of an existing one. Purchasing an aerial ladder could require the building of a new station to be postponed or, in other cases, could necessitate remodeling a station to make room for it. In all situations, management has the responsibility to establish appropriate priorities.

Computers can be of great help when fire departments respond to incidents involving hazardous materials. On-board computers tied into national databases can generate a tremendous amount of chemical-hazard information very quickly. Obviously, the use of sophisticated communications and computer equipment requires that all relevant personnel are competent with the equipment. Training and computer literacy have become especially important as equipment is more complex and assumes new functions. Computers now permeate every aspect of fire department operations, not just equipment used for emergency situations but also prefire communications and other functions.

For the competent officer who is alert to opportunities for improving operations, the advances in technology, which regularly lead to equipment with improved features, offer interesting challenges. To take advantage of the latest developments, a company officer should follow this advice:

- Stay abreast of new equipment and supplies.
- Keep informed of changes that can bring significant improvements if old equipment is exchanged for new.
- Be able to evaluate the benefits of new equipment with objectivity.
- Learn how to make the best use of newly acquired equipment.

EVALUATING THE ADEQUACY AND QUALITY
OF PHYSICAL RESOURCES

The adequacy and quality of physical resources are usually determined by fire-risk and community fire defense analyses. Such analyses not only compare the quantity of the various equipment items with what is perceived to be needed, but also evaluate their age and operational effectiveness. These assessments are performed, using NFPA standards and other recommended practices, by department personnel, private consultants, and insurance grading agencies. Fire departments use the evaluations to determine how well they can handle different types of emergencies.

The most important assessments are the analyses prepared by the Insurance Services Office (ISO), a national corporation that provides a wide range of advisory services to property casualty insurance companies. The basic goal of the ISO assessments is to prepare the *Fire Suppression Rating Schedule* (*FSRS*) [ISO, 1980a], which is also known as the Grading Schedule or just the Rating Schedule.* The *FSRS* is a tool for the insurance industry to measure, quantitatively, the major elements of a city's fire suppression system. The *FSRS* considers three basic elements: receiving and handling alarms, the fire department, and the water supply.

When ISO engineers evaluate a fire department, they compare its features (e.g., buildings, apparatus, equipment, competence) with their established standards or recommended practices. Based on these comparisons, they determine deficiency points for locations and other features of the department. For instance, for determining response capabilities, the ISO uses the recommendations shown in Exhibit 7.2 in the assessment. The number of apparatus recommended includes not only the department's but also apparatus available from mutual aid.

Based on the deficiency points assessed, the community receives a rating or grading of fire protection, in line with the *FSRS*, which defines different levels of public fire suppression capabilities. The schedule includes ten different Public Protection Classifications, with Class 1 receiving the highest recognition and Class 10 receiving no recognition.

Grading reports obtained under the schedule are used to establish base rates for fire insurance purposes for the properties under the fire department's protection. A respective community's property protection class (which determines the ISO rating) is always subject to change. Sweeping changes were made in the 1980 edition of the *FSRS* published by the ISO. Most of them liberalize the water supply components.

Although the ISO points out that its grading schedule is used only for determining commercial insurance rates in a community, chief fire executives have traditionally built their operations around the tenets of the ISO grading schedule. The grading schedule is important to fire departments because of its

---

*In the United States, the *Fire Suppression Rating Schedule* was originally developed by the National Board of Fire Underwriters (NBFU), continued by its successor, the American Insurance Association (AIA), and then became the responsibility of the Insurance Services Office (ISO).

Exhibit 7.2. Evaluation of Fire Department Response Capability

• **High-hazard occupancies** (schools, hospitals, nursing homes, explosive plants, refineries, high-rise buildings, and other high-life-hazard or large-fire-potential occupancies): At least 4 pumpers, 2 ladder trucks (or combination apparatus with equivalent capabilities), and other specialized apparatus as may be needed to cope with the combustible involved; not fewer than 24 fire fighters and 2 chief officers.

• **Medium-hazard occupancies** apartments, offices, mercantile, and industrial occupancies not normally requiring extensive rescue or fire-fighting forces): At least 3 pumpers, 1 ladder truck (or combination apparatus with equivalent capabilities), and other specialized apparatus as may be needed or available; not fewer than 16 fire fighters and 1 chief officer.

• **Low-hazard occupancies** (one-, two-, or three-family dwellings and scattered small businesses and industrial occupancies): At least 2 pumpers, 1 ladder truck (or combination apparatus with equivalent capabilities), and other specialized apparatus as may be needed or available; not fewer than 12 fire fighters and 1 chief officer.

• **Rural operations** (scattered dwellings, small businesses, and farm buildings): At least 1 pumper with a large water tank [500 gal (1.9 m^3) or more], 1 mobile water supply apparatus [1,000 gal (3.78 m^3) or larger], and such other specialized apparatus as may be necessary to perform effective initial fire-fighting operations; at least 12 fire fighters and 1 chief officer.

• **Additional Alarms**: At least the equivalent of that required for rural operations for second alarms; equipment as may be needed according to the type of emergency and capabilities of the fire department. This may involve the immediate use of mutual-aid companies until local forces can be supplemented with additional off-duty personnel.

Source: Cote, *Fire Protection Handbook*, 1997, Table 10-2A

pervasive effect throughout the community. The reports provide guidelines for municipalities to evaluate fire defenses, for decisions to augment, and for consideration by the governing bodies.

All aspects of the management of physical resources are influenced by the grading schedule, and many improvements in the fire service are a result of the grading schedule's emphasis on personnel, equipment, and water supply capability.

The insurance costs associated with ISO ratings sometimes have a significant influence on community development. In some cases, industries seeking locations for new plants have avoided a community that has excessively high insurance rates. The municipality involved thus lost tax revenue that could have helped pay for fire defense or other purposes.

Fire departments do not strictly adhere to the apparatus guidelines established by ISO because they must consider local conditions in their particular districts: the size and type of the district and the hazards in the district.

CONCLUDING REMARKS

In the management of physical resources, as in all other fire service functions, useful guidelines can be a significant help in ensuring high-level performance. If fire officers review the functional guidelines from time to time, whenever relevant decisions have to be made, and give serious consideration to the questions they raise about the department's and the company's handling of the function, they are likely to address some issues that will bring improvements. Investigations into the adequacy of physical resources may lead to evaluations of the way technological developments can enhance operations.

The aspect of each guideline segment that refers to maintenance is important not only to the entire department but also to each company. If every officer audited the performance of maintenance activities once a month or every other month, opportunities would emerge for the respective company and for the department to reach higher standards of readiness and appearance. By constantly reviewing equipment, both a department and a respective company may avoid problems and perform the necessary maintenance before further damage is done. If something is about to break, it can be fixed or replaced.

At the same time, if officers consider the 3Cs in every decision pertaining to physical resources, fire fighters are likely to gradually develop a better appreciation of the need to fully satisfy the functional guidelines. They will also be more willing to perform the sometimes onerous tasks of managing physical resources.

CHAPTER 7

STUDY QUESTIONS AND ACTIVITIES

If you are working alone, prepare your own written responses to these questions. If you are studying with a team or working in class, discuss the questions with the group and write a consensus answer.

1. Review the fires your company has been called out for, and especially fires that have occurred in your district, during the past six months or year. Discuss the appropriateness of the present location of your fire station in relation to these fires.
2. Considering your response to question 1, and the factors that should be considered when choosing a relocation site for a station, where would you recommend that the station be located? Why?

3. What are the shortcomings in the design of your present station? Make a rough sketch illustrating floor plan changes that would eliminate or reduce these shortcomings.
4. What questions should be considered when it appears that new or replacement apparatus might be advantageous?
5. Assume that some of the fires in a district are in tall buildings that have been built within the past few years. The fire department must consider the purchase of an aerial ladder or an elevating platform. Some of the streets from the station to the tall buildings are very narrow and steep, however, so such apparatus would have to move very slowly. Describe some of the options that should be considered in this situation.
6. A community has recently adjusted its zoning laws to encourage more industry. The first new major facility to be built will be an oil refinery. What changes in fire department physical resources should be considered to cope with the new fire hazard and other potential hazards?
7. What factors should be considered when deciding whether or not to purchase each of the following items:
   a. A new car for the chief
   b. A new stove for the fire station's kitchen
   c. A console desk for the communications center
   d. New uniforms for fire fighters
   e. An additional wing to the fire station
8. Review the communications system in your fire department and discuss its effectiveness for its many purposes, including emergency and routine communications with the public and intra- and interdepartment communications.
9. What are the specific responsibilities of a company officer with respect to the management of physical resources, considering the goals and guidelines?
10. In what ways can a company officer apply the 3Cs guidelines to the management of physical resources and thereby improve the performance of the company?

CHAPTER 7
REFERENCES

Cote, A. E., ed. 1997. *Fire Protection Handbook,* 18th ed. Quincy, MA: National Fire Protection Association.

Hickey, Harry E. 1985. *Fire Risk Analysis*. Emmitsburg, MD: The National Fire Academy.

Hickey, Harry E. 1993. *Fire Suppression Rating Scale Handbook*. Boston: Professional Loss Control Educational Association.

Insurance Services Office. 1980a. *Fire Suppression Rating Schedule (FSRS)*. New York: Insurance Services Office.

Insurance Services Office. 1980b. *Rating Schedule for Municipal Fire Protection*. New York: Insurance Services Office.

NFPA 10, *Standard for Portable Fire Extinguishers*, National Fire Protection Association, Quincy, MA.

NFPA 1201, *Standard for Developing Fire Protection Services for the Public*, National Fire Protection Association, Quincy, MA.

NFPA 1500, *Standard on Fire Department Occupational Safety and Health Program*, National Fire Protection Association, Quincy, MA.

NFPA 1901, *Standard for Automotive Fire Apparatus*, National Fire Protection Association, Quincy, MA.

NFPA 1906, *Standard for Wildland Fire Apparatus*, National Fire Protection Association, Quincy, MA.

NFPA 1931, *Standard on Design of and Design Verification Tests for Fire Department Ground Ladders*, National Fire Protection Association, Quincy, MA.

NFPA 1961, *Standard on Fire Hose*, National Fire Protection Association, Quincy, MA.

NFPA 1963, *Standard for Fire Hose Connections*, National Fire Protection Association, Quincy, MA.

NFPA 1971, *Standard on Protective Ensemble for Structural Fire Fighting*, National Fire Protection Association, Quincy, MA.

NFPA 1981, *Standard on Open-Circuit Self-Contained Breathing Apparatus for the Fire Service*, National Fire Protection Association, Quincy, MA.

NFPA 1982, *Standard on Personal Alert Safety Systems (PASS)*, National Fire Protection Association, Quincy, MA.

NFPA 1983, *Standard on Fire Service Life Safety Rope and System Components*, National Fire Protection Association, Quincy, MA.

CHAPTER 7
ADDITIONAL READINGS

Carter, Harry R. 1998. *Managing Fire Service Finances*. Stafford, VA: International Society of Fire Service Instructors.

Kepner, Charles H., and Tregoe, Benjamin B. 1965. *The Rational Manager*. Princeton, NJ: Kepner-Tregoe.

Kepner, Charles H., and Tregoe, Benjamin B. 1981. *The New Rational Manager*. Princeton, NJ: Kepner-Tregoe.

Rausch, Erwin, ed. 1980. *Management in Institutions of Higher Learning*. Lexington, MA: Lexington Books.

Rausch, Erwin, and Rausch, George. 1971. *Leading Groups to Better Decisions; A Business Game*. Cranford, NJ: Didactic Systems.

Wallace, Mark. 1998. *Fire Department Strategic Planning*. Tulsa, OK: Fire Engineering Books

DECISION MAKING AND PROBLEM SOLVING

Decided only to be undecided,
resolved to be irresolute,
adamant for drift,
solid for fluidity,
all powerful to be impotent.

—*Winston Churchill,* While England Slept

Of all the competencies, decision making—including problem solving and plan development—probably has the greatest influence on success or failure. Officers are well advised to ensure that they are highly competent in these activities, as they can judge from discussions during officer training sessions, and especially in their ability to apply the management/leadership and functional guidelines effectively. If no formal, ongoing program exists to ensure continuing development, officers should devote a few minutes from time to time to sharpening and honing these skills. This section provides some structure for this self-development.

Much has been written on decision-making theories and procedures, which consist of two segments: functional and managerial/leadership. Decision making as a science, with branches in mathematics and logic and even in charting, is a functional activity, just like the other fire service functional activities only in a different field. When faced with decisions that require the application of these techniques, most officers and other managers can obtain professional training in the functional considerations in which they do not have adequate specific relevant knowledge or education.

Identifying all the practical possible alternatives, including all the elements that might be their components, is, of course, the core of sound decision making. Tied to that is the competent selection of the course of action that appears to be best. We increase the chances for successful outcomes of our decisions as we sharpen the skills for finding and evaluating various choices. The better we get at the decision-making process, the more likely we will follow the procedure instinctively in minor decisions as well as in decisions that have greater potential impact. Using a plan is especially important for making decisions at fire emergency incidents. Guidelines play an important role by ensuring that we cover more bases—by providing a quality check on the decisions and by serving as comprehensive reminders to use all the appropriate resources and look at all the issues. We will not discuss the functional aspects of emergency decisions here because all training in the fire service focuses on them.

With respect to the managerial/leadership considerations in decision making, the issue boils down to the use of the 3Cs guideline questions or similarly comprehensive ones, a few relatively simple concepts, and an understanding of a sound decision-making procedure. Fire officers should develop the habit to stop and think before they act. The habit should become so much a part of one's nature that it engages automatically, together with intuitive judgments about people, the situation, the future, and other relevant factors.

DECISION-MAKING CONCEPTS

"Common Sense" and Rationality.   When thinking about the decisions we make, we have to recognize, first of all, that they rarely come alone; instead, decisions usually occur in lengthy series, such as plans. Going out for lunch, for instance, may involve several decisions, such as where to eat, what to order, how to pay for it, when to return, and so on. More complicated decisions invariably involve not only what should be done, but also how, when, and by whom. These subsidiary decisions, in effect, become the plans for implementing the initial decision that triggered them. When we ask ourselves about these related decisions early on, we can often save a lot of grief—for example, Why didn't I/you/we think of that, then . . . ? Sound decisions have a common element: rationality. Still, intuition and emotions are involved in many decisions; they need not be barriers to rational reasoning.

Making either the initial or subsidiary decisions on the basis of what "comes naturally" is called "using our common sense." We believe that we know how to make a particular decision correctly, whatever "correctly" means. The trouble is that often we are wrong. We have no quality standards against which to measure our decisions as we make them. Only with hindsight do we know whether the decision was a "good" one, and often not even then. When we consider a decision to be "good," we apply a judgment to the outcome, not to the quality of the thinking and decision making. Sometimes decisions that are carefully considered and sound, based on the available information, may not bring the desired results. In another situation perhaps we made a "poor" decision by not adequately considering the facts and possible alternatives, but then fortuitous, unforeseeable circumstances rescued us. Although our decision might have dropped us straight into dung, we came out smelling like roses.

Clearly, if we want to develop a record of making sound decisions, we need some yardsticks, quality standards if you will, that can give us some guidance as we make decisions. Common sense is a quality standard of sorts, but for most of us it is a very unreliable one. What is common sense to me isn't necessarily common sense to you. That alone tells us that common sense can be a slippery slope. Commonsense decision making is an *intuitive* process that relies heavily on accumulated learning, including some baggage we brought with us from childhood. It is based on the lessons we believe, rightly or wrongly, we have learned from what has happened to us.

Beyond trust in our "common sense," there are other good reasons we rely heavily on our intuition:

- We often don't know any better. We are simply not aware that we are making intuitive decisions because we have not trained ourselves to distinguish our inclinations from factual analysis, and we have not learned to question our assumptions.
- We like the easy reliance on intuition, and we tend to shy away from the meticulous approach that more careful decision making may require.
- We believe, with some justification, that the experiences and perspectives that we have gained have made our intuition reliable.
- Most important, we often just cannot get the information necessary to make thoroughly evaluated decisions.

The outcomes of many of our decisions depend on what will happen at some time in the future. A certain amount of intuitive judgment therefore has a place, even in very important decisions. Obviously, we should do whatever we can to carefully evaluate the pros and cons of the reasonable ways we could go.

We are biased in favor of making a decision intuitively, and we may have a tendency not to listen carefully enough to the more silent intellectual voice within us. We therefore have to adjust for that natural bias when we make important decisions—by developing the habit of asking ourselves whether we have considered at least a few alternatives. Once we have developed that habit, the red flag goes up intuitively and a voice within us screams: Stop—look at your choices.

Significance of Assumptions.   Assumptions can be silent Pied Pipers. If we assume, on the one hand, that human error caused a specific problem, then we will look for a mistake or for a person who might have committed the error. Most of the time, we'll find something and someone. On the other hand, if we assume that the problem involved systems or procedures then our search will take us in a somewhat different direction. As we start to make a decision, whether to solve a problem or to exploit an opportunity, it pays to first question what assumptions affect our thinking.

Outcome Preferences.   Another hidden obstacle to sound decision making is skipping an important step that feels redundant. It concerns clarifying what we want to get from the decision. We have a tendency to feel that the outcome is obvious; we think we know what we want from a decision. The trouble is that, sometimes, what we know just isn't so. Giving a few moments' thought to what the outcome should be can pay large dividends.

For instance, do we want to satisfy all of a series of needs, strike a good balance between full satisfaction and cost, fully satisfy one group of stakeholders, or partially satisfy several? Do we want to devote all the resources that the "best" outcome would probably need, or should we get the best we can with a fixed bundle of resources?

Even before we identify alternatives, we should ask: What do we expect from the "best" alternative? How well should it satisfy the 3Cs guidelines and the functional aspects of the desired outcome, including financial results, for the *short term* as well as the *long term*? The choice that seems most attractive for the short term might not be the best one for the long term, or vice versa. Most people have a natural, and partially justified, bias toward short-term benefits and often favor short-term benefits at the expense of more significant long-term benefits. Helping staff members become more aware of their own inclinations and of the need to consider short-term versus long-term trade-offs is important.

Postponement.   Frequently we intuitively want to or do postpone a decision. Sometimes that's pure procrastination, but sometimes there is good reason for the postponement. We may have a lot on our plate at the moment and consciously assign a low priority to the decision. Or we may feel that we will be in a better position to make a sound decision at a later time, possibly because we do not yet have enough information and we have good reason to think that we can get more.

We do have to make sure, however, that the reasons we give ourselves are valid and not just rationalizations for our failure to tackle the situation. For instance, some people never are satisfied that they have enough information. That's hardly a surprise because we rarely have all the information we would like. Still, the benefits of a timely decision often outweigh what we could gain by waiting. In some situations, such as in emergencies and in investment decisions, good timing can be more important than any other aspect of the decision.

On many occasions we do not have any reasons—good, bad, or indifferent—to postpone. As the Churchill quote at the beginning of this section pointed out, we just do it as individuals, in groups, and even as a nation, unconcerned that *not* making a positive decision is also a decision. Postponement is usually more than one decision. It can be a continuous repetition of the decision not to interfere with what is happening.

Intuitive decision making alone, including postponement, is appropriate for decisions that have inconsequential outcomes and for decisions that we have made so often that we are certain about the outcomes. For other decisions, however, we usually should take the time to think through the choices that are open to us.

Reversibility.   A decision is reversible if the selected alternative can be abandoned if it does not work out well and another one can be substituted with little loss. When no choice is clearly preferable, a reversible decision is better than an irreversible one because then we can start all over again and lose relatively little.

DECISION-MAKING AND PROBLEM-SOLVING PROCEDURES

As part of a program to improve decision-making competence of officers or staff members, it would be best for the officer leading the program to first review the following procedure with learners and then revise it on the basis of suggestions by staff members. Staff members are instructed to apply the procedure formally to their decisions until they have developed the habit to apply it "intuitively."

Decision-Making Procedure.   The steps in this procedure are represented by a series of questions:

1. What are the elements of the most desirable outcome of the decision? (The 3Cs guidelines can provide a check for completeness.)
2. Who, if anyone, should be involved in this decision? When and how should they be involved? (See Additional Insight 1.1, Participation in Decision Making and Planning.)
3. What alternatives should be considered? (At this point it is useful to consider all alternatives, even outrageous ones. Keeping an open mind may bring out a useful adaptation of an alternative from a feature of a seemingly useless one. The very act of enumerating possible choices—insisting on at least one or two more than the obvious ones—can generate better solutions. One alternative that always deserves consideration is the deliberate postponement of the decision.)
4. What information is essential for determining other alternatives and for evaluating alternatives? What information is not essential but desirable? (Include the "data" about the likely emotional reactions of stakeholders.)
5. How should the alternatives be evaluated?
6. Which alternative is the most desirable one—the one that should be implemented? (Sometimes even with careful evaluation it is difficult to determine which possible choice is "best." Still, the process of evaluation helps sharpen views on which alternatives are clearly *not* likely to achieve the desired outcomes and which appear to be among the best available.)

In seeking to identify the best alternative, we should conduct an evaluation on the basis of the 3Cs guidelines and the decision-making concepts discussed earlier in this section. Unfortunately, we often face so many uncertainties that even the most careful evaluation cannot clearly identify a "best" alternative. That's when personal, intuitive, emotional preferences, and crystal balls have to take up the slack.

Problem-Solving Procedure.   Most steps in problem solving are the same as those in other decision making. However, when we are tackling a problem—a situation in which something is not as it should be—we should ask a few questions about the problem before proceeding with the decision-making steps.

- What is the problem? (Is the problem we see only a manifestation of a more extensive or serious problem?)
- What are possible causes of the problem?
- Who has knowledge about the development of the problem?

Management of Financial Resources

INTRODUCTION

As the title and the table of contents show, this chapter discusses, as general background information, all aspects of managing financial resources. Fire officers should be familiar with these functions so they can assist the department in meeting its budgets and so they can explain related issues to the members of their staffs when the need arises.

Most fire department resources, other than those provided by certain volunteer fire departments, are purchased with public money. Public funding is supplemented by any of the following alternative means:

- Financial donations
- Equipment donations
- Fund-raising events
- In-kind donations of service by professionals
- Public-spirited citizens with special skills

In addition to justifying any special allocations of funds, a fire department must be concerned with managing its financial resources effectively and efficiently. A basic principle in municipal fire protection master planning is that the proper level of fire protection for a community is the amount for which citizens are willing to pay. Managers of progressive fire departments, however, inform their elected representatives of the probable costs of fire risk and fire protection at different levels of fire department funding. In other words, the fire department managers, as experts, provide estimates of potential fire losses at the requested budget and at other budgetary levels.

GOALS AND GUIDELINES FOR MANAGING FINANCIAL RESOURCES

In this section we will look at the goals and the guidelines for management of financial resources.

> The primary goals of management of financial resources are to ensure that the department will have the necessary funds and will administer them effectively to pay fair and equitable compensation to department staff members, maintain facilities and apparatus in full operating condition, obtain the needed equipment and supplies, procure additional facilities when needed, purchase additional apparatus, and refurbish existing apparatus to stay abreast of the needs of the community to be protected.

The guideline for these goals, like all other guidelines, is stated as questions:

> What else needs to be done to obtain funds for needed facilities, apparatus, equipment, and supplies and to ensure that the available funds are appropriately allocated to the various needs of the department?

The following are the specific questions for this guideline:

1. What else can be done to obtain needed funds for the operating budget?
2. What else can be done to obtain approval and funds for the purchase of needed facilities, apparatus, and equipment?
3. What else can be done to ensure that all purchases are made at the best possible prices?
4. What else can be done to correct wasteful practices—those that use more materials than necessary, including utilities and maintenance costs?

Please keep in mind that it is not necessary to remember the detailed wording of the guideline when making decisions. Only in the most important decisions might that be worthwhile. However, memorizing a personalized, abbreviated (summarized) version and applying it with every relevant decision are critical for developing the necessary habit that will help to ensure the highest quality performance possible under the existing circumstances.

The proper use of financial controls is essential to the responsible management of fire department funds. The primary tools of controls are budgets. Therefore, an understanding of budgets and budgeting techniques is essential to management success. Properly used budgets bring a form of discipline that forces careful planning and thereby leads to effective proposals for and judicious use of funds.

FIRE DEPARTMENT BUDGETS

A budget is a plan for future operations, expressed in financial terms. Basically, formulating a budget involves asking what a fire department intends to accomplish during a given year and how much that will cost (Carter 1986). A budget allocates financial resources to the different uses for which the money is intended. Because exact amounts cannot be known in advance, a budget reflects a forecast of what will be needed and thus a decision on how much money should be made available for each function. At the same time, the budget is a forecast of how much will be spent for a particular use.

Progressive fire departments use the budgetary process to list the goals and objectives of the department along with the financial resources necessary to achieve them. In this way, municipal officials and the general public can better understand the need for a fire department's expenditures and make more informed financial decisions. The fire department is also better able to justify its requests.

Available funds are distributed so that a balanced mix of resources is available to address community fire protection needs. Through the use of a budget, the distribution of funds is planned and controlled. A budget is used much like a road map. The department goals are set, priorities are determined, and financial plans are developed to reach those goals. The completed budget shows where the emphasis in fire department service delivery is to be placed.

The best way to make solid budget decisions is to study the actual service needs in the community. A potential service profile can be developed from the population mix and a fact-based budget can be prepared based on those data. The more that is known about existing risks, the more accurate will be the budget request.

Two basic types of budgets are used in municipal fire departments: the expense budget and the capital budget.

The Expense Budget

An expense (or operations) budget contains costs of a recurring nature, such as salaries, fringe benefits, supplies, and small-equipment purchases. This type of budget is used most often by fire department management in daily operations. The two basic operating systems used to prepare and work with recurring expenses are the line-item system and the program system (a performance-based or functional system).

Line-Item System.   In this type of budget, each expense is listed on a line (see Exhibit 8.1a). On each of the lines, the particular expense item is identified along with the amount that can be spent on it. (Expenditures are usually based on requisitions from the requesting bureau or company, which indicate the appropriate budget line from which the money is to come and are used as the basis for purchase orders to vendors.) As the money is spent, it must be accounted for by a series of matching vouchers that indicate the amount spent and the dates of the transactions.

Because the line-item system lists each expense separately, rather than incorporating many expenses under one broad category, it is difficult to determine how much of the budget is being spent on fire service functions, such as fire fighting, training, or support of community activities. An advan-

Exhibit 8.1a Sample Line-Item Expense Budget

Account number	Classification	Current year ($)	Last year ($)
001	Apparatus repairs	16,500	15,876
002	Office supplies	4,000	3,950
003	Fire hose	15,000	12,500
004	Radio repairs	10,000	9,760
005	New radios	35,000	19,500
006	Janitorial service	35,000	27,500
007	Chief's vehicle	35,000	—
008	Computer system	16,000	—
009	Software	3,000	—
010	Fire-fighting equipment	29,000	25,000
011	Breathing apparatus	21,000	12,500
012	Salaries and benefits	5,840,000	4,905,000
Total:		6,059,500	5,031,586

Note: Although some departments might place some of the expenditure items listed in this table in a capital budget, others list them as shown here.

tage of this system, however, is that it clearly shows the total expense for specific items. Line-item budgets have the added benefits of simplicity and low cost, but they have the potential negative effect of controlling resources too strictly. This may stifle initiative and limit the organization in delivering service to the public. Still, listing fire prevention week brochures in a budget line that is separate from the fire prevention literature that is stocked for daily use has its benefits. The line-item budget provides targeted budget figures that allow for better control and management of budgeted funds. Monies that are set aside for use in October are not inadvertently used during an earlier period of the year.

Budgets may also be set up by specific division. This control feature indicates where the dollars in each line-item area are going. (See Exhibit 8.1b for an example.)

Program Budget System. In this type of budget, expenditures are allocated for specific activities (see Exhibit 8.2). A department estimates the total expense—including items such as salaries and materials used—for performing a specific function over a given period of time. This estimated amount becomes the basis for measuring actual expenditures for that function. As the year progresses, the department keeps a record of the actual money spent. Officers directly involved with the budget can keep track to see how adequate the budget is for the department's needs and functions, in preparation for next year's budgetary estimates.

Exhibit 8.1b Sample Line-Item Budget Segment with Divisional Separation

Account number	Fund allocation	1999	1998
102	Apparatus repair	16,500	15,876
102-1	Suppression	12,000	11,876
102-2	Fire prevention	2,000	1,900
102-3	Public education	1,000	1,000
102-4	Administration	1,500	1,100

Many departments seek to curtail functions that are not directly concerned with emergencies to adhere to budgetary constraints. This tendency should be reviewed regularly in light of the short-term and long-term benefits of the respective function. Strict adherence to a low budget should be tempered by the future value of such functions—for example, public fire and life safety education or fire fighter training. Nonemergency functions may nonetheless bring significant long-term benefits.

As a step in controlling expenses, departments may compare budget allotments, from year to year, with actual expenditures and resources. Resources can be reallocated if necessary to allow for emergencies or to adjust for unavoidable outlays that are higher than forecast, such as the costs to repair unexpected equipment damage during storms. For example, unused salary account funds for unused overtime may be moved to the vehicle repairs account to cover a vehicle engine that might need extraordinary emergency repairs. The municipal authority responsible for budget administration needs to approve this movement of funds, which is likely to occur many times each year.

Exhibit 8.2 Sample Program Budget

Department:	Fire
Program:	Fire and Life Safety Education
Division:	Fire Prevention
Year:	1999
Goal/Objective:	To provide fire and life safety education to all residents
Measurement criteria:	Number of residents reached
Salaries (2 personnel):	$87,700
Fringe benefits (2 personnel):	$31,500
Vehicle acquisitions:	$24,800
Vehicle upkeep:	$2,400
Educational materials:	$6,500
Audiovisual materials:	$7,650
Office supplies:	$980
Total program cost:	$161,530

A program budget requires more record keeping than a line-item budget, but it permits a department to fund its work so that the activities that deserve higher priorities receive the appropriate attention. For example, if a department has decided that public fire and life safety education should receive high priority for 12 months and has allocated a certain number of hours to the project, the program budget will show how the amounts spent compare with the planned expenditures on a monthly basis. Thus, officers can see whether public education is being deprived of resources or whether it receives its share as the year progresses. This helps them to make decisions on how to perform the function in the remainder of the year.

Overseeing the budget is important for matching resources to needs. Program budgets have the distinct advantage of providing management with a control tool. Officers can make adjustments when they are desirable or necessary to achieve departmental goals.

The Capital Budget

A capital budget lists expenditures for items that have a useful life of more than one year. Communities often specify a dollar amount for an item (camera, computer, or radio equipment, for example) that, if exceeded, moves the item to the capital budget. Funding must be spread over a number of years. Examples of large capital expenditures are a new fire station and new apparatus (see Exhibit 8.3).

Because capital budgets usually involve only a few items, they are not considered to be working tools for a manager in the same sense that expense budgets are. Although capital budgets represent an authorization for a department to buy needed equipment, they do not require regular entries into the department's budget registers and they do not affect evaluation of the department's operation.

A note of caution: Including an item in the government authority's capital budget for the fire department does not mean that funds will actually be made

Exhibit 8.3 Sample Capital Budget

Account number	Classification	First year	Second year	Third year	Fourth year
001	Station 2	$550,000	$550,000	$550,000	$550,000
002	Class A pumpers	350,000	400,000	450,000	500,000
003	Aerial ladder	575,000	650,000	750,000	800,000
004	Rescue vehicle	—	—	675,000	—
005	Compressor system	—	85,000	—	—

available for that item. A capital budget reflects the intent of the government to spend the budgeted amount, but, as the year goes on, priorities change and money allocated for one purpose may instead be used for another one. For instance, the governing body may need to make major repairs to a school and take funds budgeted for a piece of fire apparatus to pay for those repairs.

Budgetary Justifications

Many times, fire department officers must justify the expense of an item. They may do this by providing a comparison with the cost of competing options. If the requested item is the least expensive, they could highlight that. If it is not the least expensive, then officers could provide a justification with the reasons for the request.

It is important to remember that restrictions exist on how public funds are spent. Bid laws require competitive bidding to assure that the lowest possible price is received. Care must be exercised in drawing up purchase specifications or requests for proposals (RFPs) so that they conform to all legal and procedural requirements. There is usually a dollar threshold above which the full set of bidding procedures must be used. Bidding and the specifications help to assure that the purchaser's intent is met by the bidder. Fire officers must be aware of the purchasing requirements that apply in their locality.

THE BUDGETARY PROCESS

"Fire administrators and city officials in general are well aware that the process of developing, presenting, justifying and monitoring a budget is . . . a turbulent, complex and sometimes chaotic process" (Burkell 1983). Mastering this process is extremely important for fire department operations. Without sufficient funds, a department cannot meet the fire protection needs of a community. Budgets thus serve many useful purposes; they are

- Guides or forecasts that make planning easier
- A communications link, joining the department with the public and officials
- Records showing how and how fast available funds are being spent

Because budgets are important management tools, every fire officer should understand the purposes they serve and the basis for formulating them. Fire officers must also understand that fiscal responsibility (management's ability to live within a budget) is a crucial element in maintaining public trust.

Except in the case of a fire district or an independent fire company, the fire department budget is a portion of the municipal budget. Personnel costs, including direct compensation and all benefits, are the most significant costs for most municipal fire departments, accounting for approximately 90% of the total expenditures of a full-time paid fire department. Personnel costs of partly paid, partly volunteer fire departments make up approximately 40%–60% of their total budgets (Cote 1997, p. 10–19). The remaining funds are allocated to cover operating expenses such as utilities for the fire stations; gas or diesel fuel for the fire apparatus and other vehicles; maintenance supplies; minor equipment such as breathing apparatus; and fire-fighting tools, office supplies, repairs to buildings and grounds, emergency medical expendable supplies, and so on.

Progressive fire departments use the budgetary process as a forecasting tool to determine the allocation of resources among the various segments of the agency. These departments list the goals or objectives of the fire department along with the financial resources necessary to achieve them. In this way, the general public and municipal officials can better understand the need for a fire department's expenditures and make better informed fiscal decisions. The fire department can also defend its requests more effectively.

Departments must give considerable time and thought to develop a budget request that will be received favorably, especially if the budget includes significant increases from the previous year. If a department solicits ideas and opinions from various groups within the department, these contributions will help to focus the budget on priorities for fire protection in the community.

Budgeting is essentially a four-part process:
1. Formulation
2. Transmittal
3. Approval
4. Management

Formulation

During the formulation phase, fire department managers review past budgets to see how well estimated outlays compared to actual expenditures. They focus on the variations to make better predictions for the future. During this time, suggestions are solicited from each division, bureau, or level of the fire department so that all useful ideas can be considered. Brainstorming by groups of fire department personnel is helpful to bring out all planning suggestions. When managers involve various groups, more people gain a better understanding of budgetary decisions and feel that they have participated in the process. Participation assists with overall acceptance of the department's budgetary request.

Using the specific knowledge of all those responsible for implementing the budget decision helps to set priorities and make more accurate estimates. Groups with special expertise can thus add detailed support and sound justification, both important elements when a budget is being considered by municipal authorities.

While a budget request is being developed, a proposed budget sheet is sometimes used. The proposed budget sheet contains several columns that show how much money was budgeted in the previous year, how much was actually spent, how much is requested for the current year, and (usually) how much greater this is than the previous year's request. Where necessary, managers attach explanations and supportive data to the budget in an appendix.

Although the layout of the budget sheet is fairly standard, many fire departments modify it to fit their individual preferences. The format provides a clear and logical approach to budgeting. Each request is based on careful reasoning and fortified by the relevant supportive information. Use of a proposed budget can prevent costly errors or omissions in budget development.

Transmittal

After the preliminary budget has been prepared, it is transmitted to the local governing body, usually to a municipal finance officer or finance committee. At this point, the budgetary request should be thorough enough to be self-supporting in the event that department personnel are not available to explain it, answer questions, or defend it. Unanswered questions can do considerable harm to the message of a preliminary budget. The message itself should, therefore, communicate all important points supporting the budget request.

Still, it is essential for the fire chief or a financial officer to be available to answer questions for the governing body. Consider your budget to be "a mute babe in arms . . . . Would you leave a baby in your out-basket on Friday, and expect its safe return on Tuesday?" (Carter 1985). Probably not. In the same way, the budget is best not left to chance. Those who will present the budget should be aware of the environment within the budget system and be ready to answer any question pertaining to the budget's contents.

Approval

Approval usually requires at least two steps. First the department budget must be approved by the town or city administration, and then it must be approved by the town or city council; in some municipalities, it must be approved by a town meeting. Some municipal charters allow the council to reduce a budget but not to increase it.

After it has been approved, the budget takes effect at the beginning of the fiscal year and becomes an instrument of control that shows a department how much it can spend during a particular time period. If the budget is not approved in time, a municipality usually permits expenditures at the same rate as the previous year (Carter 1986). This process can be altered somewhat for volunteer departments that raise funds themselves and for departments that obtain grants for special projects from a private foundation or from state, provincial, or the federal government.

Management

After the approval step, fire department administrators must, of course, attempt to manage the department within the budgeted guidelines. Excessive spending and even significant underspending are signs of shortcomings in a department's administrative competence. They can lead to credibility problems and more critical reception of future budgetary requests.

Budget worksheets provide useful guidance for staying within the budget. Worksheets usually show a record of expenditures for a current year, and they can be either distributed to officers in a particular department or kept solely for the chief's use. Although departments organize budget worksheets according to their individual needs, most worksheets are set up like the sample shown in Exhibit 8.4. Typically, one column is used for listing the monthly budget, and another column is used for recording the amount actually spent. Columns can be added to show the difference between the two or to show previous years' budgets and expenditures.

When preparing budgets for the financial management of a department's capital resources, management must consider their long-range use. To purchase all new items in one year would be impractical. Their expense would cut into a department's ability to fund other important activities

Exhibit 8.4 Sample Budget Worksheet Prepared on a Monthly Basis

				January					
Item	Budget this year	Actual this year	Difference) (+)(−)	Budget last year	Actual last year	Budget to date	Actual to date	Last year budget to date	Last year actual to date
Contractual services									
Main building and grounds									
Etc.									

Source: Rausch and Carter 1989, Table 12.4

competing for the funds, such as fire prevention work. To formulate long-range budgetary goals, officers must take into account all possible future physical resource expenditures, including renovations. They must predict problems and needs that might arise in the future. Because the entire department is affected by all budgetary decisions, management should encourage coordination among all levels of the department and should allow for the expression of individual needs and recommendations. With the cooperation of all departmental levels, the department is better able to work as a solidified unit in accomplishing its primary goal—protecting life and property.

CONCLUDING REMARKS

Typically, a fire officer has few responsibilities in the management of financial resources, except when asked to provide specific input to the planning of the next year's budget. Still, officers, as managers, should understand budgets and the budgeting process, so that they are prepared to answer any questions that arise from the members of the team.

There is one aspect of the management of financial resources in which company and battalion officers have operational responsibilities. That aspect overlaps greatly with the management of physical resources. It concerns the conservation of all resources to keep operating costs as low as possible. The guideline question asked: What else can be done to correct wasteful practices—those that use more materials than necessary, including utilities and maintenance costs? To be effective in correcting wasteful practices, officers should consider many of the specific 3Cs guidelines, especially those that pertain to goals, participation, and psychological rewards.

CHAPTER 8
STUDY QUESTIONS AND ACTIVITIES

If you are working alone, prepare your own written responses to these questions. If you are studying with a team or working in class, discuss the questions with the group and write a consensus answer.

1. Obtain a copy of the operating and capital budgets for the department or a copy of the municipal budget that includes the fire department. Figure out where equipment, supplies, maintenance, and repair costs are covered in the budget.
2. Discuss whether there might be benefits to involving company officers in budget planning (if they are not now involved) and what these benefits would be to the department and to the officers.
3. Explain how a budget system is used as a planning tool and as a control device.
4. Write a brief explanation of (a) a line-item budget system and (b) a program budget system, including a simplified example of each.
5. List and discuss the steps that must be taken to develop a budget and obtain approval for it.

CHAPTER 8
REFERENCES

Burkell, C. J. 1983. "Budgeting with Objectives." *Fire Chief,* September, p. 35.

Carter, Harry R. 1985. "Being a Good Staff Officer." *Fire Command,* April, p. 29.

Carter, Harry R. 1986. "Budget Justification, Your Fight for a Piece of the Pie." *Fire Command,* August, p. 45.

Cote, A. E., ed. 1997. *Fire Protection Handbook,* 18th ed. Quincy, MA: National Fire Protection Association.

Rausch, Erwin, and Carter, Harry R. 1989. *Management in the Fire Service,* 2nd ed. Quincy, MA: National Fire Protection Association.

CHAPTER 8

ADDITIONAL READINGS

Carter, Harry C. 1999. *Managing Fire Service Facilities*. Stillwater, OK: International Fire Service Training Association.

National Fire Academy. 1998. "Fiscal Management in the Fire Service." Budget Course. Emmitsburg, MD.

....

Fire Service Personnel Management

INTRODUCTION

Personnel administration is concerned with maintaining effective human relations within an organization. In the fire service, many specialized personnel administration functions are performed by the local government, the fire chief, and specially designated officers. Still, some aspects of personnel management are an integral part of every officer's job.

Large organizations in the private and public sectors have separate departments staffed with personnel specialists. Because most local government units cannot afford a separate personnel department for the fire service, however, the personnel function might be either in a common personnel office for all departments or one of the responsibilities of a single fire department officer. In other fire departments, the personnel function is the responsibility of the fire chief and the fire director or commissioner. Many fire departments rely on the Civil Service Commission's personnel procedures for guidance.

Department officers play an important role in implementing personnel policies, whether or not a separate personnel division exists. In departments that have a personnel office, the company officer's task is somewhat less demanding; policies and procedures are likely to be more clearly defined, and more support and advice are available. When there is no department personnel office, more responsibility for good personnel practices rests with the department's officers even though some of the personnel functions are performed by a centralized municipal or unit personnel office. In such a case, personnel services are less likely to be specifically relevant to the fire department's needs.

Whether it is the part-time responsibility of a single individual or the full-time responsibility of a fully staffed unit, personnel management involves specific responsibilities in three general areas:

1. *Routine procedures:* hiring, placing, and terminating employees; monitoring employee services; administering salaries, wages, and other compensation; administering fringe benefits; and administering career development programs, including performance evaluations

2. *Labor relations activities:* making recommendations on policy, playing a role in handling employee grievances (particularly in nonunion organizations), consulting with line managers on disciplinary actions, and conducting contract negotiations if a union represents some or all of the employees (see Chapter 10)

3. *Advisory activities:* performing an advisory role on policies pertaining to human resources, communications needs, scheduling of personnel, and

making recommendations for policy and procedural changes based on un-derstanding of the needs of personnel

SCENARIO

THE GRIEVANCE

Captain Lester Stritt was sitting at the desk when fire fighter Mark Kros entered. The captain stood up, closed the door, and motioned to Mark to sit down. Somewhat stiffly, the fire fighter complied.

"Do you know why I asked you to meet with me alone?" the captain began. When Mark shrugged his shoulders, the captain continued. "You've been in this company for many years, quite a few more than I. Still, we've been through a number of tough situations together. Isn't that right?"

Mark nodded, and then there was a brief silence.

"This isn't easy for me, Mark, because I've been satisfied with the way things have been going," said the captain. "But let me be frank. There have been a number of instances recently when it seems as though you are unhappy with your work. I don't see you putting in the same effort you used to. Your areas are sometimes not in good shape when the housekeeping time is up. You've been late for training sessions, and I've even noticed that you don't always check your equipment as thoroughly as you used to. You remember that I asked you, only the other day, to go back and check your equipment again. Last week I pointed out the poorly cleaned sink to you, when it had been your turn to do the kitchen. You remember that, don't you?"

"Yeah, I know. You're picking on me; that's what this is all about. How come you don't say anything when someone else misses up on something, like when Jim didn't clean his SCBA?

"I'm not picking on you, and I didn't ask you to come in here so we would wind up arguing. I just wanted to see whether you might share with me what's bothering you. Maybe I can be of help."

"Nothing is bothering me," replied Mark. "I don't know what you are talking about. I do my work just like everyone else does."

"Well, you may think so. You might as well know that I am not happy about what you are doing, and I would appreciate it if you would try a little harder. You know that I don't want to have to give you a warning and maybe even write you up. But unless you change your ways, you might force me to do just that."

A brief silence followed.

"Is that all? Can I leave now?" It was clear to the captain that Mark was not plan-ning to cooperate.

"Yes, that's all," the captain said sternly. Then he continued in a much softer tone.

"Come on, Mark, don't make life tough for both of us. We've known each other for a good while now, and it would be a shame if we couldn't continue the way things were."

Mark did not answer. He just got up and walked out.

During the following week, the captain didn't say anything when he had to remind Mark to fill out the apparatus checklist, but when Mark came in about 20 minutes late for the next shift without calling in, the captain decided to act.

He again asked Mark to meet him in the office. This time he got right to the point as soon as Mark sat down. "Things can't go on like this, Mark. You've got to tell me what the trouble is or shape up. I pretty well ignored it when you forgot to fill out the apparatus checklist, but this I can't ignore. Why didn't you call in when you saw that you would be more than 10 minutes late? Don't we have a strict rule here about calling in when you'll be late more than 10 minutes?"

"You and your rules. What's the big deal? I got stuck on the way in, that's all. Why don't you leave me alone?"

"That won't do. I have to have your firm promise that you will get back to doing your share the way you used to. If you don't come clean about what the problem is, you can take this as the first verbal warning. If I have to give you another one within a month, that will be the last verbal one I can give you. After that, I'll have to write you up. So, please don't let it come to that."

"Lay off, captain. You know, I always thought that they picked the wrong guy when they made you captain, and now you're proving it."

An hour later the captain received a call from the union representative, Barbara Hollister, who was off duty. She wanted to set up an appointment to discuss fire fighter Mark Kros's grievance. That annoyed Captain Stritt.

"What kind of nonsense is this? The guy doesn't do his work right, I try to talk to him and help him get straightened out, and he files a grievance. What's he grieving about? I've only given him the first verbal warning so far."

"He says you're picking on him. I guess that means you are discriminating. Let's get together and talk about this. When will you be on duty next time?"

"I'll be in Friday morning."

"OK. I'll be in to see you then, around eleven."

Barbara came in a little earlier, briefly spoke with Mark, and then asked Captain Stritt where he wanted to meet.

Barbara started, "As you both know, the purpose of this meeting is to resolve the problem, if we can, so it does not go any further. Mark is upset, captain, because he really feels that you are after him, that you are treating him differently than the others in the company. Frankly, I don't know what the facts are, but I know Mark, and he is not one to get upset easily. How do you see the problem, captain?"

The captain repeated what he had said to Mark at the first meeting and then spoke about what had happened thereafter. He concluded by reassuring both Barbara and Mark that he would like to avoid the disciplinary procedure. After all, he had given Mark only a first verbal warning and was trying to avoid having to give him the second one. He was still hoping that Mark would change his ways, or at least give the captain the opportunity to help him get over whatever was causing the problem.

Barbara turned to Mark. "What do you have to say to this, Mark?"

Mark's face had shown some inner turmoil as he had listened quietly to the captain's explanation. At Barbara's question, he turned away and there was silence. Barbara and the captain looked at each other. Their expression indicated that they had both noticed Mark's distress and were prepared to give him the time he needed to come to grips with the situation.

The next few minutes seemed like hours. Then suddenly, without turning, Mark blurted out: "Y'know that kid died in my arms, and...." Tears seemed to well up in his eyes, and he couldn't finish the thought.

There was a moment of silence again. Then the captain spoke up. "Get some rest, Mark. We'll continue at another time."

At first Mark just sat there. After a few moments he got up without facing them, said "OK," and walked out.

When he was gone, the captain and Barbara quickly agreed that they would ask for intervention by the Critical Stress Intervention Team. In the meantime, the captain would gently remind Mark whenever he noticed something that needed to be corrected. He would also keep Barbara informed of developments.

Scenario Analysis: Fire Service Function Perspective

In this section we will review the scenario with the aid of the goals for the personnel management function.

> The goals of the fire service personnel management function are to ensure that an adequate staff exists for the emergency prevention and emergency response needs of the community and to create and maintain a work environment in which all members of the department receive fair and equitable tangible rewards for their efforts, in which vacancies are filled with highly qualified candidates, and in which human resource policies and practices ensure a satisfying work climate for all members of the department.

A useful guideline for these goals follows:

> What else needs to be done to ensure that the staff is adequate for the needs of the community, that the most highly qualified candidates are selected for vacancies, that compensation and fringe benefits are fair and equitable, and that the department offers a climate that provides a high-quality work life for all members?

This guideline leads to the following specific questions:

- What else can be done to ensure that recruiting, selecting, and hiring policies are in line with all applicable laws and regulations and that the procedures used, whether administered by members of the department or by a government agency, will bring the most qualified candidates?
- What else can be done to ensure that the compensation and benefits offered to members of the department are equitable in comparison to those in other communities?
- What else can be done to ensure that policies and procedures relating to the quality of work life are appropriate to the department and that all officers have the will and the competencies to ensure a satisfying work climate.

Of all these items, only the last one applies to this scenario. It is also the one for which company officers have the greatest impact on achieving the goal. The 3Cs guidelines are particularly applicable to this item, as the analysis in the next section shows.

Scenario Analysis: Management/Leadership Perspective

The analysis of management/leadership in the scenario will be based on the guidelines for the 3Cs. We will repeat only the specific questions, however, and not the general one. (For a complete list of all the guidelines, see Appendix A, Decision Guidelines.)

Read the guideline questions and then, before reading the discussion of each, you may wish to give some thought to the strengths and weaknesses of the way Captain Stritt managed the personnel incident from the management/leadership perspective.

The specific questions that expand on the general control guideline are

1. Are the goals and objectives appropriate and effectively communicated?
2. Is the participation in decision making appropriate?
3. Are coordination, cooperation, and inter- and intra-unit communications being stimulated?
4. Is full advantage being taken of positive discipline and performance counseling?

The specific questions that expand on the general competence guideline are

1. Are changes needed in recruiting and selecting for vacancies?
2. Are management of learning concepts applied effectively?

3. Are coaching and counseling on self-development being used to their best advantage?

The specific questions that expand on the general climate guideline are

1. Are appropriate psychological and tangible rewards offered and provided effectively and efficiently to bring the highest possible level of satisfaction from the creation and use of the product/service?
2. Are policies in place to help reduce work-related stress?

As we mentioned in earlier chapters, with respect to both the functional and the management/leadership guidelines, you do not need to remember all the detailed wording of each guideline when making decisions. Only in the most important decisions might it be worthwhile to refer to it specifically. However, memorizing a personalized, abbreviated (summarized) version and applying it with every relevant decision are critical for developing the necessary habit that will help to ensure the highest quality performance possible under the existing circumstances.

All three basic guideline questions, but very few of the guideline segments, are applicable to this scenario. Of course, for issues related to personnel management, as well as for all other decisions, an officer should always consider all 3Cs. When asking the basic guideline questions pertaining to control, competence, and climate, the officer who attempts to use the guidelines with every decision is likely to think of the more detailed segments.

Discussion of the Control Guideline.

1. *Are the goals and objectives appropriate and effectively communicated?*   In this scenario, the only goal that was communicated was the one that is automatically on the list of goals, no matter how many goals pertaining to improvements are currently being worked on. That goal concerns performing all tasks that are not covered by goals or objectives at least as well as in the past. That, of course, pertains to procedures. (See Additional Insight 3.1, Goal Setting and Implementation.) We can assume that everyone was aware of the need to follow the procedures and no problems were inherent in them. Other goals and objectives were implied, such as the captain's objective to maintain positive discipline (see Additional Insight 9.1, Positive Discipline and Counseling) and his goal to be an effective leader who considers the needs of the fire fighters.

Still, in situations like the one described in the scenario, an officer might consider questions such as the following:

- What types of goals and objectives should I set with the team member, and what support should I provide?
- What types of goals and objectives should I set for myself so team members will be more willing to share their work-related concerns with me?

2. *Is the participation in decision making appropriate?*   The way the captain handled most of his contacts with Mark and Barbara indicated that he does not have an autocratic style but that he seeks participation, at least in decisions that affect the members of his team.

In general, with respect to team member performance problems, officers might ask themselves questions such as the following:

- How well are performance and related behavior norms understood and accepted? Who should be involved—and how—in providing answers to this question? (Performance and behavior norms pertain to adherence to procedures and ethical behavior. They are the kinds of things that the team expects from the individual, such as doing a fair share of the work at the understood level of quality, being polite and accommodating in one's relations with others—the golden rule, and paying attention to personal hygiene.)
- What should I do to ensure that the norms are appropriate in light of the existing needs and norms in other companies or battalions? Who should be involved—and how—in evaluating norms?
- How should norms that are no longer appropriate be changed? Who should be involved and how?

3. *Are coordination, cooperation, and inter- and intra-unit communications being stimulated?*   The captain understood that intra-unit coordination and cooperation would be in jeopardy if he allowed fire fighter Mark Kros to perform less diligently than other fire fighters. He therefore took appropriate steps to prevent damage that would result if he ignored the fire fighter's less than adequate performance.

In similar situations, officers might ask themselves these questions:

- How adequate is the level of intra-unit coordination? What procedures, if any, should be in writing or discussed from time to time?
- Are there behaviors or differences in individual performance that could lead to cooperation problems, and if any exist, what should be done about them?

4. *Is full advantage being taken of positive discipline and performance counseling?*   Captain Stritt attempted to maintain positive discipline by his approach and his reluctance to invoke the punitive steps of the formal disci-

plinary procedure. Instead, he attempted performance counseling. During his counseling sessions, he did his best to show empathy with the fire fighter's situation and tried to obtain information to help Mark overcome his problem. However, performance counseling and, even more important, maintaining a climate of positive discipline require high-level skills. Therefore, officers should ask themselves questions such as the two listed here along with the relevant ones under the competence guideline, from time to time.

- What else, besides reviewing and possibly changing norms, should I do to ensure sound positive discipline?
- What else should be done so that officers are fully competent in performance counseling?

Discussion of the Competence Guideline.

1. *Are changes needed in recruiting and selecting for vacancies?*   This question does not apply to the scenario.

2. *Are management of learning concepts applied effectively?*   This question is largely irrelevant for the scenario.

3. *Are coaching and counseling on self-development being used to their best advantage?*   It is possible, though not likely, that fire fighter Kros could benefit from coaching. It is also possible that some coaching/reviewing of standard operating guidelines would have helped him to open up on his performance problem. It might be useful, therefore, for officers to regularly ask themselves question 3 and even question 2 of this guideline.

Discussion of the Climate Guideline

1. *Are appropriate psychological and tangible rewards offered and provided effectively and efficiently to bring the highest possible level of satisfaction from the creation and use of the product/service?*   This question, on first blush, does not seem to apply to the scenario. However, if Captain Stritt had followed the suggestions for showing appreciation for fire fighter contributions, communications between him and fire fighter Kros might have been more open and Mark might have been willing to share the reason for his anxiety. The performance problem might never have occurred if the captain had been aware of the impact of the sad event on Mark. Intervention could have started much sooner to both the fire fighter's and the captain's benefit.

Officers should therefore give thought to the two critical questions from time to time:

- What else can I do to identify regularly all contributions of members of my team, not only the outstanding ones?
- What else can I do to make members of my team aware as frequently as possible that their contributions are appreciated?

2. *Are policies in place to help reduce work-related stress?*   Policies appear to be in place to provide professional counseling when a member of the department is under unusually high stress. Otherwise the scenario had no information about policies to identify or reduce work-related stress levels. The scenario did not explain why the Critical Stress Intervention Team did not help immediately after the incident. Perhaps it had not been brought in or Mark had successfully avoided counseling.

The fire service has constant potential for high stress levels. In addition to the stress of emergencies, the work load of nonemergency tasks is increasing in many departments. Increased work loads can lead to increased internal conflicts and higher levels of stress, with the resultant decrease in work satisfaction.

Officers, therefore, might ask themselves about actions to reduce stress as well as, from time to time, a question such as:

"What else can I do to ensure that increased work-related stress is monitored and that steps are taken to keep it within acceptable bounds?"

It may be difficult to determine when stress levels are beyond acceptable bounds. However, in a climate of open communications, an officer would notice various signs in behavior, in expressions, and possibly even in performance when stress levels reach the point at which one or more staff members have difficulty coping.

This scenario emphasizes the benefits and the importance of using guidelines that ensure a proactive stance by officers with respect to all matters that affect the productivity of the department and the satisfaction that staff members gain from their work.

ISSUES IN FIRE SERVICE PERSONNEL MANAGEMENT

The personnel practices, policies, and programs of a fire department are subject to many influences, including union contracts and policies; the local government; the fire commissioner, director, and/or chief; and the general public (taxpayers). These diverse interests must adhere to the federal and state laws and regulations, discussed in this section, that affect the management of fire service personnel.

Equal Employment Opportunity

Significant legislation prohibiting discrimination on the grounds of race, color, religion, gender, national origin, handicaps, or age affects all personnel functions. The relevant statutes apply not only to hiring and firing practices but also to all aspects of employment.

Civil Rights Legislation

Since the passage of the Civil Rights Act of 1964 (and specifically Title VII), many employers and labor organizations have changed their hiring practices. This legislation has had a profound impact on the fire service. Many changes have occurred voluntarily, but others have been mandated by court decisions, as supplemented by Executive Order 11246, September 24, 1965. This order specified the manner in which affirmative action compliance programs are to be structured and operate.

Civil Rights Act of 1964.   Originally, the Civil Rights Act of 1964 and its subsequent amendments and executive orders applied only to federal organizations and to those companies that do business with the federal government. Then many state legislatures, following the lead of the federal government, passed equal opportunity laws that applied the same principles to local businesses and government bodies. In summary, Title VII of the Civil Rights Act of 1964, as amended, states that it is unlawful for an employer to

> 1. Fail or refuse to hire, or to discharge any individual, or otherwise to discriminate against any individual with respect to that individual's compensation, terms, conditions, or privileges of employment, because of such an individual's race, color, religion, sex, national origin, age, physical or mental handicap, or status as a disabled or Vietnam-era veteran; or

> 2. To limit, segregate, or classify employees (or applicants for employment) in any way that would deprive or tend to deprive any individual of employment opportunities, or otherwise adversely affect that individual's status as an employee because of such individual's race, color, religion, sex, or national origin.

Age Discrimination in Employment Act of 1967.   This act added age to the list of prohibited discriminators, and in 1972 Congress amended Title VII to extend coverage to most government activities. Problems in the fire service resulted, but an exemption allowed fire personnel to be retired from employment upon reaching the age of 65. After a protracted court battle, this exclusion was affirmed in 1996. At this time, fire departments may set limits on the hiring and retirement ages of fire personnel. Recent court decisions have weakened the effect of civil rights legislation so that its impact is uneven and varies among the different federal court regions.

Americans with Disabilities Act of 1990.   On July 26, 1990, the Americans with Disabilities Act was enacted into law to protect the rights of physically challenged individuals. This new legislation prohibited inquiries into confidential matters and prohibited a number of preemployment steps to the extent that they reached beyond matters that would prevent the candidate from performing the work, such as medical examinations, inquiries about disabilities, inquiries pertaining to preexisting medical conditions, and agility testing.

For a more detailed discussion, see the section on selection of personnel later in this chapter.

Affirmative Action Programs

Executive Order 11246 encouraged employers to prepare plans to actively seek out minorities and women who are qualified to be hired and promoted. Employers that do business with the government and violate civil rights laws or do not prepare federally approved affirmative action plans are subject to loss of contracts and possibly even federal court suits. Some state and local government agencies that were in violation of affirmative action provisions have entered into consent decrees. These agreements have been monitored by the Equal Employment Opportunity Commission (EEOC) and have been found effective. In fact, the State Police in New Jersey complied with the tenets of their consent decree and have achieved the minority requirements mandated by the agreement.

An affirmative action plan contains three elements: reasonable self-analysis, reasonable basis, and reasonable action. The objective of a reasonable self-analysis is to ascertain the extent to which actual employment practices adversely affect minority hiring. The regulations indicate that no procedure is mandatory for assessing reasonableness. Organizations must, however, utilize the techniques published in Executive Order 11246.

The order uses "reasonable basis" as the criterion for indicating possible violation during self-assessment. The order discusses the indicators of adverse impact on minorities and includes the following:

- Tendency toward adverse impact on employment opportunities
- Leaving prior discriminatory practices in place

- Resulting in disparate treatment for any single individual or group of individuals

When such possible violations are identified, the employer is allowed to take reasonable corrective actions to redress the discriminatory practices.

Affirmative action programs require that criteria for selection or promotion must be essential to the job. The standards do not have to be lowered, however, to provide employment for women and minorities. For example, if a job requires a fire fighter to carry a physical load a certain distance, then a person applying for the job may be asked to demonstrate the ability to do that job (see Figure 9.1)

Tests that measure abilities not necessary for the job cannot be used. Therefore, an employer may not give intelligence tests or ask questions about marital status or personal plans unless the employer can show proof that the test or questions really measure qualities that are essential to the job.

Figure 9.1 Recruits participate in strength building to help them meet the physical requirements for full-time fire fighter positions. (Courtesy of Ken Yimm, *Palo Alto Times*, Palo Alto, CA)

Those employers who are required by law or regulation to prepare affirmative action plans must submit the following information to the federal Equal Employment Opportunity Coordinating Council:

- A statement of commitment
- A specific allocation of time and money to implement an affirmative action program
- An analysis of the present organization to determine whether discrimination exists
- Specific plans to rectify the imbalance if discrimination exists
- Goals at all employment levels and timetables for accomplishing them, with procedures for measuring how well they are being achieved

Typical goals, objectives, and specific actions to remedy any lack of compliance with affirmative action requirements include the following:

1. Achieving a work force that is representative of the population in the community, which might require:
   a. Increasing the number of women and minority employees being considered for all position levels
   b. Restructuring entry-level positions to allow qualified minority and women employees to become eligible
   c. Establishing a procedure for ensuring that recruiting materials, vacancy announcements, and so on reach minority groups and women in the community
   d. Appointing individuals to coordinate the program with minority and women's community organizations
2. Providing lower-grade employees opportunities to prepare for higher positions and giving economically or educationally disadvantaged persons opportunities to gain more marketable skills. This might require:
   a. Restructuring positions to allow movement from dead-end jobs
   b. Establishing percentage goals for enrolling employees in career development programs
   c. Identifying positions requiring bilingual ability and encouraging those with poor English language skills to study English
3. Establishing procedures to process discrimination complaints, ensuring prompt resolution, and providing corrective action procedures. This could be achieved through:
   a. Reviewing procedures to ensure that they do indeed bring improvement and revising procedures when they do not
   b. Ensuring that the appeals procedure is fair and revising it when necessary

Some of the policies and procedures recommended in the federal regulations as of the end of 1997 are listed here:

- Setting up recruitment program to attract qualified members of the group in question
- Making a systematic effort to organize work and redesign jobs in ways that provide opportunities for persons lacking "journeyman" level knowledge of skills
- Revamping selection instruments or procedures that have not yet been validated to reduce or eliminate exclusionary effects on particular groups
- Training members of the affected pool of potential employees to arm them with the skills necessary to secure employment
- Using classroom training to improve the promotional skills of the members of affected groups
- Creating a system that monitors the effect of programs to improve the representation of affected groups

The agency that carries out fire department hiring practices should know the requirements of the latest applicable laws and regulations. Reasonable accommodations might be necessary under certain circumstances. Personnel officials have to consider certain questions (Brannigan 1993):

- Who is an individual with a disability?
- What is a reasonable accommodation?
- What are the essential functions of the job to be performed?

Further information on relevant laws and regulations may be obtained from the International Association of Fire Chiefs and the International Association of Fire Fighters.*

The enactment of equal employment opportunity laws led to the formation of two minority associations: the International Association of Black Professional Fire Fighters (IABPFF) and Women in the Fire Service.

International Association of Black Professional Fire Fighters.[†] The IABPFF, which is a member of the National Association for the Advancement of Colored People (NAACP), was organized in 1970 with the following goals:

- Create a liaison among black fire fighters across the United States.
- Compile information on unjust working conditions in the fire service and implement corrective action.
- Collect and evaluate data on all deleterious conditions under which minorities work.
- Ensure that competent blacks are recruited and employed as fire fighters.
- Promote interracial progress throughout the fire service.
- Motivate black fire fighters to seek promotions.

Women in the Fire Service.* This nonprofit organization provides networking, advocacy, and peer-group support for women in the fire service. It maintains an information service, provides consulting services, and holds national conferences on topics relating to the gender integration of the fire service.

Occupational Safety and Health Regulations

Beginning in 1970, a whole new set of regulations began to have an impact on fire department operations. The Williams-Steiger Act authorized the federal government to set and enforce regulations to protect private-sector workers from job-related injuries. Enforcement is the responsibility of the then newly created Occupational Safety and Health Administration (OSHA) of the U.S. Department of Labor. Over time, these regulations were broadened to cover public-sector organizations such as fire departments.

Interest in safety-related issues increased during the 1970s and 1980s. In 1987 the National Fire Protection Association's membership voted on a new standard to address safety and health issues specifically. NFPA 1500, *Standard on Fire Department Occupational Safety and Health Program*, addresses such issues as these:

- Organization
- Training and education
- Vehicles, equipment, and drivers
- Protective clothing and protective equipment
- Emergency operations
- Facility safety
- Medical and physical requirements
- Member assistance and wellness program
- Critical incident stress program

---

*The address for Women in the Fire Service is P.O. Box 5446, Madison, WI 53705.

This standard is adopted by OSHA, and all fire departments should conform to it to reduce the incidence of safety and health problems and the liability in case of lawsuits. NFPA 1500 is OSHA's factual source for procedures. The latest OSHA regulations state that any fire department complying fully with NFPA 1500 will be considered OSHA compliant. A companion document, NFPA 1521, *Standard for Fire Department Safety Officer*, addresses the requirements for individuals who function in safety officer-related positions.

PERSONNEL MANAGEMENT FUNCTIONS

Even though some personnel functions are solely the responsibility of personnel professionals, we will discuss the following functions to provide an overview of personnel administration:

- Staffing: selecting, hiring, and promoting personnel
- Salary administration
- Fringe benefits and employee services
- Training and development programs
- Record keeping
- Administration of disciplinary action in union and nonunion departments

Staffing: Selecting, Hiring, Placing, and Promoting Personnel

The personnel function must determine a fire department's staffing needs and to fill openings as they occur, keeping in mind the federal laws outlined earlier. Job openings may be the result of planned expansion, retirement, disability, voluntary terminations, or dismissals.

Specific Tasks in Selecting and Hiring.   In addition to the antidiscrimination statutes, one statute of importance to selecting, hiring, and placement is the Fair Labor Standards Act of 1938. This act was long used to determine the work week of private-sector employees. A 1985 U.S. Supreme Court decision broadened the coverage of that law to include public-sector employees.

> Under the *Fair Labor Standards Act*, employees of local government are eligible for time-and-one-half overtime for hours worked in excess of a statutory maximum (53-hour average work week for fire fighters working 24-hour shifts; 40 hours per week for fire employees who work a five-day week). But some employees are considered exempt from FLSA's time-and-one-half requirement if their jobs are professional, administrative, or executive.

This law and subsequent decisions have had an impact on the staffing of fire departments. Officers must consider the implication when making staffing decisions.

When a department is subject to civil service regulations, it uses the civil service roster to obtain qualified candidates. Otherwise, a department recruits through normal procedures, such as advertising, using a community network, maintaining a list of interested persons, and so on. Once applicants are identified, the selection usually proceeds along these steps:

1. Reviewing applications
2. Interviewing applicants
3. Testing to determine capabilities and physical fitness
4. Analyzing interview results
5. Checking references
6. Repeating interviews where necessary
7. Selecting the most qualified individuals

Because human life and safety are at stake, competency in the fire service is of major importance; therefore, the interviewer must also try to establish how an applicant is likely to perform under emergency conditions.

In many states, the Civil Service Commission maintains a roster of individuals who want to join a paid fire department. To become eligible for this roster, a candidate usually must pass a written examination, a practical examination (in some locales), and a physical checkup. Qualifying candidates are usually listed on the roster in the order of their test results. The commission submits several names from the top of the list to a fire department for selection. A personnel officer, the company officer, or a small selection team then chooses from these candidates the one who is best qualified to fill the position.

Considerations for Effective Selection Interviews. Planning interviews is of great importance. It provides an opportunity to review thoroughly what needs to be discovered. With respect to any one position, this plan should include some questions that are similar or essentially the same for all applicants, especially with respect to assessing skills according to comparable criteria. The plan should also include specific questions unique to each applicant, based on issues that the application or resume raises. During the interview, when the interviewer's mind is on many different subjects, a plan will ensure that no important questions are overlooked. Although the interviewer may develop new questions during the interview, the basic structure of the interview is determined beforehand.

Planning of each interview can best be done by preparing a checklist to include the following items for all applicants:

1. Ideas for starting the interview informally so that the candidate will feel comfortable
2. A few notes to help explain the requirements, difficulties, and opportunities of the position
3. Reminders of specific information to get a full picture of the applicant
   a. Work background
   b. Feelings about new experiences
   c. Interest in learning and self-development
   d. Career aspirations
   e. Attitudes about people, work, and responsibility

To obtain such information, some of which is delicate and may be partly concealed by candidates, an interviewer should encourage a candidate to speak as freely as possible about background and experiences. Closed questions that can be answered with "yes," "no," or a single word or brief phrase rarely elicit such information directly. When attempting to get the candidate to speak freely, an interviewer should avoid questions such as: "Did you have any difficulties with . . . ?" or "You used (special equipment), did you?" or "Did you ever rescue a resident from an apartment on a third or fourth floor?" or "Do you like to learn new things?" or "Did you ever have to paint or touch up deteriorating paint?" *Use open-ended questions*

More useful approaches that encourage a candidate to speak freely are questions or statements such as: "Could you tell me about any difficulties you encountered with . . . ," "Tell me about any difficulties you encountered with...," "What kind of special equipment did you use?," "Can you describe any incidents during which you rescued residents from higher floors?," "What did you find most interesting or useful to learn?," or "Under what circumstances were you called on to do some painting or repairing of quarters?" Such open-ended questions usually elicit more information than the simple closed questions.

Many interviewers have a tendency to accept factual information about experience as being equal in importance to the more subjective judgments about knowledge and abilities. Experience is much easier to measure than the potential of an applicant; it is therefore understandable to give experience considerable weight. However, the extra effort it takes to explore beyond experience usually proves to be worthwhile, even though such exploration requires searching questions and continual, careful evaluation while interviewing. (For more suggestions on interviewing, see Additional Insight 9.2, Interviewing.)

During the interview, personnel professionals often develop a list of questions about specific dates, accomplishments, and names of people with whom the applicant worked closely for later use when checking references. If the references prove satisfactory, interviewers often call one, two, or three applicants for a second interview before making the final selection.

Interviewing for volunteer fire departments is similar to interviewing for paid staff positions. The primary difference is that applicants for paid positions are screened to determine whether they are suitable for a particular position. Applicants for volunteer positions usually are screened to see how their services or skills might be used, possibly in activities other than the ones for which they are applying. Often volunteers are interviewed when they apply, rather than when an opening exists.

The following are some specific suggestions for interviewing volunteers:

- Clearly and honestly describe the position, its requirements, and the amount of time it will take to do the job.
- Explain training requirements in detail.
- Outline clearly the psychological impact of the fire service on a volunteer's family.
- Discuss policies and procedures of the department.
- Describe the support the volunteer can expect from the paid staff, if one exists, and the financial assistance to be provided.
- Never use high-pressure tactics or continue to persuade a volunteer who shows reluctance.
- Inform volunteers that the first few months of affiliation are considered probationary.
- Describe any realistic opportunities for obtaining prestigious or paid positions.
- Allow volunteers a few days to think about the position and its requirements before accepting. Ask them to call back when they have decided.

If accepted, new volunteers should be assigned as soon as possible. The fire department should confirm such assignments by letter. A letter is also an appropriate way to confirm training dates, uniform requirements, and other job specifications.

Acceptability of Preemployment Interview Questions.   Questions that have been proven to have a discriminatory effect on the selection process are explicitly prohibited by court rulings under equal employment opportunity legislation and cannot be asked. The following considerations apply not only to interviews but also to any part of the selection procedure, including the tests that are used to differentiate qualified from unqualified candidates. The basic principle that should be applied is that all selection must be based on criteria that actually are job related—not only in the mind of the interviewer but also factually.

To ascertain whether or not a question is discriminatory, an interviewer should ask the following questions:

- Does the question have a disproportionate effect in screening out minorities and women?
- Is the information necessary to judge an individual's competence for performance of this particular job?
- Are there alternative, nondiscriminatory ways to secure the necessary information?

Exhibit 9.1 is a checklist of the topics to avoid during preemployment interviews to ensure that the interview is job related and nondiscriminatory.

Placement of New Fire Fighters. After an individual has been selected, the personnel office or officer must handle the routine duties involved in hiring the candidate and providing him or her with information on the department's organization, insurance and health benefits, duty hours, vacation, sick leave, holiday policies, educational benefits, retirement and social security plans, and any other rules and regulations necessary for adherence to the department's policies.

When new fire fighters are assigned to duty stations, they usually are subject to a probationary period of several months so that their capabilities and attitudes can be evaluated before they are given permanent status. When a preemployment training procedure is not used, some of those who can pass the physical and written tests for candidates may not meet the actual performance requirements of department work. Many candidates voluntarily drop out when they find that they cannot handle the work during the probationary period.

Promoting Personnel. Fire departments are moving away from using the traditional written examination as the sole criterion for promotion. Assessment centers, formally structured interviews, and practical performance tests assist in evaluating a candidate's ability to perform by providing information that is more extensive and more relevant than a written examination.

Exhibit 9.1 Checklist of Possible Discriminatory Topics to Be Avoided During Preemployment Interviews and on Application Forms

- **Age and date of birth:** Date of birth or age cannot be asked.
- **Maiden name or previous name, if name was legally changed:** Inquiries concerning whether the applicant has worked or been educated under another name are allowable only when the data are needed to verify the applicant's qualifications.
- **Birthplace:** It is discriminatory to inquire about the birthplaces of applicants and their parents.
- **Race, national origin, and religion:** Federal civil rights laws specifically outlaw questions on race, national origin, or religion. In the absence of any logical explanation for the questions, they will be viewed as evidence of discrimination because the race, national origin, or religion of an applicant rarely has anything to do with job performance.
- **Gender, marital, and family status:** Questions considered to be discriminatory toward women include those inquiring whether a candidate is married or single and those requesting information about the number and age(s) of children. Such questions do not relate to job performance. Although an employer might believe that married women with young children have more absenteeism or higher job turnover, actual studies show little difference in the absentee rates of men and women. Job turnover is more related to type of job and pay level than to gender or family status. Investigation of an applicant's previous work record is a valid method of evaluating employee stability. It is a violation of the law for employers to require preemployment information on child-care arrangements. The Supreme Court has ruled that an employer cannot use different hiring policies for women and men.
- **Education:** Educational requirements that are not job related and that have a disparate effect on some groups are considered to be discriminatory. The Supreme Court has explicitly affirmed the Equal Employment Opportunity Council guidelines prohibiting the requirement of a high school education as a condition of employment or promotion where this requirement disqualifies minorities at a substantially higher rate than others and where there is no evidence that it is a significant predictor of job performance.
- **Experience:** Requirements for specific experience should be reviewed and reevaluated periodically. Requirements should be eliminated for jobs that can be learned quickly or reduced if they are not essential for job performance.
- **Physical characteristics:** Questions related to height, weight, and other physical requirements can be asked only if necessary for the performance of a particular job.
- **Credit rating:** Negative employment decisions based on an applicant's poor credit rating have been found unlawful when credit policies have a disproportionately negative effect on minorities and the employer cannot show "business necessity" for such rejection. For example, inquiries about charge accounts or home or car ownership (unless the latter is required for the job) have been found to have an adverse effect on minorities and are unlawful unless required by "business necessity."
- **Arrest and conviction record:** An individual's arrest record has been ruled by the courts to be an unlawful basis for refusal to employ. An arrest is not an indication of guilt. Courts have found that where minorities are subject to disproportionately higher arrest rates than whites, refusal to hire on this basis has a disproportionate effect on minority employment opportunities. Also, a federal court has ruled that conviction of a felony or misdemeanor should not, by itself, constitute an absolute bar to employment and that the employer should give fair consideration to the relationship between the nature of the act resulting in conviction and the applicant's qualifications for the job in question.

Exhibit 9.1 (continued)

- **Appearance:** Employment decisions involving hiring, promotion, or discharge that are based on factors such as length or style of hair, type of apparel, and other aspects of appearance have been found to violate the law if they disproportionately affect employment on the basis of race, national origin, or gender. Some courts have ruled it illegal to refuse to hire or to discharge men with long hair without imposing similar restrictions on women. Hairstyle requirements also can be racially discriminatory.
- **Availability for Saturday or Sunday work:** Although it might be necessary for an employer to have this information, the law requires that employers make reasonable accommodations for an "employee's or prospective employee's religious observance or practice without undue hardship on the conduct of the employer's business."
- **Friends or relatives working for the department:** Questions surrounding this issue can reflect a preference for friends and relatives of present employees. These questions are unlawful if they have the effect of reducing employment opportunities for women or minorities—for example, if the makeup of the present work force differs significantly from the proportion of women or minorities in the relevant population area. Questions on this topic can also reflect a practice of allowing only one partner in a marriage to work for the employer. There is growing recognition that such rules have a disproportionate, discriminatory effect on the employment of women and that they serve no necessary business purpose.

Assessment centers gauge candidates' ability to perform within the parameters of the position they seek. Performance is assessed through a series of written and oral interactions relevant to the position with other candidates and with assessment center staff members.

NFPA 1201, *Standard for Developing Fire Protection Services for the Public,* contains requirements for the selection and promotion of fire service personnel. Exhibit 9.2 presents selected requirements from NFPA on these topics.

NFPA 1201 also requires internal and lateral entry candidates for promotion to positions requiring special qualifications to have education and experience that meet the requirements for the effective performance of the mandated duties of those positions. Candidates should be able to meet the professional qualifications required in accordance with NFPA 1021, *Standard for Fire Officer Professional Qualifications;* NFPA 1031, *Standard for Professional Qualifications for Fire Inspector and Plan Examiner;* NFPA 1033, *Standard for Professional Qualifications for Fire Investigator;* NFPA 1035, *Standard for Professional Qualifications for Public Fire and Life Safety Educator;* and NFPA 1041, *Standard for Fire Service Instructor Professional Qualifications.*

Exhibit 9.2 Selected Requirements for Selection and Promotion of Personnel

Age	A minimum age limit that is consistent with state, provincial, and federal labor laws shall be specified to ensure that members possess sufficient physical and mental maturity to perform fire-fighting duties.
Education	A high school education or state-recognized equivalent shall be required as a minimum to ensure the candidate has the educational background to accommodate the wide variety of activities in which fire fighters now participate.
Physical performance	Job-related physical performance requirements shall be used to select candidates who are physically qualified.
Medical examination	All persons offered a fire fighter's position shall meet the medical requirements as outlined in NFPA 1582, *Standard on Medical Requirements for Fire Fighters.* The confidentiality of personal health records shall be maintained.
Probationary period	For a specified period of at least 12 months before permanent appointment to the department, candidates shall be assigned to a probationary training program and supervision. Candidates shall meet the requirements for Fire Fighter I and, preferably, Fire Fighter II of NFPA 1001, *Standard for Fire Fighter Professional Qualifications,* before permanent appointment.
Probationary status	Candidates shall be kept in probationary status until all phases of the selection process are completed, including the period of probationary training. The chief shall dismiss any candidate at any point during the period of probation for unsatisfactory performance after reasonable written warning and notice.*
Promotion program	The fire department shall establish a documented job-related personnel evaluation program for internal and lateral entry promotion to the various ranks. The program shall be coordinated with the procedures of municipal or other personnel or civil service agencies having jurisdiction.
Rank requirements	Internal and lateral entry candidates for officer shall meet the requirements for that officer rank for which they have applied in accordance with NFPA 1021, *Standard for Fire Officer Professional Qualifications.* Promotion examination questions, whether administered by the fire department or a personnel agency, shall be related to the principal duties to be performed.

*The fire chief's authority may be limited to recommending action when a personnel agency outside the fire department has jurisdiction over probationers or when another agency makes the actual appointments.

Source: Requirements are from NFPA 1201, *Standard for Developing Fire Protection Services for the Public,* 1994, p. 9.

Salary Administration

Another major function of a personnel department is to administer salaries and recommend appropriate pay scales to higher authorities (if they are not already set by state civil service regulations). Even though fire service officers rarely become directly involved in salary administration, we will present some of the general principles here. Officers who understand the principles can explain the salary structure to new fire fighters and discuss salary-related problems. The key elements of a good salary administration program include the following elements: job analysis, job description, job evaluation, job pricing, increase determination, frequency and timing of increases, communication and control, and other considerations.

Job Analysis. Information about a position must be available or determined for job descriptions and training and sometimes for the purposes of compensation. Job postings usually state the purpose of the job, define the job to be performed, and identify working relationships and inherent authority. The essential fire-fighting functions shown in Exhibit 9.3 may be used as part of the fire fighters' job analysis.

Exhibit 9.3 Essential Job Functions for Fire Fighters

- Operate both as a member of a team and independently at incidents of uncertain duration.
- Spend extensive time outside exposed to the elements.
- Tolerate extreme fluctuations in temperature while performing duties. Must perform physically demanding work in hot (up to 400°F), humid (up to 100%) atmospheres while wearing equipment that significantly impairs body-cooling mechanisms.
- Experience frequent transition from hot to cold and from humid to dry atmospheres.
- Work in wet, icy, or muddy areas.
- Perform a variety of tasks on slippery, hazardous surfaces such as on rooftops or ladders.
- Work in areas where sustaining traumatic or thermal injuries is possible.
- Face exposure to carcinogenic dusts such as asbestos, toxic substances such as hydrogen cyanide, acids, carbon monoxide, or organic solvents through either inhalation or skin contact.
- Face exposure to infectious agents such as hepatitis B or HIV.
- Wear personal protective equipment that weighs approximately 50 pounds while performing fire-fighting tasks.
- Perform physically demanding work while wearing positive pressure breathing equipment with 1.5 inches of water column resistance to exhalation at a flow of 40 liters per minute.
- Perform complex tasks during life-threatening emergencies.
- Work for long periods of time, requiring sustained physical activity and intense concentration.
- Face life-or-death decisions during emergency conditions.
- Be exposed to grotesque sights and smells associated with major trauma and burn victims.
- Make rapid transitions from rest to near-maximal exertion without warm-up periods.
- Operate in environments of high noise, poor visibility, limited mobility, at heights, and in enclosed or confined spaces.
- Use manual and power tools in the performance of duties.
- Rely on senses of sight, hearing, smell, and touch to help determine the nature of the emergency, maintain personal safety, and make critical decisions in a confused, chaotic, and potentially life-threatening environment throughout the duration of the operation.

Source: From NFPA 1582, *Standard on Medical Requirements for Fire Fighters*, 1997, Appendix C.

Job Description. Job descriptions are used to record the facts and information obtained during the job analysis process. They can be used to thoroughly explain the job to applicants and sometimes to reporters who write articles on the fire service. Job descriptions provide a written record of job duties and responsibilities. They usually are written in a uniform manner in a standard format. In addition, the descriptions frequently outline the education, experience, special knowledge, and desirable qualities necessary to perform the duties. Exhibit 9.4 provides a sample job description for an entry-level fire fighter position.

Job Evaluation. The relative worth of the job within an organization must sometimes be determined, primarily for setting compensation levels. Four basic formal job evaluation methods are in use, with many variations of these basic systems:

1. *Ranking* is a comparison and placement of jobs in the order from most difficult to least difficult. Ranking is a nonquantitative method and, because it is subjective in nature, it is often difficult to explain and justify.

2. *Classification* establishes predetermined definitions for each salary class or grade. Jobs are compared against these predetermined definitions and are then slotted into the classification that best describes the characteristics and difficulty of the job. This method, which also is a nonquantitative system, is used by the federal government and some state and local governments.

3. *Point system* defines factors that are present in all jobs. Different degrees of each factor are identified, and point values are assigned to each degree in order of relative importance. Jobs are evaluated and points compiled based on this method. The total point value then determines the relative worth of the job. This quantitative system is used widely today.

4. *Factor comparison* is similar to the point system, except that factors are selected and assigned values. Jobs are compared to one another, one factor at a time, and are ranked accordingly. This quantitative system is also widely used.

Job Pricing. Based on the job evaluation results, jobs are grouped and grades assigned. The next step is to price these grades, which is normally done by conducting an outside salary survey utilizing key or benchmark jobs to compare salaries being paid by other agencies. As a result of this survey, base salary and pay ranges are established consistent with the organization's wage policies (above, below, or comparable to what is being paid elsewhere). Salary ranges (minimum to maximum) within each classification differ among different organizations. Those most commonly used are based on a scale with the top salary 30% to 50% higher than the minimum salary.

Exhibit 9.4 Entry-Level Fire Fighter Job Description

- **General overview:** Fire fighters play a major role in the protection of life and property. Therefore, a fire fighter must possess the knowledge, skills, abilities, and other characteristics (KSAOs) necessary to be ready to react instantaneously and effectively in all emergency situations.

A fire fighter must be familiar with safety policies and procedures, fire-fighting equipment and methods, and first aid techniques. In addition, a fire fighter must be able to carry, secure, and climb ladders; carry and/or drag victims; and use equipment to gain access and ventilate buildings. Therefore, physical strength, endurance, and agility are also required. Finally, a fire fighter must be able to interact and communicate with the public during periods of crisis and in standard community settings.

- **Educational requirements:** Must meet the minimum educational requirements established by the authority having jurisdiction.

- **Age requirements:** Must meet the age requirements established by the authority having jurisdiction.

- **Medical requirements:** Must meet requirements for entry-level personnel developed and validated by the authority having jurisdiction and in compliance with applicable legal requirements (e.g., Equal Opportunity and Americans with Disabilities Act regulations).

- **Work behaviors:** What follows is a list of work behaviors required for success as a fire fighter:

(a) Understands the organization and mission of the fire department as well as the applicable rules and regulations of the position
(b) Performs hose evolutions; controls fire with water, maintaining orientation in fire building
(c) Operates pump, determining appropriate water pressure to adjust equipment properly
(d) Recognizes hazardous conditions such as backdraft, presence of noxious fumes, or possible structural collapse of building
(e) Raises and secures appropriate ladders in safe areas and manner
(f) Performs lifesaving and rescue operations applying proper search and rescue techniques
(g) Performs overhaul and salvage of buildings
(h) Drives or tillers fire apparatus safely and properly, adapting to changing conditions while en route (blocked streets, etc.)
(i) Stands station watch using all equipment appropriately, receiving and transmitting information clearly and accurately
(j) Inspects residences or structures for possible safety hazards
(k) Participates in drills, fire strategy sessions, and hydrant inspection
(l) Maintains all fire department tools and equipment, inspecting these for defects and performing proper maintenance procedures

Note: This list must be tailored to each jurisdiction. Some behaviors might or might not be required in all jurisdictions.

Source: NFPA 1582, *Standard on Medical Requirements for Fire Fighters*, 1997, Appendix E; Copyright © 1996. Psychological Services, Inc. All rights reserved.

Determination of Increases. Individuals can move within a salary or wage grade either automatically or on a merit (pay for performance) basis, or a combination of both. The automatic system provides increases at a fixed rate based primarily on longevity or seniority. Under this system, an individual receives either a predetermined fixed amount if performance is satisfactory or no increase at all if performance is less than satisfactory. The merit system, in contrast, compensates employees on the basis of individual performance and output. An employee receives an increase that is directly related to performance. In general, automatic increases are lower than possible increases under a merit system. However, an automatic system combined with a merit increase can provide a capable and competent person with more compensation, while a poor performer receives less, sometimes much less.

Frequency and Timing of Increases.   Generally, increases under a merit system can occur more frequently than automatic increases, depending on the individual's performance rating and location in the salary range. Each department must decide on its own method for awarding increases. In addition to individual pay increases, most organizations raise all ranges simultaneously from time to time to maintain a comparable relationship with salary ranges in the community and to keep salaries compatible with increases in the cost of living.

Communication and Control.   The salary administration plan must be communicated to officers and fire department personnel so that they can carry out and support the program. In addition, for purposes of information and guidance, administrative practices and procedures relating to the program should be in writing. Large departments need not only budgetary controls for salary increases but also a system to ensure that proper approvals are obtained for individual, promotional, and special salary increases.

Other Considerations.   The entire salary administration program should be based on basic principles for a sound system.

- Equal work receives equal pay.
- Appropriate pay differentials are applied for work that requires different levels of knowledge, skill, and physical exertion.
- Pay scales have a reasonable relationship to the salaries and wages paid in the job market in which the department competes, taking into consideration all the tangible and psychological benefits a position offers.
- Each position has a pay range that allows for raises over a number of years. These pay scales should have an appropriate relationship to each other with respect to the principles in the preceding item.
- Capabilities of the individual that are beyond the requirements of the position the person holds are not considered when a pay scale for a position is established.

- Employees with several years of service receive somewhat higher pay than new employees in identical positions. These differentials, however, might become very small between employees with several years of service and those with many years of service.
- Salaries reflect in some way an individual's contribution to the mission of the organization.
- As much as possible, pay scales are known to employees.
- Pay scales are in line with the organization's ability to pay.

Some of these principles are in obvious conflict with others. For example, awarding fire fighters appropriate merit raises may conflict with the fire department's ability to pay. These inherent conceptual conflicts make it difficult to develop a salary system that all employees consider fair and acceptable.

Fringe Benefits and Employee Services

Members of most paid fire departments are civil service employees. As such, their benefits usually are determined by state or municipal legislation. For example, the pension plan might be under the control of a legislative commission, and health insurance might be under the jurisdiction of a local labor organization. When fringe benefits are not subject to outside control, the department must administer these benefits:

- Workmen's compensation, which is an insurance coverage for job-related injuries and is a statutory requirement
- Life insurance
- Hospitalization and medical (surgical) insurance
- Accidental death and dismemberment insurance
- Major medical expense (nonoccupational) insurance
- Disability insurance (weekly payments in case of nonoccupational illness or accident)
- Retirement income (pension)

Other important fringe benefits that require administration include the following:

- Annual leave or vacations
- Holidays
- Sick leave
- Other leave with pay, including time for a death in the immediate family, time to vote, maternity/paternity administrative leave, military training, and short-term educational leave

Some personnel functions that usually are considered fringe benefits can contribute toward satisfying the tangible and psychological needs of people at work by encouraging fire fighters and officers to participate in various activities. The following are some programs that meet tangible needs:

- Credit unions
- Educational incentive programs
- Employee assistance programs

Programs that satisfy psychological needs include the ones listed here:

- Donations to blood banks
- Car pools
- Bowling or baseball teams
- Special commendations and awards for such activities as training and educational achievements, special services, transfers and travel, and action above and beyond the call of duty [see Additional Insight 5.1, Enhancing Work Satisfaction (Providing Recognition) and Performance Evaluation]

Training and Development Programs

Training and development programs for fire fighters are necessary not only for operational effectiveness but also to prevent injury and possible death. Even though many aspects of training are personnel functions, in most fire departments the basic job training is the responsibility of the line officer. The personnel division or officer is in charge of analyzing and advising on training needs in other areas, such as personnel work or officer competencies. In addition, representatives from personnel often assist line officers in improving their on-the-job training techniques. These techniques are an important function because many fire departments do not hire several people at one time, and therefore they must train new fire fighters one at a time. The entire subject of training and development is so important to the fire service that Chapter 11 is devoted entirely to it.

Safety and Physical Fitness.   One of the functions of personnel management is to ensure safety and physical fitness of department staff. For that reason, the personnel function should go beyond the requirements of OSHA and ensure that both safety and physical fitness needs are fully met.

Fire fighting has always been recognized as a dangerous occupation. "According to 1997 data from the NFPA, more than one-half of the 94 fire fighter deaths were stress-related" (Leblanc, Washburn, Fahey 1998). Although the total number of deaths has declined slightly over the past decade, the incidence of stress-related problems is still high. Approximately 50% of fire fighter deaths were directly related to heart attacks.

Historically, the fire service took pride in being a dangerous occupation. However, during the early 1980s attitudes began to change. Fire department administrators recognized people as a resource that must be protected and conserved. Officers developed programs that sought to improve the physical fitness of fire fighters and raise the level of safety in fire departments.

Several National Fire Protection Association standards deal with health and safety issues. Among them are NFPA 1500, *Standard on Fire Department Occupational Safety and Health Program*, which was approved in 1987 (see earlier discussion); a companion document, NFPA 1521, *Standard for Fire Department Safety Officer*, which contains the minimum requirements for the assignment, duties, and responsibilities for fire department safety officers; and NFPA 1403, *Standard on Live Fire Training Evolutions*, which provides guidance in developing safe training.

Suggestions for developing and managing safety programs are available in NFPA's *Fire Department Occupational Health and Safety Standards Handbook* (Foley 1998). Fire department officials can call on local health and fitness experts for help in developing physical fitness programs. Municipal risk managers can also provide assistance.

Career Development.   Career development is a personnel function that often does not receive the attention it deserves. Yet, for effective operation, hiring, promotion, and, of course, training, it must consider all aspects of career development, from ensuring that competence standards of all positions are met to career counseling (see Additional Insight 9.1, Positive Discipline and Counseling).

On December 14, 1972, the National Professional Qualifications Board for the Fire Service directed four technical committees to develop minimum standards for each of the following areas: fire fighter, fire instructor, fire investigator, and fire officer. These standards were planned to accomplish major objectives:

- To identify and define levels for an effective organization so that positions exist to ensure that each company has staff with the skills to accomplish its mission
- To provide for comprehensive training programs and testing of competence
- To provide career steps for individuals

The intent of the committees was to develop clear and concise performance standards that could be used to determine whether a person possesses all the necessary skills for a fire service job. Performance standards, such as

NFPA 1001, *Standard for Fire Fighter Professional Qualifications*, can be used in any fire department in any city, town, or private organization throughout North America.

The performance standards for Fire Fighter I require the following level of achievement concerning water supplies:

- Connect a supply hose to a hydrant and fully open and close the hydrant.
- Demonstrate hydrant-to-pumper hose connections for forward and reverse hose lays.
- Assemble and connect the equipment necessary for drafting from a static water supply source.
- Describe the deployment of a portable water tank.
- Describe the assembling of equipment necessary for the transfer of water between portable tanks.
- Describe loading and off-loading of tanks on mobile water supply apparatus.

The performance standards for Fire Fighter II require the following level of achievement concerning water supplies:

- Identify the water distribution system and other water sources in the local community.
- Identify the parts of a water distribution system: distributors, primary feeders, and secondary feeders.
- Identify the operation of a dry-barrel hydrant and wet-barrel hydrant.
- Define the following terms as they relate to water supply: static pressure, normal operating pressure, residual pressure, and flow pressure.
- Identify indicating and nonindicating types of water main valves.
- Describe how the following conditions reduce hydrant effectiveness: obstructions to use of hydrant, direction of hydrant outlets to suitability for use, mechanical damage, rust and corrosion, failure to open hydrant fully, and susceptibility to freezing.
- Identify the apparatus, equipment, and appliances required to provide water at rural locations by relay pumping or a mobile water supply apparatus shuttle.
- Identify and explain the four fundamental components of a modern water supply system.
- Given a Pitot tube and gauge, read and record flow pressures from three different sized orifices.
- Identify the pipe sizes used in water distribution systems for residential, business, and industrial districts.
- Identify two causes of increased resistance, or friction loss in water mains.

Record Keeping

An important function of personnel officers is to maintain complete and accurate records. Records are necessary for volunteer departments as well as for paid and partially paid departments. Such records provide the following information:

- A detailed history of an individual's association with the department
- Data for determining personnel availability and qualifications for planning purposes, such as sick leaves and unauthorized leaves that were taken, and percentages of volunteers who report to emergencies in various months
- Information necessary or useful for future contract negotiations, such as deferred compensation plan enrollment and costs of the various benefits
- Records for tax and insurance purposes, such as salaries, taxes withheld, insurance premiums paid, and claims filed

Personnel Records for Paid Employees.   If records are not kept by a personnel office in the local government, the fire department must maintain a complete file on each employee. Besides containing the data necessary to provide reports to government agencies such as tax information and workers compensation injuries, the file enables management to select individuals objectively for promotion. The file should include the following items:

- The completed application form, interviewer's notes, reference checks, and results of all tests taken
- Payroll records, including salary or grade status; federal, state, and municipal income tax deduction authorizations as well as those for savings bond, credit union, insurance, union, and pension deductions; annual leave and sick leave taken; and insurance claims filed
- A work record that includes complete information pertaining to on-the-job training programs and courses or schools attended; performance evaluations; commendations and special awards; grievances, complaints, and disciplinary actions; and positions held (including dates)

In addition to keeping records on individuals, a personnel officer in a unionized department should maintain a record of significant factors that could affect future negotiations with the union. The records include this information:

- Complaints about unilateral policy changes by management
- Complaints about changes in job assignments

- Complaints about promotions or transfers
- Problems encountered in job assignments
- Numbers and types of disciplinary actions by types of infraction
- Contract settlements by other departments

Personnel Records for Volunteers. A volunteer fire department is responsible for maintaining records related to the overall relationship of the volunteers to the organization. This responsibility includes keeping the following items for each volunteer:

- A master file containing complete data related to the selection of each individual, such as the application, interview report, committee investigation report, and physical examination results
- A record of training received
- Insurance information
- A record of participation in various areas of activity, including leadership positions and information regarding interest in and availability for other assignments
- Performance evaluations and recognition of achievement, such as service awards, expressions of appreciation, and honor awards
- A record of the reason for discontinuing service
- A record of individual responses to calls and routine assignments

Administration of Disciplinary Action in Union and Nonunion Departments

By its very nature, the fire service is a paramilitary organization. Members must adhere to rules and regulations, particularly during emergencies. As in similar organizations, because an individual or group of individuals may fail to adhere to reasonable rules, some form of disciplinary action may become necessary, even when there is a successful positive discipline environment. Disciplinary action is a serious matter and a significant source of dissatisfaction, not only for the individuals involved but also for their sympathizers. Therefore, disciplinary actions should be taken only on infracted rules that have been thoroughly communicated and understood by all. Furthermore, every officer is responsible for identifying and helping to change rules that are no longer appropriate. Although seemingly inappropriate rules may be in effect, officers must enforce them fairly and impartially.

In many fire departments, the rules are established by the chief or by volunteer committees. The penalties for infractions often are established in the same way. In some states and municipalities, the penalties for infractions are stipulated by legislative action. Typical penalties for various offenses are listed in Exhibit 9.5.

Exhibit 9.5 Typical Penalties for Various Offenses

Infraction	First penalty	Second penalty	Third penalty
Consuming alcoholic beverages or using nonprescription drugs while on duty	Warning and suspension of 1–5 days; mandatory enrollment in counseling and random testing	Suspension of 4–10 days	Dismissal
Reporting for duty while under the influence of alcohol or drugs	Warning and suspension of 1–5 days	Suspension of 4–10 days; mandatory enrollment in counseling and random testing	Dismissal
Violating a safety regulation	Warning	Warning and suspension of 1–5 days	Dismissal
Fighting while on duty	Warning and suspension of 1–30 days	Dismissal	
Stealing from fellow workers	Warning or dismissal	Dismissal	
Stealing from the department	Warning or dismissal	Dismissal	

Whether or not a disciplinary procedure is mandated by higher authority, employees usually can appeal any action through the grievance procedure and ultimately to the courts. Thus, the procedure must be thorough and must effectively protect an employee against unfair and arbitrary disciplinary action so that any action sustained through the procedure will be upheld in the courts.

When grievances are subject to the civil service procedure, an appeal from a disciplinary action is usually heard by a board (usually consisting of three members) that has been established by law or appointed by the head of the fire department. One member of the board is often elected by the employees or approved by the employee making the appeal. Even in volunteer fire departments, a board can be convened to hear and settle the dispute.

Whatever the appeals procedure, it might result in an officer's action being overturned. Because this possibility always exists, officers should be careful when resorting to formal disciplinary action. Nevertheless, to preserve the validity of rules and policies, disciplinary action must be taken from time to time. On such occasions, the officer involved should observe the following precautions.

- The fire fighter or other person involved should be clearly informed that a rule violation is the issue and that disciplinary action might result. At this point, it generally is wise for an officer not to be specific about the consequences to avoid committing to a specific course of action.
- If the employee persists in violating the rule or if the violation has already occurred, it usually is best for the officer to move deliberately and slowly by consulting with either a personnel officer or a superior officer before imposing any penalty greater than a verbal warning.
- To avoid possible future embarrassment, an officer should always point immediately to the appeals procedure and suggest that the offender utilize this procedure if he or she thinks the disciplinary action is not warranted.

CONCLUDING REMARKS

The personnel function involves issues that are primarily related to the third of the 3Cs guidelines, climate. The administration of the personnel function, the point where policies for compensation, benefits, and labor relations are determined, of course, has a major impact on the climate. In every organization, however, though all other conditions are the same, some supervisors are able to ensure that members of their team enjoy greater satisfaction from their work than members of other teams. The same is true in fire departments. The difference lies in the way the officer makes decisions, either intuitively or as a result of rational thought, that affect climate. Using 3Cs guidelines on climate, or similar ones, with every decision can be of help.

CHAPTER 9
STUDY QUESTIONS AND ACTIVITIES

If you are working alone, prepare your own written responses to these questions. If you are studying with a team or working in class, discuss the questions with the group and write a consensus answer.

1. Describe the organization of the personnel function in the fire department where you work.
2. What are some ways to assemble a work force that is representative of the population of a community?
3. What advances has your fire department made in meeting the requirements of affirmative action programs?

4. Explain the usual procedure for hiring people in paid fire departments that are subject to civil service regulations and in those that are not subject to civil service regulations.

5. One or more of four formal job evaluation methods are used by most supervisors of salary administration programs when determining the worth of particular jobs within an organization. In outline form, describe the method you think is the most objective. Discuss your choice with the class. Use your outline to defend your choice.

CHAPTER 9
REFERENCES

Affirmative Action Program, as per Title VII, *Civil Rights Act of 1964* (as amended in 1979). 29 CFR 1608. Washington, DC: U.S. Government Printing Office, pp. 1–10.

Brannigan, Vincent. 1993. "Lex de Incendiis." *Fire Chief*, October, p. 17.

Foley, Stephen, ed. 1998. *Fire Department Occupational Health and Safety Standards Handbook*. Quincy, MA: National Fire Protection Association.

Leblanc, Paul R., Arthur E. Washburn, and Rita F. Fahey. 1998. "1997 Fire Fighter Fatalities," *NFPA Journal*, vol. 92, no. 4, pp. 50–62.

NFPA 1001, *Standard for Fire Fighter Professional Qualifications*, National Fire Protection Association, Quincy, MA.

NFPA 1021, *Standard for Fire Officer Professional Qualifications*, National Fire Protection Association, Quincy, MA.

NFPA 1031, *Standard for Professional Qualifications for Fire Inspector and Plan Examiner*, National Fire Protection Association, Quincy, MA.

NFPA 1033, *Standard for Professional Qualifications for Fire Investigator*, National Fire Protection Association, Quincy, MA.

NFPA 1035, *Standard for Professional Qualifications for Public Fire and Life Safety Educator*, National Fire Protection Association, Quincy, MA.

NFPA 1041, *Standard for Fire Service Instructor Professional Qualifications*, National Fire Protection Association, Quincy, MA.

NFPA 1201, *Standard for Developing Fire Protection Services for the Public*, National Fire Protection Association, Quincy, MA.

NFPA 1403, *Standard on Live Fire Training Evolutions*, National Fire Protection Association, Quincy, MA.

NFPA 1500, *Standard on Fire Department Occupational Safety and Health Program*, National Fire Protection Association, Quincy, MA.

NFPA 1521, *Standard for Fire Department Safety Officer*, National Fire Protection Association, Quincy, MA.

NFPA 1582, *Standard on Medical Requirements for Fire Fighters*, National Fire Protection Association, Quincy, MA.

CHAPTER 9

ADDITIONAL READINGS

Carkhoff, Robert R. 1977. *The Art of Helping III*. Amherst, MA: Human Resource Development Press.

Drake, John D. 1982. *Effective Interviewing: A Guide for Managers*. New York: Amacom.

Fear, Richard A. 1984. *The Evaluation Interview*, 3rd ed. New York: McGraw-Hill.

Fisher, Roger, and Ury, William. 1981. *Getting to Yes*. Boston: Houghton Mifflin.

Rausch, Erwin, ed. 1970. *Interviewing*. In Simulation Series for Business and Industry. Chicago: Science Research Associates.

Rausch, Erwin, and Washbush, John B. 1998. *High Quality Leadership: Practical Guidelines to Becoming a More Effective Manager*. Milwaukee: ASQ Quality Press.

Yate, Martin J. 1987. *Hiring the Best: A Manager's Guide to Effective Interviewing*. Boston: Bob Adams.

POSITIVE DISCIPLINE AND COUNSELING

This section covers two topics—positive discipline and counseling—that are related, yet not often perceived that way. Positive discipline is the type of discipline that individuals accept voluntarily because they believe in the benefits that will come from subordinating their personal interests to some larger cause. Fundamental to such acceptance is an understanding of these benefits. That is where counseling comes in. In its several forms, competent counseling of staff members can be a major contributor to a climate of positive discipline.

POSITIVE DISCIPLINE

Most people are uncomfortable when they hear the word *discipline*. To many, the word is almost a synonym for punishment, as when it is used to describe tight authoritarian controls, criticisms, or penalties for individuals who do not meet standards or adhere to rules and accepted practices.

Yet *discipline* also has another, more positive meaning: the discipline that unites a successful ball team as it works in well-coordinated fashion, or the discipline in teams fighting a fire. This type of discipline is largely self-generated. Each member is aware of the team's objectives, strategies, and tactics and voluntarily, even eagerly, subordinates personal interests to those of the team. In this environment, trust between leader and team is strengthened.

Positive discipline encompasses these values:

- A common understanding of the rules of the game and the standards of performance
- An awareness of the personal and team benefits of the rules and standards
- A willingness by individual team members to make personal sacrifices, if necessary, to help the team achieve its goals

Positive discipline rests on five foundations:

1. The culture supports open, two-way communications.
2. The same standards and norms apply to all.
3. Department members who deserve commendation and privileges receive them.
4. Those who violate accepted rules or fail to adhere to reasonable norms and standards, cooperatively determined, receive help at first and then gradually more stringent warnings until their behavior conforms or disciplinary steps have to be taken.

5. Counseling is used competently to reduce, to a minimum, the use of the organization's disciplinary procedure.

Positive discipline thus supports all 3Cs—control, competence, and climate—and has no negative implications. It is closely related to coaching and to counseling. Through coaching, members of the team come to thoroughly understand their roles as well as *what* has to be done and *how* it is to be done. Counseling helps them see *why* something is necessary and to their benefit, so they will be fully motivated to devote maximum effort, when needed.

Effective positive discipline demands these behaviors from the fire officer:

- Create, with appropriate participation, fair standards and norms
- Review standards and norms regularly with members to ensure that they continue to be fair and equitable
- Expect similar if not equal adherence to standards and norms from all staff members
- Demonstrate a willingness to discuss standards and norms when there are complaints of unfairness
- Ensure that all officers:
  - Coach where necessary
  - Provide recognition and psychological awards fairly and equitably
  - Communicate the standards and norms clearly during performance evaluations and coaching and counseling sessions
- Criticize constructively by making sure that suggestions to improve are given in ways that will bring positive reactions
  - Provide counseling when appropriate
  - Administer the disciplinary procedure in a consistent, equitable way if the other steps do not achieve positive discipline and full cooperation

If you are skillful in establishing an atmosphere of positive discipline, you have little need to apply disciplinary procedures. Still, situations will come up from time to time when, no matter how skillful you are, someone will not accept your suggestions or instructions to change inappropriate behavior. Under those conditions, after initial counseling rounds have failed, you have to apply the penalties of the organization's disciplinary procedure.

You must be thoroughly familiar with disciplinary procedures. It is equally important to the positive climate in your organizational unit that you apply those procedures with compassion. That means reasonable concern for any personal hardship that a disciplinary step may impose on an individual.

At the same time, avoiding disciplinary procedures is unwise. Ignoring infractions is unfair to all other members of a team and violates two of the five foundations of positive discipline.

COUNSELING

Counseling has four uses:

1. Performance-related opportunities, challenges, or problems and the related self-development and learning challenges
2. Career decisions
3. Personal issues affecting work performance
4. Personal issues possibly not affecting work performance

All four uses affect the climate of positive discipline. Higher-level officers, therefore, need to ensure that counseling is performed competently in each segment of their organizational units. Counseling is not a comfortable task for most officers. Conflict may result, especially in performance counseling and possibly in self-development and learning issues. Career development and personal problem counseling may present significant risks.

For counseling on performance problems, officers need guidance to become aware of and adopt appropriate behaviors or courses of action. If they are counseling staff members on career decisions, they have a more limited, role. Finally, if counseling applies to personal issues that do or may affect performance, the officers' role is even more sharply limited.

Recognizing when counseling is appropriate and being adequately proactive without being intrusive require greater knowledge and skills than the average officer possesses. Lack of skills, however, is not necessarily obvious. Even in crisis situations, the problem is often seen to lie with the staff member rather than with counseling competence that might have prevented it. Self-development and ongoing development of counseling competence are responsibilities that must be undertaken without any clear symptoms of organizational need.

Developing competence for counseling, and especially for performance counseling, revolves around several questions that also can serve as guidelines for counseling:

- Will counseling intervention in this situation enhance positive discipline?
- If a counseling interview is called for, which steps should be taken prior to the counseling interview and which during the counseling interview?
- Should an action plan be developed and formalized?
- Will follow-up be desirable?

Counseling on Performance-Related and Self-Development/Learning Challenges

Staff members may fail to do what is appropriate in a situation because of these weaknesses:

ADDITIONAL INSIGHT 9.1

- *Unaware*—not be aware that the respective behavior is required or desirable at that point
- *Unable*—not have the knowledge and/or the skill to perform the task at the expected level of competence
- *Unwilling*—have reservations about performing the task at that point or simply do not want to do it

The third reason is different from the first two, because staff members have the skill to perform the task as expected and know that they are expected to do it. Still, they do not do it because their norms do not match those of the organization.

Different managerial actions are appropriate for these three situations. In the first two, coaching can remedy the behavior discrepancies by helping the staff members become aware that certain behavior is appropriate or by helping them achieve the desired competence level. In the third case, the appropriate first remedy is counseling. Shaking out whether a staff member has the competence to do something, *cannot* do it, or is actually *unwilling* to do it is an essential early step in counseling. The unwillingness may be due to apprehension about admitting lack of competence. Counseling may have to be used to overcome that concern and possibly to overcome resistance to learning. With competent counseling, reliance on the disciplinary procedure is held to a minimum and positive discipline is strengthened.

Counseling on Career Decisions

Effective officers consider the career aspirations of their staff members. At appropriate times, they share with their staff members the broader perspective that their position and experience are likely to have given them. They thus help each individual gain a more realistic view of the potential career paths that may be open to them within the department and sometimes even elsewhere.

Competent career counseling can add to the effectiveness of performance evaluations. Career guidance can relieve two major problems:

1. Officers have little or no opportunities for providing compensation increases to staff members. These are determined almost exclusively by the government agency that has jurisdiction.

2. It appears to staff members that performance evaluations are geared primarily toward improving staff member performance because on the surface, advantages to staff members are clear only if there is a visible relationship between the evaluation and tangible rewards. When the emphasis of performance evaluation is on a

performance improvement plan, career counseling can change that impression by showing that the plan can help to better prepare a staff member who seeks career advancement or a career change.

Most officers are not involved in career counseling situations very often. However, performance evaluations can sometimes lead to discussions of career possibilities. There are also occasions when a staff member takes the initiative, seeking advice on how to progress both financially and in position.

Requests for information on opportunities for higher earnings are not necessarily requests for career counseling. They could be introductions to requests for counseling on personal issues. Requests for information on how to advance to higher or better paying positions, on the other hand, are requests for career counseling.

Officers need to be aware of the potential sources of risk—financial and ethical—that can accompany discussions about career opportunities. When officers give advice, they assume a certain amount of responsibility for the outcome. Instead of giving advice, it is wiser for officers to point out that they cannot provide any specific advice, only general guidelines, for enhancing the staff member's competitive position in relation to others who may seek a position and that there is no assurance that these guidelines will actually lead to the staff member finding and obtaining a desired position.

In career counseling, officers should clarify what they may be able to do. Following are some possibilities:

- Provide assistance with clarifying the facts that will help staff members evaluate the advantages and disadvantages of the position to which they aspire.
- Provide information about the competencies that are necessary for success in that position.
- When personally qualified, provide assistance, guidance, or suggestions for acquiring those competencies.
- Consider arranging for staff members to speak with incumbents in the positions to which they aspire (to obtain information on the pluses, minuses, and requirements from that person's point of view).

When a department helps officers understand their role in providing career planning assistance is not intended to make them career counselors. They need to understand that their objective is limited—to help their staff members gain perspective on the knowledge, skill, and other requirements for those positions in which they may be interested. Often it is best to suggest that the staff member also seek other sources of information or guidance, such as the municipal or other government personnel office.

ADDITIONAL INSIGHT 9.1

Counseling on Personal Issues Affecting Work Performance

Officers are sometimes called on to provide either advice or assistance to staff members with personal problems. In these situations they need to have a clear view of their responsibilities.

Some of the problems directly affect work performance. Acrimonious divorce proceedings, serious illness of close relatives, and abuse of alcohol and drugs are examples of personal situations that have considerable potential for affecting performance. Another example is interpersonal relationships with other staff members. These situations may even first manifest themselves in performance problems that require counseling. Unfortunately, sometimes they may be so serious that they cannot be resolved with counseling.

Laws and regulations, such as the Americans with Disabilities Act and those pertaining to ethnic, age, and gender discrimination, have to be considered when deciding whether counseling has a chance to do some good or might contain too many potential risks. Counseling may be provided to decide whether more than counseling is necessary, to decide whether the disciplinary procedure is applicable or whether the employee's problems are beyond his or her control and therefore call for professional help or involvement of social services agencies. In any case, the organization's policy on these matters should, of course, be followed. The policy probably is to send the staff member to the human resources department. If there is no such department or if no policy is in place, the 3Cs guidelines call for serious consideration of developing a policy. Without a policy, control may suffer, competence is threatened, and various stakeholders may be affected negatively in various ways, thus jeopardizing the climate.

Counseling on Personal Issues Possibly Not Affecting Work Performance

Staff members may also have personal problems that may not affect their work, such as illnesses, relationships with family or friends, or difficulties with children.

Staff members often look to their officer as an experienced, knowledgeable friend, to whom they will turn for advice in difficult situations. Before starting a counseling session, officers are wise to clarify appropriate limits for helping in such situations. Organizational policy is likely to call for obtaining guidance from the human resources department, if one is available. Policies should also make it clear that officers need to keep in mind the possible liability that could lead to problems for the organization if well-meaning suggestions or advice does not work out well.

Counseling on personal problems should be totally nondirective, intended solely to help the staff member see issues in better perspective. Even explaining what the officer might do if he or she were in that situation might be risky.

INTERVIEWING

I keep six honest men (they taught me all I knew)
Their names are What and Why and When
And How and Where and Who.

—Kipling, *The Serving-men*

The term *interviewing* usually conjures up employment interviewing and that is the use on which this section focuses. However, interviewing is also used by health-care professionals to obtain a patient's history or to zero in on a diagnosis. Police interview a suspect; media professionals interview people of interest to their audiences and those who are witnesses to events. In organizations, interviewing is used not only for candidate selection but also for promotion decisions and for performance assessment.

Interviewing is at the heart of one of two major components of competence, the second of the 3Cs: seeing to it that capable people are recruited and selected for open positions. Recruiting and selection is covered in this chapter; the other component, competence development, is the subject of Chapter 11.

Interviewing candidates and selecting finalists based on the interviews are key activities in providing fire departments with the "raw material" on which to build required competencies. Interviewing skills are also important for management of learning and coaching. We have already discussed some aspects of interviewing earlier in this chapter.

Although the fire chief or other high-level officer and the human resources department may perform most of the interviewing and selecting, individual officers sometimes play a role in the selection process. Unfortunately, most managers and officers have had little, if any, training in how to obtain information during interviews, and they have to rely heavily on their common sense.

Company officers need interviewing skills for the rare instances when they are involved in job applicant selection, for discussion and counseling on performance problems (possibly as part of the disciplinary procedure), for management of learning and coaching to gain insights into competence strengths and weaknesses, for other counseling sessions, and for performance evaluation interviews. Interviewing is primarily an art, and considerable practice is necessary for mastery. Effective interviewers seek and obtain information relevant to the purpose of the interview.

Communicating During Interviews

Success in interviewing depends on competent interpersonal skills so the interviewer will freely share information. These skills include creating a positive climate that reduces stress on the interviewee and using questions and silence effectively for general and in-depth probing.

All interviews, whatever their purpose, should be conducted in a professional atmosphere that puts the interviewee at ease so that he or she will answer questions truthfully, without apprehension. Interviews are highly stressful for interviewees. An interviewer can help reduce the stress level by using a conversational tone rather than making the session seem like an interrogation.

Empathy

Empathy is the ability to see a situation as another person perceives it. Empathy is not sympathy. When you are empathetic, you understand the other person's situation and position. You need not agree with it or even share the feelings. Sympathy adds feelings to understanding.

Conveying honest empathy with the interviewee's situations and needs is an effective way to reduce the tension for an interviewee and to enhance mutual understanding and trust. Allowing candidates to first ask questions about the position shows interest.

In-depth Probing

An interviewer may use probing to conduct an in-depth examination of subjects in which candidates claim to be knowledgeable. The technique consists of asking very general questions first and then increasingly more specific questions on smaller and smaller segments of the topic, until the limits of the candidate's knowledge have been reached. For instance, questioning about pump operation could start with a question on what the applicant knows about water flow rates and pressures from hydrants and then probe more specifically about the various steps to take after hydrant connection in all types of circumstances.

Even an interviewer who is not familiar with the subject under examination can use in-depth probing to gain insight into qualifications of the candidates:

- Depth of knowledge of the specific subject
- Perception of their depth of knowledge
- Credibility of explanations
- Ability to organize
- Ability to verbalize

In-depth probing is primarily useful to gain information on depth of *knowledge* of a topic. Probing on details of job history or personal characteristics can quickly become overly inquisitive and be resented by applicants. When the technique is used skillfully to gain information on a technical topic, an interviewer can make the applicant feel that he or she is providing information of interest to the interviewer.

Even when a series of questions is softened through careful choice of words, if the interviewer asks too many questions during the probing process without intervening discussion, candidates may perceive probing as an interrogation and feel increasing stress. If interviewers show genuine interest, however, and shares thoughts between some of the questions, this interest will be reflected in various ways, including nonverbal signals and the candidate will not perceive a large number of questions as an interrogation, but rather as evidence of real interest in the subject.

Applicable Laws

Interviewers should be aware of the relevant details of federal and state equal employment opportunity laws and regulations to ensure compliance with them, not only when the interview is for candidate selection purposes but also when its purpose is for counseling.

ADDITIONAL INSIGHT 9.2

Labor Relations in the Fire Service

INTRODUCTION

Labor relations is the term used to describe the relations between an employer and a group of employees who have joined together to present their views formally, based on the rules of their organization. The employees' organization is usually a union. Labor relations in the fire service are similar to labor relations in industry and government. They involve negotiation of a contract on the terms and conditions of employment and a procedure for settling grievances.

Although nonunion labor relations (sometimes referred to as human relations) provide the same functions for employees who are not organized, labor relations are distinguished by being far more formal. The term *human relations* generally pertains to the policies that an employer establishes, often in consultation with employees, to retain employees and provide a satisfying work environment. Labor relations involve a more bilateral decision-making process in these matters, called collective bargaining, with formal negotiations on all issues.

Antidiscrimination laws apply to both labor and human relations. Labor relations, however, are also governed by laws specifically designed to regulate the relationships between unions and employers, to keep the balance of power as equal as possible and not to inconvenience the public any more than is necessary to protect the interests of the parties.

The goal of a fire department in labor relations is to achieve the best relations possible within budgetary limitations (not the limits of the current budget, but what resources can realistically be obtained) and to provide as satisfying a climate as possible for the department's staff. For top management of the department this goal involves negotiations with financial authorities. It also requirees officers who are competent in establishing a positive discipline environment with open communications and mutual trust and who know how to show their appreciation for the contributions of staff members. In short, labor relations involve all the 3Cs guidelines, especially competence and climate.

LABOR RELATIONS

Labor relations, defined broadly, is management's relationship with the non-managerial people in the organization. For both management and labor to meet their goals (that is, job performance and job satisfaction), both sides must be committed to a course of cooperative action. Achieving cooperation depends on the attitudes and approaches of all parties involved. Management, however, has the primary responsibility for establishing and maintaining good relations with labor, and every officer in the fire service plays an important role in that endeavor. In the fire service, good labor relations are actually good human relations because an officer who establishes and maintains rapport with the team is likely to have good labor relations. The human resource office or officer for paid fire fighters is an active participant in labor relations. The office or officer provides advice and guidance about labor relations along with performing the following additional functions:

- Keeping a personnel policy manual with written statements of personnel practices
- Determining whether personnel policies are fair and equitably administered throughout the department
- Determining whether personnel policies are current with court rulings and statute changes
- Investigating and recommending changes in policies that are unfair, inequitable, or no longer applicable
- Arranging for discussions and conclusions on any significant incidents and problems between officers and fire fighters, or their representatives, that might affect the labor relations climate or the labor–management agreement, if one exists

Labor Relations Laws

Before 1932, labor relations problems were resolved in the courts without specific labor laws. Judges, through the use of common law, decided the rights of both management and labor. Since then, however, four pieces of federal legislation have established the rules and regulations for the present collective bargaining system: the Norris-LaGuardia Act of 1932, the Wagner-Connery Act of 1935 (often referred to as the Wagner Act), the Taft-Hartley Act of 1947, and the Landrum-Griffin Act of 1959. These laws, together with the Railway Labor Act of 1926 and some antitrust legislation, are the basis for all labor negotiations in the United States.

The Norris-LaGuardia Act. This act set the stage for additional labor relations laws by specifying that an employee cannot be forced into a contract by the employer in order to obtain and keep a job. Before this act, many employers made workers sign a pledge that they would not join a union as long as they were employed by the company. Workers who violated the pledge were fired. Unions called those who signed the pledge "yellow dogs," and the contracts were so named.

Partly because the "yellow dog" contracts were legal and partly for other reasons, courts were likely to side with management in labor disputes and issue an injunction that prohibited striking or, if a strike was in effect, prohibited picketing. These injunctions were enforceable by the police.

The Norris-LaGuardia Act made two stipulations:

- It said that "yellow dog" contracts were not enforceable in any court in the United States.
- It made getting an injunction to prevent a strike almost impossible.

In 1932, the only way a union could gain recognition was by striking or by threatening to strike. In effect, the employer had to be forced to recognize the union. Even with the passage of the Norris-LaGuardia Act, employers could threaten and discharge workers engaged in union activity. The act, however, gave unions the right to use their major weapons to gain recognition—striking, picketing, and boycotting—without interference from the courts.

When Franklin Roosevelt became president in 1933, the Great Depression was three years old. In an attempt to bolster the faltering economy, the Roosevelt administration took many steps, including instituting the National Industrial Recovery Act (NIRA). Section 7a of the NIRA guaranteed unions the right to collective bargaining in order to keep wages up and thus maintain the purchasing power of the workers. This guarantee was the "shot in the arm" that unions needed. Workers flocked to join both the American Federation of Labor (AFL) and the new Congress of Industrial Organizations (CIO).

The Wagner-Connery Act. In 1935, the Supreme Court struck down the NIRA as unconstitutional. Senator Robert Wagner of New York then introduced the Wagner-Connery Act, which was quickly passed by Congress. In 1936, a strike in the automobile industry brought the act before the Supreme Court, where it was upheld. The Wagner-Connery Act included the following provisions:

- Workers were allowed to decide, by a majority vote, who would represent them at the bargaining table.
- The National Labor Relations Board (NLRB) was established.
- Unfair labor practices were defined, and the NLRB was given the power to hold hearings, investigate unfair practices, and issue decisions and orders concerning them.
- Management was prohibited from interfering or coercing employees when they tried to organize.
- Management was required to bargain with a union, although management was under no obligation to agree to any of the union's terms.
- So-called "yellow dog" contracts were outlawed (the Norris-LaGuardia Act had only made them unenforceable).

The entire area of unfair labor practices as covered by the Wagner-Connery Act restrained management. The act, in effect, was an attempt to equalize the positions of management and labor. However, the act imposed no penalties for any violations, and it did not provide the NLRB with any real power to enforce its decisions or orders. Not until the matter went to the courts did the act become effective. For example, when the NLRB decided that some employees had been fired in violation of the Wagner-Connery Act, the courts upheld the decision, and the employees were reinstated with back pay.

Since the Great Depression and especially since 1936, the unions (under the protection of the Wagner-Connery Act and favorable court decisions) continued to grow. With their growth came increasing strength.

The Taft-Hartley Act. Shortly after World War II, a series of industrywide strikes threatened the smooth return of the economy to civilian production. During these strikes, it became apparent that the power of unions had grown to such an extent that they were now substantially stronger than their management adversaries. Congress, in an attempt to restore the balance, passed the Taft-Hartley Act of 1947 over the veto of President Harry Truman. Besides spelling out specific penalties, including fines and imprisonment, for violations, the Taft-Hartley Act modified the Wagner-Connery Act in five major areas.

• *Union representation.* The act gave workers the right to refuse to join a union and outlawed the "closed shop." The act specified that only one election a year can be held to determine whether a union, and which union, should represent the employees. The act gave employers the right to express "any views, argument, or opinion" about union representation, provided "such expression contains no threat of reprisal, or force, or promise of benefit."

• *Unfair labor practices for unions.* The act protected employees from coercion by unions. Employees were protected from paying exorbitant dues and initiation fees. If nonunion employees refused to join, they were protected against possible union reprisals because unions no longer could force employers to fire antiunion people. The act required unions to "bargain in good faith," as the employers had been previously forced to do by the Wagner-Connery Act.

• *Bargaining procedures.* The Taft-Hartley Act provided for a 60-day cooling-off period when a labor agreement ends. Section 8(d) of the act stipulated that written notice must be served if one party to the agreement is terminating the agreement. The written notice must be given to the other party 60 days before the contract ends. Thirty days later the Federal Mediation and Conciliation Service must be notified of the dispute.

• *Regulation of the union's internal affairs.* Union rules regarding membership requirements, dues, initiation fees, elections, and so on must be made available to the government and to the union membership.

- *Strikes during a national emergency.* In the event an imminent strike affects an entire industry or a major part of an industry and imperils the health and safety of the nation, the president was granted certain powers to help settle the dispute.

The Landrum-Griffin Act.   In 1955, the AFL and the CIO merged. Two years after the merger, Senator John McClellan of Arkansas conducted committee hearings that revealed evidence of crime and corruption in some older local unions. At the height of the resulting furor, Congress passed the Landrum-Griffin Act, which provided the following:

- Established a bill of rights for members of labor organizations so that unions would be run in a more democratic manner
- Required that labor unions file an annual report with the government, listing the assets of the union and the names and assets of every officer and employee of the union, and required every employer to report on any financial relationship with a union or union representative
- Established minimum requirement guidelines for the election, responsibilities, and duties of all union officers and officials
- Amended portions of the Taft-Hartley Act concerning secondary boycotts, union security, and the rights of some workers to strike, and imposed additional restrictions on the rights of unions to picket for recognition

Growth of Unions

Early in this century, joining a union was dangerous because employers were quick to fire the people they believed engaged in union activities. The conflict caused by the firings first resulted in lawsuits and later led to the enactment of labor laws. Today an employee has extensive rights to join a union or to participate in union organization activities without fear of retaliation. The federal legislation allowed unions to grow in the private sector, but public-sector growth was not so rapid; fewer than 1 million government employees were members of unions in 1956.

In January 1962, President John Kennedy, consistent with the growing civil rights movement, issued an order that, for the first time, gave federal employees the right to bargain collectively under restricted rules. The order stipulated the rights of management, grievance procedures, and rules for union recognition. In 1969, President Richard Nixon further expanded the rights of the government employee unions. He established a Federal Labor Relations Council that is similar to the NLRB for unions in the private sector. By 1970, membership in public-employee unions had grown to 4.5 million persons.

Until 1970, government employees were forbidden to strike under existing legislation. The unions' most effective weapon, therefore, was denied to them. The first rumblings of discontent and rebellion to this disenfranchisement began in the late 1960s. Several major cities, notably New York and Baltimore, suffered strikes by sanitation workers, teachers, police, and fire fighters. The federal government was no longer immune to strikes. At one time, air traffic controllers called in sick in such large numbers that commercial flight operations were severely hampered.

In 1970, post office employees went on strike and thereby set the tone for all government employees. Although the strike was illegal and the postmaster general was, by law, forbidden to negotiate with the strikers, he nevertheless did negotiate. The result was that the strikers were reinstated without penalty and received raises, and Congress recognized the union as representing the employees for the purpose of collective bargaining.

power of unions

Unlike the 1970s, the 1980s were difficult times for unions in the public sector. In a 1981 strike by air traffic controllers, President Ronald Reagan took a hard line with the Public Air Traffic Controllers Organization. When the air traffic controllers failed to return to work as ordered, he fired all striking members and decertified the union. This action set a negative tone for employer–employee relations during the remainder of the Reagan administration. The growing militancy of public sector workers was met by a rising tide of right-to-work activity by employers.

limits of unions (power)

Although the 1990s have seen some fairly large strikes, in general labor strife has been subdued. Strikes have been milder than those in the past because of the low rate of inflation, the reduced power of unions in light of the shift to service and knowledge jobs, and the generally high level of employment and prosperity.

International Association of Fire Fighters (IAFF)

As the other government employee unions grew, so did the IAFF. In early 1998, the IAFF represented more than 225,000 members in 2,316 local union chapters. Unlike most unions, the IAFF enrolls and represents supervisory personnel. Although not every employee is a member of the union in every unionized department, some departments are completely unionized, including the chief. In many of the communities where the IAFF represents the fire fighters at the collective bargaining session, some members also hold membership in benevolent associations. Some benevolent associations provide health and insurance benefits, whereas others are more concerned with professional activities (e.g., continuing education) and social events.

In 1968, IAFF dropped a 50-year-old rule prohibiting strikes. Subsequently, like unions in the private sector, fire departments in several communities have gone on strike when they could not reach agreement with the municipality.

COLLECTIVE BARGAINING IN THE PUBLIC SECTOR

To comply with labor relations laws, management and labor must bargain collectively and in good faith on the subjects of wages, hours, and working conditions. In unionized organizations, the result of bargaining is a formal agreement or contract between management and an agent (the union) that represents the employees. The contract specifies the rights and obligations of all three parties: management, the union, and employees. In nonunionized organizations, an implied contract exists in the form of a verbal agreement, customary practices, statutes, and possibly employee handbooks, supervisory manuals, and community personnel policies.

The collective bargaining procedures at the local level in the public sector are different from the methods used by private industry. In the private sector, management and labor leaders generally "hammer out" an agreement and, after ratification by the employees, the agreement becomes the contract. State and local employees generally go through a two-stage process: (1) the labor leaders and a government-appointed representative or committee negotiate the terms of the contract, and (2) once the employees have ratified the agreement, the contract goes before the governing body for final approval.

Several states have adopted legislation that sets up specific steps for working out difficult negotiations. Impasses are often referred to a fact-finding process during which both the employer and the unions can present their cases. Agreements are sometimes settled this way. Some states allow binding arbitration, in which an outside arbitrator rules on the merits of positions taken by the union and the employer. Because legislative approval is customary, however, a contract is often subject to further modifications and negotiation.

The Collective Bargaining Agreement

The final contract outlines the conditions to which both parties agreed and the duration of the agreement. One-year agreements are the most common, although recently the trend has been toward three-year agreements, with two-year agreements often being compromises. The typical clauses in all labor–management agreements can be grouped into five types:

- Routine clauses, which contain the preamble and purpose, term(s) of the agreement, reopening condition(s), and amendment(s)
- Union security clauses, which contain the bargaining unit definition and union recognition
- Management rights clause, which reserves for management the right to make decisions in any area not specifically covered by the agreement and stipulates in detail those areas that are solely the rights of management
- Grievance procedure clauses, which spell out the steps in the grievance and arbitration procedures
- Conditions of employment clauses, which cover details not included in the other clauses, such as wages, work hours, strikes and lockouts, holidays, vacations, leaves, reporting, shift differentials, discharge, benefits, safety, apprenticeship training, and so on

In the following sections we will look more closely at the union security, management rights, and grievance procedures clauses.

Union Security Clauses.   A union must develop and maintain a secure organization so that it can speak with assurance for its members. A union must have strength to be able to bargain effectively with management. Unions insist on a clause that defines the bargaining unit and recognizes the union as the agent (third party) that represents the employees in that union.

The bargaining unit is an important concept during organizing drives. The union prefers to define the bargaining unit to assure a majority during an election vote. Management, on the other hand, usually prefers to define a different, wider bargaining unit. In the event of a dispute, the final decision on the bargaining unit is made by the National Labor Relations Board (NLRB) when it certifies the election. A bargaining unit can be selected by function, by craft, by location, or by some other logical grouping. For example, in some locals of the IAFF, officers are members of the bargaining unit; in other locals, officers are excluded.

A municipality can have agreements either with separate unions that represent police, fire, and municipal employees or with one union that represents all municipal employees. When several unions represent the employees of a department, jurisdictional disputes (disputes over who represents whom) are possible. In such an event, the personnel officer has the responsibility of cautioning all supervisory personnel to remain neutral. This caution is most important because the local or state agency that might become involved can consider any comment or action by a supervisor that implies favoritism toward one union as an unfair labor practice.

In addition to defining the bargaining unit, union security clauses might contain one or several of the following provisions:

- *The preferential shop.* Management agrees to give the first chance for employment to union members (not usually applicable to the fire service).
- *Maintenance of membership.* New employees do not have to join the union to gain employment; however, any employees who voluntarily join the union must maintain their membership for the duration of the contract.
- *Union shop.* New employees must join the union to retain their jobs after the probationary period.
- *Agency shop.* Employees do not have to join a union, but they must pay dues to the union.
- *The check-off.* Employers must deduct dues from the wages of employees and remit them to the union.

Management Rights Clauses.   Initially, unions came into being because employees believed, often with good justification, that management misused its power to gain economic advantages at the expense of others. Employees formed and joined unions for mutual protection. After the formation of the unions, capricious and arbitrary decisions by management forced unions to spell out more and more detail in their contracts to protect the employees against unjustified detrimental actions. Management frequently objected that some areas, such as work assignment or overtime, were not subject to bargaining. Arbitration awards and court decisions, however, gradually have established that any subjects not specifically reserved for management must be negotiated. To protect its rights against encroachment, management has always endeavored to maintain and strengthen management rights clauses in contracts.

Management rights clauses reserve for management the right to make decisions in any area not specifically covered by the agreement. In addition, such clauses stipulate in detail those areas that are strictly the rights of management. Management rights usually include the following:

- To direct the work force
- To hire, promote, transfer, and assign without interference
- To suspend, demote, discharge for cause, or take other disciplinary action
- To take action necessary to maintain a department's efficiency
- To make reasonable rules and regulations

When management is weak, arbitrary, or inept, a union often has cause to demand that one or the other of these rights be removed from the management rights clause and be made the subject of separate contract clauses

[handwritten margin note: given rise of Labour Laws of the past 60 yrs, management cannot be faulted for this]

that specifically delineate what managers can and cannot do. To avoid such additional restrictions on their freedom to manage, officers must be careful not to abuse any of their rights, and management spokespersons at the bargaining table must be skillful in protecting their rights clauses.

Grievance Procedure Clauses.   In negotiating an agreement, both management and labor realize that it is impossible to anticipate every conceivable problem. Both parties also realize that some problems will occur that are not directly covered by the agreement. Therefore, every contract provides for a grievance procedure that is intended to serve as a mechanism for resolving problems. The grievance procedure is so important to contracts that any agreement with a federal agency that does not contain such a clause is in violation of President Nixon's order. Although actual grievance procedures might vary somewhat from contract to contract, most of them are comparable to the following example (all of the steps have time limits):

- The aggrieved fire fighter discusses the problem with an immediate supervisor.
- If the problem is not settled at that time, the fire fighter discusses the problem with a union representative, who then submits the grievance in writing either to the same officer or to the officer at the next higher level.
- If the grievance is not settled, the union can appeal to the chief, the fire commissioner, or the governing body.
- If the grievance is not answered or resolved to the satisfaction of the employee or the union, it can be submitted to an arbitrator or a panel of arbitrators.

Arbitrators usually are agreed on by both parties or, depending on the contract, can be appointed by an impartial group such as the American Arbitration Association, the Federal Mediation and Conciliation Service, or a similar state agency. The decision of an arbitrator is final. (For further discussion of arbitration, see Additional Insight 4.1, Management of Potentially Damaging Conflict.)

Either party can, of course, challenge the decision in court. However, judges rarely overturn the rulings of arbitrators, and do so only when an arbitrator has clearly gone beyond the authority granted by a contract.

Sometimes either management or the union deliberately uses the grievance procedure to obtain a ruling on an issue in which the wording in the contract is not clear. For example, if no agreement can be reached on a difference of opinion, either party can act on its interpretation. The union or employee would file a grievance, or management would take some disciplinary action to bring the issue to an arbitrator.

Arbitration can be a tool

Using an arbitrator occurs rarely, however, because unions and management generally try to avoid referring disputes to an outsider. Both unions and management consider it a failure of their ability to maintain good working relationships with each other, and both are concerned that the outsider (unfamiliar with all the subtle relationships that exist) could inadvertently resolve the issue in a way that is unsatisfactory to both sides.

An officer can take several steps to ensure that misunderstandings do not lead to grievances and that those grievances that reach arbitration will be resolved in favor of management. An officer should follow these suggestions:

- Ensure that all facts are available to both sides and that either side did not misunderstand the situation
- Maintain accurate and complete records of every occurrence related to a given problem
- Maintain as cordial a relationship as possible with the aggrieved employee(s)

[handwritten margin note: management is not democratic in nature]

Contract Negotiations

[handwritten margin note: it may have an advantage over democratic union. union leadership goes with what is popular not what is right or most beneficial]

Management and union representatives meet to negotiate the terms of a new agreement either immediately after a union has been recognized as the collective bargaining agent or before a contract approaches the expiration date. This process usually is initiated by the union, which submits of a list of requested changes to the previous agreement. Because the union is a political body operating under a constitution, bylaws, and democratic procedures, the requests for changes are the result of meetings either by the union members or by a large group of union officials who have been elected for that purpose.

During contract negotiations, management representatives listen to the proposed changes and then, after discussion, submit counterproposals. The counterproposals might be just responses to the items brought up by the union, or they might contain contract changes that management would like to make. Union representatives answer these counterproposals, and either side (or both sides) gradually makes concessions that narrow the issues separating them.

A new contract gradually emerges from these discussions and compromises. Often, the tension grows as the contract renewal date approaches or even passes. Threats of job actions, such as a strike or refusal to perform nonessential duties, are common. In the event that talks break down, management, union, or both might call on the services of a third party (a mediator) to break the deadlock. The mediator has no specific powers but acts strictly as an impartial go-between in reconciling differences.

The attitude of both parties is extremely important during negotiations. To achieve an amicable agreement, each side must exercise restraint. On the management side, company officers face heavy and often difficult responsibilities, especially if they are also union members. As managers, they must see that the work of the unit continues to be performed without interruption. At the same time, as human beings, union members, and members of the team, fire department managers are under great pressure to help their people achieve their negotiating goals. Officers need to demonstrate considerable competence and good judgment during negotiating periods to retain the team enthusiasm and devotion that required so much effort to build.

In addition to being a member of the negotiating team, the major function of a fire department's personnel officer (if one exists) during negotiations is to prepare for the negotiations and to serve as a source of information and technical expertise for the other members of management's negotiating team. The data that personnel officers should have available during negotiations include

- Information about settlements by other fire departments or local governments and by industrial facilities in the area
- A list of contract changes that would improve the effectiveness of operations, obtained from the officers in the department during the life of the union contract
- Basic statistics about numbers of fire fighters, hours worked, costs of benefits, benefits used, and other information to estimate the probable costs of the union's requests

UNION–MANAGEMENT RELATIONS

The relationship between union leaders and managers contains some curious contradictions. On the surface, each represents a threat to the other, and the general public as well as union members perceive their roles that way. During negotiations, union leaders usually see management representatives as their antagonists because management is the force that prevents them from obtaining the legitimate improvements in working conditions, salary, and benefits that they believe their members deserve. At the same time, managers often see union leaders as irresponsible opportunists who would do thoughtless damage merely for the sake of getting their way. Managers and union leaders can plot strategy in such a way that any confrontation will result in

there are issues which unions/management will never agree on

their respective side "winning" in the negotiations. Also, both of them can often claim victory by favorably comparing a particular feature of the settlement with initial demands or offers or with other settlements.

At any single moment, with respect to any single issue, the union leaders and the organization's managers are clearly antagonists with opposing interests. On closer examination, however, many of these differences are not fundamental or serious. Strategic, long-term thinking about labor–management relations reveals that the three major issues are (1) wages and fringe benefits, (2) working conditions (including satisfaction of psychological needs), and (3) the job security of the union leaders and of the organization's managers. As a result of this broader picture and for practical day-to-day working needs, managers and union leaders often work tacitly, without mutual understanding and in different ways, on the same side of a problem. This cooperation occurs most frequently when they are dealing with unreasonable employees and with excessive demands by irresponsible groups within the union. (See also the section on managing conflict in Additional Insight 4.1, Management of Potentially Damaging Conflict.)

Wages and Fringe Benefits

The union leaders' objectives in negotiating for wages and fringe benefits are to obtain the best possible package as defined by their membership. The requests that the union places on the table early in the negotiations have been arrived at by a political procedure, at one or many union meetings, and have been established through a democratic process. All the union members involved are clearly aware that, by asking for more, it is likely they will receive more than if they started by asking for less or for what they really want. During negotiations, the union can concede some demands and still obtain at least the minimum they are willing to accept. Obviously, the union will not come in with requests that are foolish; nevertheless, it generally requests substantially more than the union leaders or members expect.

each side must present a fair case to an arbitrator

Management's immediate interests are to hold cost increases to a minimum. In a profit-making enterprise, amounts that are not paid out as additional wages are available as profit to the owners. In a government organization, the demands on the funds available are so great that managers are under pressure to avoid giving greater increases or benefits than necessary. All managers are aware that, to some extent, their performance is evaluated on the basis of their ability to achieve favorable settlements.

What we have described is only the immediate picture, however. In the long run, no organization can remain healthy and effective if it fails to adhere to reasonable principles of salary administration. One of the most important principles is that an organization should pay wages and fringe benefits that are equal, or possibly even superior, to those for similar positions in the community. Every manager is aware that, if the organization can pay higher wages and provide a richer benefit package, it has a greater opportunity to retain employees with exceptional competence and high abilities than if its wages are low or lower than the average. For this reason, competent managers do not oppose a union's legitimate and reasonable demands when they are based on good salary and fringe-package administration principles. Competent managers will not try to obtain the very lowest settlements they are able to force on the union when they are in a position to do so. They will, instead, start to bargain at a figure that is lower than what they believe the membership should receive. At the same time, however, they will work diligently to obtain agreement either from stockholders or from higher-level government officials for amounts that will establish fair salaries and fringe benefits.

Working Conditions

The term *working conditions* includes more than just physical facilities and amenities; it also concerns the extent to which officers can help satisfy employees' psychological needs. To some extent, what is true of salary and fringe benefits is true of working conditions. Organizations that have substandard working conditions have greater difficulty retaining qualified and competent people.

Officers, like all other managers, know that good working conditions, including reasonable work rules, provide a more desirable work climate and lead to higher productivity. Similarly, union leaders prefer to work with organizations where little conflict exists and employees have few complaints. Such organizations demand little attention, thus permitting union leaders the freedom to pursue activities they consider more important than settling grievances or processing complaints. Because it is in the best interests of the unions that their members have good working conditions and adhere to reasonable rules, competent union leaders willingly help management establish such environments.

Job Security of Union Leaders and Managers

Because unions are political bodies, union leaders advance in their careers or remain in office only if they are competent leaders or if they do what the membership wants them to do, as long as the membership is not too badly split in its opinions. Managers advance their careers, in part, on the basis of their competence in working with union officials. If union leaders and managers have difficulty establishing a good relationship and if grievances mount, their career goals can become more difficult to achieve.

Consequently, having competent leaders in both management and unions is mutually beneficial. Capable managers do not try to use their powers to undermine the strength of capable union opponents. Enlightened managers are aware that, ultimately, competent union leadership is in everybody's best interest because only capable union leaders are able to control those members who make unreasonable demands or who attempt to undermine a fair settlement. Similarly, competent union leaders are aware that they should not use their power so as to endanger the security or career advancement opportunities of competent managers. Competent managers are difficult to take advantage of, and sometimes less competent management might allow a union to obtain more favorable short-term settlements. Over the long term, however, a lack of management competence, combined with the repercussion of excessive settlements, can lead to difficulties for the union. It might receive either fewer increases or no increases at all or even worse, the organization might lose the strength to survive.

The Role of the Steward

Management's actions are restricted by a union agreement and by the Wagner-Connery Act of 1935. The union agreement and the Taft-Hartley Act of 1947 similarly restrict the activities of a union. Officers can still run their organizations by insisting on adherence to safety rules; attendance regulations; maintenance standards for equipment, apparatus, and facilities; and regulations that apply to fire fighting and prevention. The major effect of the union agreement for the first-line officer is that day-to-day operations now involve another individual—the union's shop steward.

Establishing good relations is the joint responsibility of the line officer and the shop steward. New contract clauses, in particular, are subject to varied interpretations. When a question arises, competent officers consult with the officers to whom they report, who often seek advice from the personnel officer to clarify the intent of a particular clause. Only after having obtained such clarification of the management position should an officer take a firm stand with a shop steward.

A steward occupies the same position, relative to the union, that a supervisor enjoys with an organization. However, a steward is elected by the

[handwritten note: in a perfect world union would not exist because there would be no need for them]

employees and is never paid; the job is voluntary. Therefore, in dealings with a steward, a supervisor must keep in mind that a steward, whose job is awarded by constituents, might conceivably lose the position at the next election if constituents become displeased. A supervisor must also remember that, although a steward may be well informed on union matters, stewards rarely have any formal training in settling disputes.

NONUNION LABOR RELATIONS (HUMAN RELATIONS)

In a nonunion environment, serious dissatisfaction on the part of employees can lead to performance problems and can bring employees to seek outside representation. Major sources of dissatisfaction include the following:

- *Job or security of position.* Employees may feel threatened because of inattention to their complaints or grievances, unsympathetic management, harsh or erratic disciplinary steps, or other high-handed actions on the part of officers. Such actions can be viewed as favoritism or unfair treatment.
- *Wages, salaries, benefits, and promotions.* If compensation is not competitive with that of comparable community positions or other communities, if pay raises are unfair, or if promotion procedures are unknown or unfair, dissatisfaction is likely.
- *Employees not knowing where they stand with management.* Uncertainly generally is the result of inadequate, improper, or irregular appraisals of work performance or managers' failure to communicate to employees the established criteria used in making evaluations.
- *Lack of involvement in policy and decision making.* Employees may feel that they are not involved in decisions affecting their jobs or that they do not know what policies are in effect. Employees like to know not only what is happening but also why it is happening.
- *Poor working conditions.* A poor physical environment has an unfavorable effect on employees.
- *Discrimination.* Discrimination, in any way and on any basis, is a detractor from good relations.

Because many employers, both in industry and in government, have failed to create a climate that reduces these six major sources of dissatisfaction, employees have sought the protection of unions and the safety that labor–management contracts are intended to provide.

[handwritten note: if an employer wants to keep a union out, they must follow these guidelines.]

Administration of Grievances in Departments Without a Union

Every organization has occasions when an individual or group of individuals is dissatisfied. What sometimes starts as a minor complaint can lead to a more serious grievance if a supervisor does not give it fair consideration. If a department is unionized, the contract spells out the procedure a fire fighter can follow to obtain satisfaction. If no union exists, an established procedure must specify the way in which fire fighters can air their complaints. The way grievances are handled is important to the atmosphere of a department. Effective handling of a complaint at the earliest possible time prevents dissatisfaction from building up. Officers certainly do not need to accede to all requests from fire fighters or lower-level officers, but they must give everyone a serious and fair hearing.

If an officer must deny a request, thorough explanations are necessary to ensure that, at the least, the petitioner clearly understands why a personally favorable resolution of the request is not possible. Employees are often dissatisfied because they believe that a particular decision is unfair, but their opinion may be based on incomplete knowledge of the facts. Often, therefore, all an officer has to do to resolve the dissatisfaction is to provide a thorough explanation.

Sometimes the problem is primarily an emotional one for which the employee does not really expect any action on the part of the officer. In such cases, a friendly, empathetic listener may be all that is necessary. As a fire fighter discusses the situation, he or she might realize that the problem is not as important as it first seemed to be, that nothing can be done about the problem, or that something has already been done to avoid a similar problem in the future.

Discussion and explanation will not always resolve the problem that an employee presents. In such instances, considerable dissatisfaction can build up and spread if a fire fighter cannot go to higher authority to file an appeal against a first officer's decision. Therefore, a grievance procedure that is clearly communicated to all fire fighters and officers functions as an important and necessary safety valve. The procedure must be a fair one that does not subject the fire fighters to possible reprisals for using it.

In departments with officers who are alert to identifying practices and policies that could lead to complaints, few complaints are likely to become serious. Managers at all levels, in the fire service as well as in all organizations, should strive to maintain good communications with their teams so that all actions that affect people can at least be explained if a decision cannot be made through some form of participative process. Complaints can stem from many sources, including the following:

- Inadequate or unsafe parking facilities
- Favoritism in the assignment of jobs
- A suggestion that is given little consideration
- A privilege that is not granted
- Physical conditions in the station
- Inadequate or old uniforms
- Lack of facilities for personal belongings
- A rule that is dislike
- Disciplinary actions

Labor Relations in Volunteer Fire Departments

Volunteer fire departments have no clear management–employee relationships because every member of the department can be, potentially, a member of the managing team during the next year. No clear line is drawn between manager and employee because department managers are selected by the members at an election.

Many volunteer organizations have two lines of command: the administrative line and the tactical, or fire-fighting, organization. The administrative organization usually is headed by elected officials and operates like any democratically run organization. Members can bring up any suggestions for policy or procedural changes at a regular meeting. Any changes that are made must be in accordance with the constitution and the bylaws of the fire department or volunteer organization.

At a fire or fire drill, the line of command is clear. The elected chief is in charge, and officers are in command of their respective units. If members are dissatisfied with the way the chief sets strategy or determines tactics or with the way an individual officer performs, they can bring up the matter at a regular administrative meeting. Officers can, however, discipline member fire fighters in a manner similar to that in paid fire-fighting companies.

CONCLUDING REMARKS

Contract negotiations occur once a year or less frequently. In the day-to-day work of a fire officer, labor relations involve anticipating and preventing as many grievances as possible and considering climate and personal competence issues, as well as staff competence, in all significant decisions.

CHAPTER 10

STUDY QUESTIONS AND ACTIVITIES

If you are working alone, prepare your own written responses to these questions. If you are studying with a team or working in class, discuss the questions with the group and write a consensus answer.

1. Before 1932, judges determined the rights of both management and labor in labor relations disputes. After 1932, four federal laws established the rules and regulations for the present collective bargaining system. Explain how each of these laws contributed to establishing the rules and regulations.
2. How do collective bargaining procedures at the local level in the public sector differ from the methods used by private industry?
3. Collective bargaining agreements contain a series of clauses, each relating to a particular area.
   (a) Name four typical clauses in labor–management agreements.
   (b) Write brief descriptions of the purposes of at least two of these clauses.
4. Although grievance procedures vary somewhat from contract to contract, most of them follow steps comparable to the example presented in this chapter. Review the example, and then describe a more satisfactory procedure for settling management–labor problems by either rewriting the procedure in the example or adding other steps to the example.

Training As a Management Function

INTRODUCTION

Realistic fire department training is essential if fire fighters are to perform safely and successfully on the fireground. Because of the need for realism, safety must be an integral part of every aspect of fire training. For department personnel to conduct training operations properly, they must be familiar with the appropriate training standards. Chapter 3 of NFPA 1500, *Standard on Fire Department Occupational Safety and Health Program*, contains extensive information and recommendations for education and training. Other resources are NFPA 1403, *Standard on Live Fire Training Evolutions*, and NFPA 1404, *Standard for a Fire Department Self-Contained Breathing Apparatus Program*.

After a scenario and its analyses, this chapter provides more detailed information on fire service training and relevant standards. The bulk of the chapter, however, addresses the more general issues pertaining to learning and training that can be useful to officers as they work to make their various training and coaching activities more interesting and effective.

SCENARIO

LADDER DRILL

"OK, let's get going," Captain Joe Streak called into the day room. He was becoming a little impatient. He had scheduled ladder training for 10 o'clock. It was already almost ten minutes after the hour, and only one of the fire fighters was at the engines. Pete hadn't finished his mid-morning coffee yet, and Sam was reading the morning paper.

The captain turned to the fire fighter who was putting on his coat. "Jack, get ready to pull out."

While Jack climbed into the driver's seat and started the engine, Pete came into the apparatus room and, without rushing, began to put on his boots and turn out clothing. By the time Sam finally came through the door, the captain was clearly annoyed. Still, he kept quiet when he heard Sam's loud whisper to Pete about ladder drill being just a dumb waste of time. Pete's response, "I agree," was too much. The captain snapped, "You guys know better than to make stupid remarks like that. Sam, grab your protective gear."

When Sam joined them, Jack and Pete had already taken the 14-foot roof ladder off and the captain was standing at the top end of the 24-foot ground ladder, waiting for Sam to pick up the foot.

"OK, first Sam and I will do a two-person ladder raise to the second-story window. Then Jack and Pete will do the same thing. Finally we'll do three person raises in rotation."

After that part of the drill, the captain wanted to go through a beam raise drill, but the men demurred: "One ladder drill a day is enough. Let's do it some other time," Jack expressed the sentiment of the three. The captain didn't like it, but, rather than forcing the issue, he instructed Jack to take the engine back into the station house.

Immediately after they returned the apparatus, the alarm sounded. "Dispatch to Station Three. Reported working fire, 61 Bloomdorn Street, second floor of four-family wood frame tenement.

On the way, the team reviewed what they knew about the location. Jack had lived nearby when he was in high school. "It might be the house where there's an alley on the north side. These houses have hallways running down the middle, from front to rear. The alley is a really narrow one. I never understood why they've got windows on the side of the building facing a solid wall."

"Let's hope there are no people in there, if that's the place," the captain added.

When the team reached the scene, they could see heavy smoke billowing out from the second-story windows in front. It was indeed the house with the narrow alley. Looking into it, they could see a woman and a child in the last window in the rear, screaming for help.

"Pete, grab the 3 length preconnect and attack the fire. Jack, work with Pete. Sam, help me with the ladder; we'll have to do a beam raise—too many wires in the way to take a chance. You take the foot end," the captain shouted as he jumped off the engine.

As soon as the woman could hear them, they called to ask her whether others were in the building. Much relieved at seeing rescue on the way, she responded that they were all out at work.

Meanwhile the captain had begun to lift the ladder. It slipped a few feet down the alley, falling out of his and Sam's hands and landing flat on the ground. Sam realized that he had not braced it with his foot. By the time they were able to get it back in position, a few moments were lost that could have been precious if the emergency had been more serious. The ladder then went up smoothly. Sam extended it, and the captain asked him to go two more rungs before leaning it against the wall where it just reached the window sill.

Sam climbed into the room to check whether the hallway might be passable so the woman and child could get to the fire escape in the rear of the building. It wasn't.

While Sam was doing that, Jack stuck his head into the alley and called out, "The line is in, captain."

The captain, suddenly thinking about the extinguishment part of the operation, called back, "Did you vent?"

"No." Then Jack rushed back.

Meanwhile, Sam had started to take the child down the ladder and handed the child to the captain, who had climbed up to meet him. Sam then helped the woman, while the captain rushed forward to see what was happening there.

When he came up the stairs, Pete had forced entry into the front apartment and was directing the stream at the fire in the room facing the street. Fortunately, Jack had knocked out one window and was just hitting the other one. When Jack saw the captain, he could not help expressing his annoyance that Pete and he had to begin the fire attack before the second crew was in place. Despite the noise, the captain reminded him that they were shorthanded and that they would discuss this again during the debriefing. [In an earlier discussion, the captain had pointed out that minimal staffing is allowed under the emergency rescue exception of the Two In/Two Out Rule as listed in both NFPA 1500 and OSHA 1910.134(g)(4).]

The fire was soon under control. After salvage and overhaul, the captain turned the location over to the police, who had notified the tenants and were waiting for them to arrive.

Back at the station, at the beginning of the critique, the captain asked how everyone felt now about training, about doing much more of it and more often. No one answered. All three looked sheepishly to the ground.

Scenario Analysis: Fire Service Function Perspective

Before reading the analysis, you may wish to give some thought to the strengths and weaknesses of the way Captain Streak conducted the drill and managed the incident from the fire service function perspective.

This scenario involves primarily the goals and guidelines of training. The goals are to ensure a high level of competence in all fire department members for all functions in which they may be involved, and particularly for those that affect the safety and health of the fire fighters themselves and of all others involved in an emergency incident. The goals of officers are to be competent in management of training and in coaching fire fighters.

A guideline follows from these goals:

> What else needs to done to achieve a high level of competence in all members of the fire department for all functions in which they may be involved?

This guideline leads to the following specific questions:

1. What can be done to achieve a reasonably clear picture of the knowledge and skill levels of each fire fighter and officer with respect to each of the major tasks to be performed?
2. What can be done to ensure that all fire fighters and officers are skilled in performing all tasks, and especially those that require close and detailed coordination between two or more fire fighters?

3. What can be done to motivate fire fighters and officers toward willing participation in training, including drills, that will maintain a peak level of competence?

4. What can be done to motivate fire fighters and officers toward self-identification and self-development on knowledge and skills that need sharpening?

5. What can be done to monitor the knowledge and skill levels of all fire fighters and officers?

6. What can be done to ensure that officers are competent in the management of training and in the coaching of fire fighters?

You do not need to remember the detailed wording or specific subsections guideline when you are making decisions. Only in the most important decisions might that be worthwhile. However, memorizing a personalized, abbreviated (summarized) version and applying it with every relevant decision are critical for developing the necessary habit that will help to ensure the highest quality performance possible under the existing circumstances.

We will consider the questions one at a time to analyze Captain Streak's scenario.

1. *What can be done to achieve a reasonably clear picture of the knowledge and skill levels of each fire fighter and officer with respect to each of the major tasks to be performed?* The scenario contained no evidence that Captain Streak had anything but a very general view of the specific knowledge or skills of any of his fire fighters. He might ask himself questions like these:

- If I do not have easy access to a list of the knowledge items and skills that the fire fighters should possess, should I create such a list, and how? Should I work in cooperation with other company commanders, with some or all the members of my company, or with both?
- How can I best go about gaining a clear picture of what each of the fire fighters can do?
- How do the knowledge and skills in my department compare to the standards in NFPA 1001, *Standard for Fire Fighter Professional Qualifications?*
- What do I need to do to clarify my own skills and weaknesses?

2. *What can be done to ensure that all fire fighters and officers are skilled in performing all tasks, and especially those that require close and detailed coordination between two or more fire fighters?* The scenario was silent on what other training activities were done in Captain Streak's company. Still, the questions he might ask himself include the following:

- How can I identity the tasks that require close and detailed coordination between fire fighters? How or where should I record the information about them?
- Should I prepare a realistic training schedule, including drills? What priorities of topics and skills are appropriate for it? Who should participate in the preparation of such a schedule, and how?
- What learning/training methods are best for each item on the schedule?
- What do I need to do to ensure that my own skills and knowledge are as sharp as necessary?

3. *What can be done to motivate fire fighters and officers to be willing to participate in training, including drills, that will maintain a peak level of competence?* There is little evidence that Captain Streak made any attempt to motivate his fire fighters to enhance their competence. He might ask questions like these:

- What techniques can I use to make training exercises more interesting for fire fighters so they can gain greater satisfaction from them?
- Are there ways in which I can package some training for individuals or the entire company with daily activities?
- How can I combine the least interesting drills with interesting learning activities so the drills are easier for the fire fighters to do without boredom or resentment?
- What can I do to make effective use of the competence strengths of individual fire fighters to lift the skills of others in those areas? When and how will it be advantageous to assign training responsibility for specific topics to individual company members?

4. *What can be done to motivate fire fighters and officers toward self-identification and self-development on skills and knowledge that need sharpening?* This guideline question is one of the toughest because the desire for continuing self-development is an intensely personal trait. Still, the captain might ask questions such as

- How can I challenge fire fighters to assume a greater role in keeping up with new developments in the community and technology, achieving continuing competence improvement, and maintaining skills at peak levels?
- How can I best use the knowledge or skill strengths of each fire fighter to demonstrate the benefits of such knowledge and skill and to create the desire for matching those strengths?

5. *What can be done to monitor the knowledge and skill levels of all fire fighters and officers?* Here the captain might ask questions such as

- How can I observe the competence of the fire fighters besides during training and drill sessions?
- How can the fire fighters contribute to an ongoing assessment of knowledge and skill levels?
- What kind of activities should I schedule strictly for assessing competence, and what should the role of the fire fighters be in such activities?

6. *What can be done to ensure that officers are competent in the management of training and in the coaching of fire fighters?* This issue involves Captain Streak only with respect to his own self-development. The chief of the department should ask the question with respect to all officers. It speaks for itself and does not need additional questions to clarify or augment it.

Scenario Analysis: Management/Leadership Perspective

We will analyze the management/leadership in the scenario based on the guidelines for the 3Cs. We restate only the specific questions that expand on the basic guidelines and then use them in the analysis.

The control guideline questions are as follows:

1. Are the goals and objectives appropriate and effectively communicated?
2. Is the participation in decision making appropriate?
3. Are coordination, cooperation, and inter- and intra-unit communications being stimulated?
4. Is full advantage being taken of positive discipline and performance counseling?

The competence guideline questions follow:

1. Are changes needed in recruiting and selecting for vacancies?
2. Are management of learning concepts applied effectively?
3. Are coaching and counseling on self-development being used to their best advantage?

The climate guideline questions are listed next:

1. Are appropriate psychological and tangible rewards offered and provided effectively and efficiently to bring the highest possible level of satisfaction from the creation and use of the product/service?
2. Are policies in place to help reduce work-related stress?

Before reading the analyses, you may wish to give some thought to the strengths and weaknesses of the way Captain Streak conducted the drill and managed the incident from the management/leadership perspective.

Discussion of the Control Guideline.   Two elements are especially important to this discussion:

• *Coordination.* Learning activities and learning needs must be well coordinated. Confusion or lack of clear direction has a demotivating impact. An officer should build the sequence of topics in training logically and carefully to ensure that all new concepts are understood before they are used to explain other matters that will serve as a basis for drawing conclusions.

• *Norms and behavior rules.* Doubt, confusion, and distractions are reduced when members are expected to follow clear norms or rules. Simple rules, such as learning assignments must be completed on time and learning sessions must start and stop at the appointed times (when emergencies do not interrupt), can significantly increase confidence in a training program.

1. *Are the goals and objectives appropriate and effectively communicated?*   The scenario contained no evidence that Captain Streak communicated learning goals or objectives to the fire fighters. He set up a single learning experience, but he did not refer to other sessions when he admonished the two fire fighters who were late. Similarly, at the incident, the captain's failure to give immediate attention to ventilation suggests that objectives for tasks at a fire either were not set or were poorly communicated. Jack overlooked his obvious role in ventilation. He seemed to be the most motivated of the fire fighters, and still he did not immediately realize that he should knock out the windows.

The captain might ask himself questions like these:

• What types of guidelines should I set, clearly communicate as behavioral objectives at a fire incident, and drill so that fire fighters do not overlook obvious tasks and can perform them without specific instructions?

• What can I do to review these guidelines regularly and gradually augment them with new ideas, so they will become automatic behavior at an incident? For example, a guideline might say that, in the absence of specific commands by the incident commander, the pump operator should attend to ventilation or, if ventilation has been assigned, assist with the attack.

2. *Is the participation in decision making appropriate?*   We can assume from the scenario that the training plan and the schedule, the primary decisions, were not determined participatively. In contrast, the decision to end the training session as planned by the captain was highly participative, but too much authority was allowed the fire fighters. These issues raise questions such as these that the captain might ask himself:

- What level of participation is appropriate for decisions pertaining to training content and schedule? Should training proceed according to a comprehensive schedule that is followed as closely as possible? Or should the training schedule be developed jointly by the team, with the content of individual sessions specified in the schedule?
- What should the procedure be for continuing a training session that is interrupted by an emergency?

3. *Are coordination, cooperation, and inter- and intra-unit communications being stimulated?* We know too little about the company and the department to make meaningful comments on this question. However, with respect to this guideline issue, the captain might ask himself questions such as these:

- What else should I do to ensure that coordination procedures for emergencies and nonemergency tasks are clearly communicated and understood?
- What obstacles might interfere with full cooperation (such as norms that are not fully shared or insufficiently resolved conflicts)?

4. *Is full advantage being taken of positive discipline and performance counseling?* The captain did not make preparations to counsel the fire fighters on their reluctance to attend training sessions. We can assume that it would be useful if he asked himself questions such as these:

- On which of the issues raised by the training session and by the performance at the fire should I consider specific performance counseling (as distinct from the necessary training and coaching)?
- What else do I need to do so I can effectively counsel a fire fighter on issues that are not performance problems but rather opportunities for improving performance?

Discussion of the Competence Guideline.

1. *Are changes needed in recruiting and selecting for vacancies?* This question is not relevant to the scenario.

2. *Are management of learning concepts applied effectively?* We covered this question extensively in the fire service function analysis.

3. *Are coaching and counseling on self-development being used to their best advantage?* We also covered this question.

Discussion of the Climate Guideline.

1. *Are appropriate psychological and tangible rewards offered and provided effectively and efficiently to bring the highest possible level of satisfaction from the*

creation and use of the product/service? The captain took no steps to reinforce positive behaviors by acknowledging them. Even though Jack was on time and cooperative at the start of the training session, the captain did not say to him, "Thanks for being on time" or "It's nice to know that I can count on you to be punctual." Depending on the relationship, the comment could be made privately or publicly. Captain Streak might consider asking questions such as these:

- What can I do to better recognize opportunities for showing that I notice and appreciate cooperative or effective actions?
- In what ways can I communicate that I have noticed such actions?

2. *Are policies in place to help reduce work-related stress?* The scenario contained no hint about policies on this subject. Lack of adequate competence and preparedness contributes to stress. Over the long haul, a sound approach to honing skills and knowledge can do much to bring about a higher confidence level and lower stress for all members of the company, including the captain.

The analysis of the scenario confirmed the synergism between the fire service function guideline on training and the competence guideline, the second one of the 3Cs. The fire service function guidelines raise some specific questions for training. Officers who review those guidelines, and the 3Cs guidelines, at least from time to time and with every relevant decision, will help to ensure that all aspects of the training function receive the attention and emphasis they deserve.

TRAINING IN THE FIRE SERVICE

Many fire departments consider training to be among the most important functions because it is so central to operational effectiveness.* One reason is the need to maintain skills at peak levels; another concerns the training of new recruits, both paid and volunteer.

In the selection of fire fighters, departments consider many factors, such as physical condition, mechanical ability, personality, educational background, and ability to learn. Experience is rarely a major consideration because few people have fire-fighting experience.

---

* The International Fire Service Training Association (IFSTA) book, *Essentials of Fire Fighting* (1998), contains more than 500 pages of basic skills that every fire fighter should possess, and it does not even include more advanced training topics such as hazardous materials, specialized rescue, or aircraft and petroleum fire fighting.

Fire service personnel require training to perform effectively. Only with a comprehensive training program is a fire department able to establish and maintain a staff that is prepared and fully skilled. New fire fighters usually go through a course of basic training that provides the foundation for more thorough on-the-job training. Some of the material in the basic training course is repeated in the refresher sessions of in-service programs.

Continuing training and in-service training programs are necessary for fire fighters because unused skills fade. Precious minutes may be lost in fumbling or accidents, and near-accidents may be caused when fire fighters are not familiar with equipment or its proper use. Lapses can be avoided only by continuous training and repeated drill.

Unfortunately, fire fighters perceive drills as redundant. No one wants to be forced to relearn something that is already well understood and practiced. Therein lies the challenge to the officer: how to gain willing cooperation with drills on routine evolutions or other tasks, and not to drill too frequently on skills that are adequately honed.

When routine drills are part of an interesting learning experience on advanced skills or part of a simulated exercise on one aspect of a serious emergency, fire fighters will accept them much more readily than when they stand alone. Often training and learning can also be woven into the daily contacts between officer and fire fighters, where they are more readily accepted.

Training Standards and Professional Qualifications

At present, no single national, mandatory formula for training fire fighters exists to which local fire departments are compelled to conform. However, the most formal set of basic requirements—a series of model fire training standards—was developed through the auspices of the Joint Council of National Fire Service Organizations (JCNFSO). In 1972, the JCNFSO created the National Professional Qualifications Board for the Fire Service (NPQB) to facilitate the development of nationally applicable performance standards for uniformed fire service personnel. On December 14, 1972, the board established four technical committees to develop those standards using the National Fire Protection Association (NFPA) standards-making system. The initial committees addressed the following career areas: fire fighter, fire officer, fire service instructor, and fire inspector and investigator.

The original concept of the professional qualifications standards, as directed by the JCNFSO and the NPQB, was to develop an interrelated set of performance standards specifically for the fire service. The various levels of achievement in the standards were to build upon one another within a strictly defined career ladder. In the late 1980s, revisions of the standards recognized that the documents should stand on their own merit in terms of job performance requirements for a given field. Accordingly, the strict career ladder concept was abandoned, except for the progression from fire fighter to fire officer. The later revisions, therefore, facilitated the use of the documents by other than uniformed fire services.

In 1990, the NFPA assumed responsibility for the appointment of professional qualifications committees and the development of the professional qualifications standards. The Correlating Committee for Professional Qualifications Standards was appointed by the NFPA Standards Council in 1990 and assumed the responsibility for coordinating the requirements of all the professional qualifications documents. Professional qualification standards have been developed to cover the following functional areas:

- Fire Fighter (NFPA 1001, *Standard for Fire Fighter Professional Qualifications*)
- Fire Apparatus Driver/Operator (NFPA 1002, *Standard for Fire Apparatus Driver/Operator Professional Qualifications*)
- Airport Fire Fighter (NFPA 1003, *Standard for Airport Fire Fighter Professional Qualifications*)
- Fire Officer (NFPA 1021, *Standard for Fire Officer Professional Qualifications*)
- Fire Inspector (NFPA 1031, *Standard for Professional Qualifications for Fire Inspector and Plan Examiner*)
- Fire Investigator (NFPA 1033, *Standard for Professional Qualifications for Fire Investigator*)
- Public Fire and Life Safety Educator (NFPA 1035, *Standard for Professional Qualifications for Public Fire and Life Safety Educator*)
- Fire Service Instructor (NFPA 1041, *Standard for Fire Service Instructor Professional Qualifications*)
- Wildland Fire Fighter (NFPA 1051, *Standard for Wildland Fire Fighter Professional Qualifications*)
- Public Safety Telecommunicator (NFPA 1061, *Standard for Professional Qualifications for Public Safety Telecommunicator*)

These standards, along with others pertaining to rescue and emergency vehicle technicians, provide in-depth coverage of the professional competencies required for the functional areas. The levels of competency that have been developed in each of the classifications allow fire personnel to progress in an orderly manner.

Another set of standards provides guidance on fire department training programs. In many places, these standards are adopted by reference and become operating requirements. In New Jersey, all live fire training is mandated to conform to NFPA 1403. Because of the need for safety, it is important for fire departments to base their training operations on these standards:

- NFPA 1401, *Recommended Practice for Fire Service Training Reports and Records*
- NFPA 1402, *Guide to Building Fire Service Training Centers*
- NFPA 1403, *Standard on Live Fire Training Evolutions*
- NFPA 1404, *Standard for a Fire Department Self-Contained Breathing Apparatus Program*
- NFPA 1410, *Standard on Training for Initial Fire Attack*
- NFPA 1451, *Standard for a Fire Service Vehicle Operators Training Program*

Training Programs

How fire departments organize to meet the training requirements is as varied as the information in training programs. The construction of fire department training programs—their priorities, content, and methods—is ultimately left to individual localities, chiefs, and training instructors. As listed in Exhibit 11.1, many organizations in the public and private sectors provide training materials that fire departments can use in developing training programs.

In large departments, a training officer, usually a high-ranking officer appointed by the fire chief, coordinates all training activities. The training officer in medium or large municipal fire departments is responsible for selecting textbooks, writing department manuals, presenting materials in the classroom, and ensuring that training records are kept on probationary and in-service fire fighters throughout their terms of employment. The training officer and the company officer to whom the trainee is assigned share on-the-job training and record-keeping responsibilities.

Typically, training of new fire fighters begins with a basic training course, which can last from 1 to 18 weeks. The training officer is responsible for the basic training program and will assign the trainee to a company for on-the-job training. The company officer then assumes the responsibility for arranging training schedules to cover evolutions of every type as soon as possible.

An example of the way one fairly small department schedules its training is illustrated in Exhibit 11.2. Although not all departments have schedules like

Exhibit 11.1 Sources of Materials for Fire Department Training Programs

Federal Emergency Management Agency
(FEMA)
Federal Center Plaza
500 C Street SW
Washington, DC 20472

International Association of Fire Chiefs
(IAFC)
4025 Fair Ridge Drive
Fairfax, VA 22033-2868

International Association of Fire Fighters
(IAFF)
1750 New York Avenue NW
Washington, DC 20006

International Fire Service Training
Association (IFSTA)
930 N. Willis
Stillwater, OK 74078-8045

International Society of Fire Service
Instructors (ISFSI)
U.S. Fire Administration
16825 S. Seton Avenue
Emmitsburg, MD 21747

National Fire Academy (NFA)*
16825 S. Seton Avenue
Emmitsburg, MD 21727

National Fire Protection Association
(NFPA)
1 Batterymarch Park
P.O. Box 9101
Quincy, MA 02269-9101

United States Fire Administration (USFA)*
16825 S. Seton Avenue
Emmitsburg, MD 21747

* These organizations are under the umbrella of the Federal Emergency Management Agency (FEMA).

Note: Various insurance companies and trade associations also provide specialized training materials in their areas of expertise.

the one shown, schedules provide the company officer with a plan for ensuring that training is continuous and thorough for the needs of the company. At the same time, the plan gives training somewhat higher priority in the work schedule.

In-service training for fire fighters is an ongoing process. Fire fighters cannot maintain their skills in all essential fire-fighting evolutions because many of these operations are performed at only large or special types of fires. Each member of a department is therefore required to put in enough training work not only to retain skills in performing standard evolutions but also to keep abreast of current technical developments in particular fields. Good fire departments devote part of every day to drills and training work.

Through regular drill procedures at a training school, methods of performing all operations are standardized so that a department can transfer personnel to various companies without compromising efficiency. In well-run departments, companies periodically are assigned to drill at their fire training academy.

Company training activities usually are recorded by the company officer and presented to the training officer for inspection. Exhibit 11.3 is an example

Exhibit 11.2 Sample Training Schedule

January
- 911 answering
- Dispatch duties
- Streets/numbers/map study
- Records and reports
- General orders/SOPs/directives/notices
- Rules and regulations
- Alarm response patterns
- Strategy and tactics
- Prefire planning
- Tools and equipment
- Engine operations

March
- Advancement, placement, and use of handlines/nozzles and appliances
- Power-operated tools
- Ground ladders
- Ropes and knots
- Infectious disease control/bloodborne pathogens*
- Defibrillation/rescue 1 equipment
- Public relations

May
- Second due—all evolutions
- Foam operations
- Standpipe and sprinkler systems
- Hose test/maintenance
- Tools and equipment
- Aero-medical procedures/interagency operations
- Mutual aid procedures
- Unusual emergencies/special operations and rescues
- SCBA training*

July
- Incident critique
- Fireground operations
- Multi-alarm fires
- Safety/backdraft/flashover, etc.
- Fire cause and origin/arson investigation procedures
- Relay pumping
- Rope rescue

September
- Second due—all evolutions
- 911—answering/*refresher*
- Dispatch duties/*refresher*
- SOPs

February
- Sizeup/building construction
- Driving and spotting—all apparatus
- Forcible entry procedures
- Search and rescue
- Communication/incident command system
- Salvage
- CPR/first aid
- Hazardous materials*

April
- Pump operations
- Deck gun
- Ladder pipe
- Hose
- Thief
- Hydraulics
- First due—all evolutions
- Extrication tools and equipment
- Amkus tool
- Live fire training drill
- Confined space rescue

June
- Ventilation
- Power tools
- Tower/truck—all evolutions
- Hydrant inspection
- Hose test
- Deck gun operations
- Overhaul
- Incident command system/*refresher*
- SOPs/*refresher*
- Legal considerations

August
- Prefire planning
- Driving and spotting—all apparatus/ *refresher*
- Tower 1—all evolutions/*refresher*

October
- *All refreshers*
- Amkus tool
- Extrication tools
- First due—all evolutions
- Hydraulics
- Ropes/knots
- Rope rescue

(continued)

Exhibit 11.2 (continued)

November	December
• All Refreshers	• Winter fire fighting
• Tower 1	• Placement of hoselines/*refresher*
• Power-operated tools	• Overhaul/*refresher*
• Ventilation	• Search and rescue/*refresher*
• Forcible entry	• Any training area not covered as
• Confined space rescue	suggested in the previous months
• Hazardous materials*	

*Indicates 8 hours mandatory training as required by the Division of Fire Safety per year. All training must be documented and records maintained.

Source: Courtesy of Captain Patrick Kelleher, Roselle, NJ, Fire Department (Robert Hill, Chief).

of a form that can be used to record training activities in a fire department. These reports, usually prepared by company officers throughout a department, supply the training officer with a consolidated monthly training activity report that is reviewed periodically by the chief. An annual training report can be compiled easily from these monthly reports. Another form should be maintained to track the training activities of each individual (see Exhibit 11.4).

Small departments that lack budgets for full-time training staffs usually train new fire fighters by sending them to state or regional fire fighter training programs. These programs operate in most states and include state or regional fire schools; courses are conducted annually, with sessions held either for a few days or for a full week of intensive training. The state courses include officer and leadership conferences, instructor training classes for fire department instructors, industrial fire brigade training courses, and training conferences for fire prevention officers. Some states have traveling training instructors who work year-round with local fire departments that are too small to have full-time training officers.

TRAINING DESIGN

In the educational world, the traditional perception of an instructor as one who lectures and provides explanations is changing to one in which the instructor pays equal attention to both the process of learning and the content of the material presented. *Process* refers to the steps necessary to ensure that learners gain the greatest benefits from the learning experiences. *Content* is the topic itself, the knowledge or skills to be acquired. In the fire service, involving

Exhibit 11.3 Sample Form for Recording Departmental Training Activities

Date _____

Officer in charge of drill _____

Drill subject _____

Audiovisuals used _____

Number of training hours _____

Members attending

1. _____   11. _____

2. _____   12. _____

3. _____   13. _____

4. _____   14. _____

5. _____   15. _____

6. _____   16. _____

7. _____   17. _____

8. _____   18. _____

9. _____   19. _____

10. _____   20. _____

_____ **Officer in charge**

_____ **Training officer**

_____ **Fire chief**

Exhibit 11.4 Sample Form for Recording Individual Fire Fighter's Training Activities

| Name _____ | Date joined company _____ |

Subject	Location	Date	Hours
1. _____
2. _____
3. _____
4. _____
5. _____
6. _____

•
•
•

learners by letting them practice has always been an important aspect of training. Paying attention to the process of learning is therefore important for training officers who design training programs for new and in-service fire fighters. We cannot overemphasize the importance of paying attention to both process and content in developing fire service training programs.

Perspectives on the Learning Process

A clear perspective on the learning process is essential for officers who design training programs and for those who teach training courses for the fire service. Familiarity with three concepts of learning can help assure the success of the training effort: the paradox of clear explanations, the level of aspirations, and the volume of explanation.

The Paradox of Clear Explanations or Lectures. A carefully organized and clearly delivered explanation or lecture can actually be self-defeating. Although a clear, logical presentation should lead learners step by step from the fundamentals to a complete picture of the topic, this approach may actually conflict with learner comprehension. A well-organized presentation may not make learners aware of the practical limits of their new knowledge. Because they easily follow a presentation, the learners assume that they fully understand the message. Few questions come to their minds and, even though they have mastered some difficult thoughts, they leave the learning environment with only a shallow comprehension of the subject—a comprehension that is inadequate for application to complex, real-life situations.

The circle (A) in Figure 11.1 graphically depicts this kind of learning experience. The learners have little contact with the unknown that surrounds the small island of knowledge they have gained. The learners believe that they have developed an adequate understanding of a subject, and therefore they are not likely to probe for more information. At some later time, however, when they attempt to apply what they have learned, they discover their lack of understanding. By this time, the instructor is no longer readily available to supply guidance, and the learner makes mistakes or at best must either devote more effort to learning the necessary material or forgo a clear comprehension of the subject.

The star-shaped drawing (B) in Figure 11.1 represents information given in a rugged, uneven format. Material presented this way gives learners only

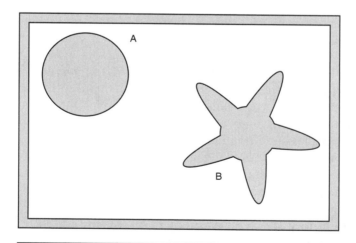

Figure 11.1 A circle (A) depicts a learning experience in which learners are not likely to probe beyond the small circle of new knowledge for more information. The starshape (B) depicts a learning experience in which the information is presented with techniques that give learners greater contact with application and the complexities of the subject.

enough information to serve as a basis for exploration and discovery of more knowledge. In this type of learning situation, learners probe for understanding in terms most meaningful to each individual. If such a learning experience is designed carefully, it involves some form of problem solving and provides individualized answers to questions that arise in the process—as the answers are needed, and asked for, by the learner. Many of the outreach programs offered by the National Fire Academy use this problem-solving approach to learning. It allows students to practice skills learned in the classroom to solve real-world problems and scenarios.

The Level of Aspirations.   An effective officer/instructor is concerned with the level of difficulty that a topic or learning goal or objective presents to the learner. Figure 11.2 shows the relationship between the learner's level of aspiration (how strong the learner's desire is to learn) and the difficulty of the task (the complexity of the task level). Only the sufficiently motivated learner—that is, one who has the necessary level of aspiration to overcome the difficulty the task presents—will devote maximum attention to the task of learning and lift the zone of ego involvement. The diagram suggests that high motivation is possible only if the task level is set so that it is perceived to be within the experience, courage, and competence of the learner (even if some failures are expected). Aspirations influence the way a task is seen and the attitudes with which it is approached.

If a task is perceived as much too difficult and has never been attempted, it is outside the zone of ego involvement shown in Figure 11.2. For example, a Russian language literacy test would fall into this category for most Americans. At the other extreme, any task that is immediately recognized as much too easy also fails to engage ego involvement and therefore presents little challenge and will quickly be considered a waste of time. The most satisfying learning experiences are those that present a modest threat of failure and are therefore challenging. Here, of course, the payoff in satisfaction of accomplishment is high, and motivation to tackle the task is greatest. According to Wallace Wohlking's reports on research with children, which he has confirmed informally with his adult students at Cornell, those who have a history of success usually will set realistic goals and objectives, whereas those who fail regularly tend to set unrealistic ones—either too high or too low. Those who set goals or objectives that are too high seem to do so because they usually are rewarded for trying, whereas those who set them too low do so as a defense against possible failure.*

_____
*Based on work by Wallace Wohlking.

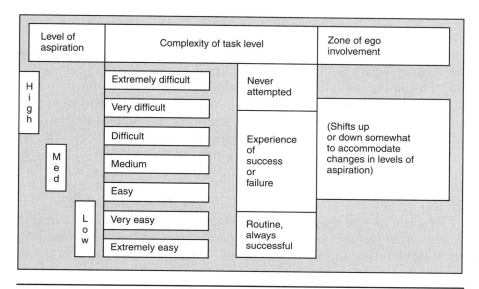

Figure 11.2 The level of aspiration, the complexity of task level, and the amount of ego involvement are compared here. (*Source:* Based on work by Wallace Wohlking)

The conclusions for training design are clear:

- For all complex tasks, the designer of a training experience must set goals.
- Goals and objectives must be attainable so that successful experiences are developed.
- Specific effort must be made to teach tolerance for making mistakes.

To create an environment in which learners can find maximum motivation requires consideration of all elements of the 3Cs, especially the guideline concerned with setting goals and objectives (see Additional Insight 3.1, Goal Setting and Implementation) and the one concerned with competence (see Chapter 1). In addition, the manager of learning must understand what learners already know and the extent of their knowledge or skill deficiencies.

The Volume of Explanation.   The four diagrams in Figure 11.3 emphasize the importance of motivation and show the limitations of teaching presentations from the perspective of volume. In part A of Figure 11.3, the larger rectangle (1) represents all currently available knowledge about a specific subject, and the smaller rectangle (2) represents what the instructor knows about the subject. Faced with limited class time, the instructor prepares a presentation that covers the portion of knowledge he or she considers most important. However, during class, the instructor encounters various interruptions and delays and thus covers less material than planned. The prepared material compared to what is actually discussed is represented by the two smaller rectangles (3 and 4) shown in part B of Figure 11.3.

The amount of knowledge dwindles even more as it is transmitted to learners, as shown in part C. Because of lapses of attention, screens of personal biases, misunderstandings, misinterpretations, and false impressions of concept clarity, what reaches the student is far less than what was presented. What is retained a few weeks, months, or even years later is still another matter. The three small rectangles (5, 6, and 7) in part C of Figure 11.3 represent what the student hears (5), what penetrates the student's consciousness (6), and what the student remembers at some later date (7).

The intention of this somewhat exaggerated analogy is to illustrate the point that *how* a subject is presented is often far more important than *how much* of the subject is presented. The instructor can arouse the curiosity of learners by presenting only that portion of the material that they can readily absorb. If the material is presented in an interesting, stimulating format, learners usually will seek out more information (D of Figure 11.3). Thus, they can fully understand and readily absorb more information, and their ability to apply the newly acquired knowledge will be enhanced substantially.

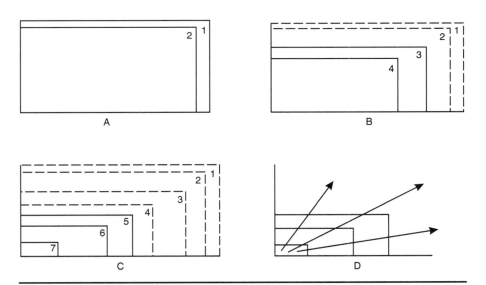

Figure 11.3 The volume of explanation in a presentation is represented pictorially.

Conditions Influencing the Effectiveness of Learning

Many factors influence the effectiveness with which learners can absorb and then apply new skills and knowledge. We will list and discuss the more important factors that training designers and instructors should keep in mind.

Pace. Instructors should adjust the learning pace to the learner's ability to absorb new concepts if the learning process is to proceed without interruption. All learners have different capabilities for learning. Some understand quickly, others have good memories, and still others have long attention spans. An explanation or lecture that is fairly fast paced can lose some learners, and a slow-paced one usually holds the attention of only the least capable learners.

Absorption. Learning will be absorbed better and retained longer if it is directly related to work and can be applied on the job. Most subjects covered in fire service training are work related; however, some subjects (such as complex rescue operations or those involving a specific type of hazardous material) are rarely applied. If the task might be very important during an emergency, learners should be given the opportunity to practice it regularly in a simulated environment to ensure adequate skills when they are needed.

Understanding. Skills that are learned and thoroughly understood or practiced until they become habits are retained better than skills that are learned by rote. Learners should be tested to determine that they thoroughly understand a subject and have not just memorized some facts.

Practice. Practice distributed over several periods of time is more valuable to the learning process than the same amount of practice concentrated into a single period. If learners need to practice an important skill extensively to achieve full mastery, practice sessions do not have to be continued until they reach this goal or objective. Pauses of several days between relatively short sessions lead to better retention.

Order. The order of presenting materials is important. Although this learning principle might seem obvious in many programs, instructors sometimes present complex topics before before they establish an adequate foundation for them.

Recognition/Recall.   Recognizing something is easier than recalling it. This principle is of great importance in selecting ways to test the knowledge of learners. Testing should emphasize recall, not recognition. For example, requiring fire fighters to draw their own prefire plans and insert appropriate symbols and tables correctly tests the learning process more accurately than presenting a ready-made prefire plan and requiring only that a given list of symbols be placed correctly on the plan. Fire fighters must recall much more information to draw their own plans than to simply recognize proper symbol placement on a plan drawn by someone else.

New Knowledge.   Learning something new can interfere with remembering something that has been learned earlier. Instructors often overlook this principle. While helping learners acquire new skills or knowledge, instructors may fail to reinforce previously acquired skills or subject matter. This principle is especially important when the new subject overlaps, or is similar to, a previous one. For example, learning a new set of code numbers can hinder retention of an earlier group of numbers.

Feedback.   If learners are unsure whether they are acquiring knowledge correctly, or if they are wondering whether what they are learning is correct, they will be hesitant to devote their full efforts to learning. Therefore, observation with feedback, testing, and participative activities in which learners receive feedback on the correctness of what they are learning are important segments of satisfactory learning experiences.

Although the preceding factors in learning effectiveness are not the only ones to be considered when planning an instructional session, they are helpful techniques. The officer/instructor who successfully incorporates these principles into a teaching plan will motivate learners to seek greater competence.

MANAGEMENT OF LEARNING

Management of learning strives to shift the responsibility for learning and acquiring competence from the instructor to the learners. The instructor who acts as a manager of learning does not concentrate solely on the topic and how it can best be presented. Instead, the manager of learning seeks to identify learning needs and finds ways to help the learner satisfy those needs. The manager of learning stimulates curiosity and motivation to learn by presenting only that portion of the material that learners can readily absorb and by presenting it in an interesting format so learners will reach out for more information and greater competence.

Basically the concept of management of learning is simple. As with the 3Cs guidelines, a number of questions go to the heart of the challenge:

- What does the learner need to learn?
- Which goals and objectives will lead to the most improvement in competence?
- What learning experiences will be most effective in reaching the goals and objectives?
- How can progress best be measured and ensured?

Using a Knowledge/Skills Profile: What Does the Learner Need to Learn?

To understand what learners need to learn, a manager of learning must first have a clear view of what they need to know. A knowledge/skill profile depicting all the topics and skills that are needed for fully competent performance provides background for such a view. A manager of learning can then use the profile to identify competence strengths and deficiencies in an effort to gain the benefit of the strengths and reduce or eliminate the weaknesses.

A knowledge/skill profile is a list of topics that defines the knowledge and skills—the competencies—required by a position or by a major segment/function of a position. Each line on such a profile represents either (1) a limited amount of knowledge that can be learned from a short presentation or discussion or by studying some written material or (2) a skill that can be enhanced with a limited amount of practice. Most lines on a profile are likely to represent both—that is, some knowledge and a skill.

Beyond providing step-by-step guidance in analyzing strengths and weaknesses, a knowledge/skill profile can serve as the foundation for setting goals, objectives, and priorities and also for recommending learning assignments. Exhibit 11.5 lists the topics and skills that can be included in a knowledge/skill profile for a fire fighter. Each of the items in Exhibit 11.5 consists of many subsegments. For the purposes of the profile, however, this broader listing is adequate. After specific learning experiences are determined and scheduled, a manager of learning may consider the subsegments.

The officer who is an effective manager of learning will develop a knowledge/skill profile in many ways, including the three alternatives presented here:

1. The profile may be based on a detailed job description, even though it usually does not match the job description exactly. For example, two lines on an officer's job description might read as follows:

Exhibit 11.5 Sample Knowledge/Skill Profile for Fire Fighters

Knowledge	Skills
Organization of fire department	Hose evolutions
Scope of fire department operation	Ladder evolutions
Standard operation procedures	Breathing apparatus use
Fire department rules and regulations	Forcible entry
Safety policies	Ventilation operations
Fire behavior—chemistry of fire, types of	Hydrant operation and connection
fire, etc.	Salvage operations
Basic physiology of body systems	Rope use
Fire streams and use of nozzles and	Basic apparatus maintenance operations
couplings	Cleaning, maintaining, and inspecting
Use and types of equipment, such as	equipment, such as breathing apparatus,
breathing apparatus	ropes, salvage equipment, and ladders
Life-threatening injuries	Care of hoses and nozzles
Ventilation methods	EMT skills
Salvage process	Recognizing, identifying, and working with
Inspection procedures and standards	hazardous materials
Reporting	Use of chemical protective equipment
Safety	
Basic chemistry	
Hazardous materials	

- Supervise the activities of the company's fire fighters during overhaul operations.
- Supervise the activities of the company's fire fighters during salvage operations.

These two tasks cover many knowledge and skill items. An officer must know the technical aspects of each operation, communicate clearly, and apply the knowledge and skills involved in supervision. Another job description task for an officer might concern a single topic, such as the following:

- Must be thoroughly familiar with the hydraulic principles related to fire streams.

2. The profile can be derived from a list of all goals and objectives that apply to a particular position at one time. The goals and objectives may be converted to knowledge and skills in the same way that a job description is converted to a profile. Some goals and objectives require more than one skill, but others overlap closely with a single skill or specific knowledge.

3. A profile can be developed cooperatively with a fire fighter or training officer through the direct listing of knowledge and skill requirements. A jointly developed profile will probably lead to learning goals and objectives

that are challenging and to which a fire fighter or officer will devote significant effort.

Analyzing Team Learning Needs

In the fire service, an analysis of learning needs cannot concentrate exclusively on the knowledge and skills that individuals must acquire. Team needs go beyond the individual needs of the team members. Even when every person on a team has mastered all of the knowledge for performing specific tasks, the team must learn how to coordinate its efforts. Joint team practice can gradually develop the necessary coordination.

First, team skill needs must be identified. For example, each fire fighter in a specific company must have clear knowledge of the company's standard attack evolution guidelines, which are represented as follows:

1. Officer and fire fighter 1 advance one preconnected 1¾-inch line to fire for immediate fire attack.
2. Operator charges 1¾-inch line from engine tank.
3. Fire fighter 2 pulls supply hose from body, connects to hydrant, and then prepares to advance 2½-inch line.
4. Fire fighter 3 ventilates and assists fire fighter 2.
5. Operator switches from tank to hydrant supply as soon as connection is made.
6. Operator connects 2½-inch hose to pump discharge gate and charges it.

Although each fire fighter might practice these steps alone, without joint drills to develop team skills, fire fighters 2 and 3 will not perform their tasks as fast as they would if they practiced together. Joint drills can also be used to prepare for contingencies, such as snags in switching from tank to hydrant.

Setting Learning Goals and Objectives

Little needs to be said about learning goals and objectives. They are no different from other goals and objectives. All the relevant points apply from Additional Insight 3.1, Goal Setting and Implementation.

To be most useful, learning goals and objectives should be developed jointly by the person charged with assisting the learner (the company or training officer) and the learner. Learners know better what they know and can do, but often they do not know what they have to learn. The officer knows more about the competencies that have to be acquired.

After the knowledge and skills needs of a person or team have been identified, goals and objectives can be set and developmental activities can be planned to stimulate the greatest desire for achieving the goals or objectives. Goals and objectives define new behavior that is to result from new learning. When properly set, each goal represents a contract to oneself (or to others) to achieve such new behavior. For this reason, when educators discuss the goals and objectives learners have agreed to achieve, they often speak of a learning contract.

As discussed in previous chapters, goals and objectives define what is expected of a fire fighter and of the officer and what support the latter is expected to provide. Similarly, a learning contract defines the goals, objectives, and respective responsibilities of learner and instructor and thereby places these responsibilities into clear perspective.

Delivering Learning Programs

Delivery systems that allow managers of learning to adapt the learning process to individual needs achieve learning goals and objectives effectively. A wide range of systems for delivering information and competency training systems, to be used singly or in combination, are available to training designers and instructors. With modern technology, more choices exist than ever before, and many are available with and without multimedia support. Following are some delivery systems:

- Traditional classes consisting of lectures, case studies, simulations, and role-playing
- Field assignments between classes
- Traditional self-study with appropriate materials
- Learning-topic discussion and hypothetical or real decision-making sessions at meetings, with reflection or other evaluation of learning
- Distance instruction (learning) on electronic networks or with individual PC programs (including lectures, case studies, and simulations)
- Supervised individual work projects
- "Action learning" techniques in which learning is drawn from work projects
- Developmental work assignments
- Supervised cooperative team work projects
- Coaching (traditional on-the-job training)

An extensive body of literature exists on all these methods. Some will be discussed briefly later in this chapter, but space limitations do not permit a detailed discussion of all of them.

Measuring Learning

Many different techniques are used to measure learning, including tests, self-evaluation by the learner, observation during simulated and on-the-job application, and task performance. In the fire service, careful observation during drills is probably the best way to evaluate competence. Unfortunately, drills usually show only one condition and give little insight into a fire fighter's ability to apply knowledge to other situations. An officer who wants to expand on a drill to gain insight into knowledge and skills in related situations can discuss other situations and address specific questions to individual learners.

Another somewhat more precise, but also cumbersome way to obtain information about a learner's knowledge and thereby his or her progress is the in-depth probing technique discussed in Additional Insight 9.2, Interviewing. Gaining insight into a learner's progress is, of course, critical to ensuring that desired progress has been achieved. When instructors can identity knowledge and skill inadequacies, they can eliminate them.

THE LEARNING EXPERIENCE

To be most effective, learning experiences need to involve four phases: acquisition, demonstration, personal application, and feedback. The first three phases are shown in Figure 11.4.

1. *Acquisition.* The learner receives new knowledge through the written word, from live presentations, and in group or one-on-one discussions.

2. *Demonstration.* The instructor makes new material more meaningful or more explicit by demonstrating its practical applications through dramatic presentations, films, slides, audio- or videotapes, or actual demonstrations with props or actual equipment.

3. *Personal application.* Learners come face to face with their individual ability to apply the material to various situations. Learning experiences that are designed properly will be similar to those the learners deal with in their jobs.

Figure 11.4 The first three phases of a complete learning experience are acquisition, demonstration, and personal application.

Managers of learning consider the third phase to be the most important one. The personal application phase is the foundation for the fourth phase of a complete learning experience: feedback and correction. Personal application gives both the learner and the instructor an opportunity to evaluate the success of the learning experience. In so doing, two identical feedback loops emerge (only one is shown in Figure 11.5):

- A loop for the instructor to observe the progress of the learner or the learning group in meaningful terms
- A loop for the learner to diagnose areas of particular knowledge deficiencies

The two feedback loops become the basis for a continuing process that managers of learning use to achieve any learning goal or objective that is not blocked by attitudinal or emotional obstacles.

As Figure 11.5 shows, the process of achieving a learning goal or objective is a spiral-type repetition of the three phases of learning (acquisition,

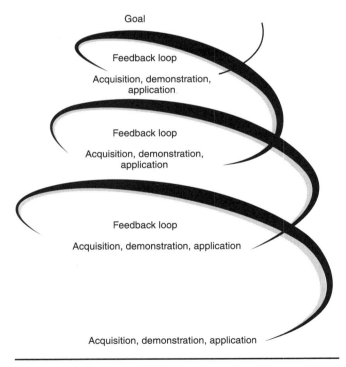

Figure 11.5 This model represents all four phases of a complete learning experience.

demonstration, and personal application) and the feedback loops; less and less new knowledge has to be acquired in each turn. This model becomes especially apparent to instructors when they experiment with various types of simulated experiences. Serious-minded learners inevitably ask more precise and penetrating questions during and after the personal application than they are able to ask beforehand.

For new knowledge to be meaningful, instructors must present it in terms that are familiar to the learner. We cannot overemphasize this point because words and images often have different meanings to different people. Therefore, instructors must apply new knowledge in a group or individual training session in situations that are similar to those that learners normally experience. Also, learning must proceed at a learner's own pace, must be adapted to personal ability, and should be related to previous experience.

For optimum learning an instructional strategy should consider two interrelated components: topic sequence (content) and processes used for each segment. Managers of learning should present segments of the subject in a logical progression. They should introduce simple concepts or skills first, followed by more complex concepts and skills. In the next sections we will discuss specific learning techniques for each of the three phases of the learning experience shown in Figure 11.4: acquisition, demonstration, and personal application of new knowledge or skills.

Instructional Techniques for Acquisition of New Knowledge or Skills

The learning experience for acquiring new knowledge or skills usually relies on three basic instructional techniques: lectures and explanations, reading (self-study), and conferences.

Lectures and Explanations.   Lectures and explanations to groups and to individuals are the most widely used instructional techniques for transmitting information. Often they are the least expensive and the most readily available instructional techniques. They are flexible, they can be changed on the spot by an instructor who perceives a previously hidden learning need, and they can be used to teach one person or many. Lectures can be given by one lecturer or many lecturers.

Like all instructional techniques, however, lectures and explanations have disadvantages as well as advantages. The primary disadvantage of lectures is that they are essentially one-way communication with little or no opportunity for practice, reinforcement, or knowledge of results. An instructor can reduce the disadvantages of one-sided explanations, however, by handling questions effectively. Instructors who encourage questions and answer them profession-ally can often clarify subjects that were confusing to learners.

Reading/Self-Study.   As an alternative to lectures and explanations, the learning experience can rely on independent study through reading and programmed instructional materials. The learner who is reading has the op-portunity to stop when necessary, to make notes, to clarify a point by refer-ring back to earlier material, and to consult other sources for more detailed information. The learner may find it more difficult to obtain answers to questions while reading than during lectures when a question can be ex-plored with the instructor.

Programmed instruction is a special form of guided reading used in an in-dependent learning situation. Programmed instruction contains frequent questions that the learner must answer before proceeding. The learner is told whether a response is correct by comparing it with the answer in the program. Following are several advantages to the use of programmed learn-ing materials:

- Knowledge is acquired at an individual's own pace.
- The learner follows a logical progression of thought.
- The learner must respond to questions after each presentation of mater-ial, which encourages learner acceptance and retention of the material.
- A wide variety of supplementary materials and audiovisuals can be pre-sented as supplements to the programmed instruction.
- Programmed instruction is self-administered (participation occurs whenever the learner is ready to learn).

Despite the benefits, however, programmed materials often are con-sidered to be tedious. Programmed courses of instruction are currently in limited use. However, use of computers for a more sophisticated form of programmed learning does have some advantages over paper-and-pencil learning material. Computer assisted instruction (CAI), in its crudest form, was a form of programmed instruction that made it easier for learn-ers who were familiar with a topic to branch out to more sophisticated topics. More modern versions of CAI use simulations, speaking heads, or video demonstrations and may even link several learners in a group learning experience.

Conferences.   A conference is a planned learning experience in which the instructor is not a lecturer, but a facilitator. The instructor encourages group thinking and, when necessary, provides background information that enables all learners to take part in the discussion. At the end of each discussion topic, the instructor should summarize what the participants should have learned from the experience.

Conferences or group discussions can offer immediate reinforcement of newly acquired knowledge to small groups. For this reason, group discussions often follow a lecture. Unlike a lecture, the discussion provides for two-way communication, which gives the learner an opportunity to compare personal reactions to the subject matter with the reactions of other learners and with the views of the instructor. Discussions may be combined with brief lectures to provide a mixture of acquisition, demonstration, and even some personal application.

Conferences tend to encourage learners or trainees to give thought to the subject matter, and they show each conferee how others think about a subject or problem. The facilitator adapts the presentation of the subject matter to the needs of the group. By exposing learners to group opinions, conferences help to change attitudes that are detrimental to learning. The development of satellite technology has spurred the use of teleconferences, through which people, at a number of sites in widely scattered locations, can participate in a common learning effort.

One major disadvantage of the conference technique is that it requires a highly skilled instructor—one who can carefully plan for a variety of eventualities that might occur during the conference. Another disadvantage is that a conference has limited usefulness: It requires that all participants have adequate prior knowledge of the subject for discussion. Here too, technology has broadened the application of conferences. Computer networks, the Internet and videoconferences are being used to assist learning programs. In fact, more and more universities offer courses in cyberspace, some without any live contact with a faculty member.

Instructional Techniques for Demonstrating New Knowledge or Skills

Demonstrations show learners how to apply the information, concepts, or materials presented during the acquisition phase of the learning experience. Illustrating how to use a piece of equipment and showing how to don breathing apparatus are demonstrations, as is a role-playing situation in which an instructor takes the part of a building owner while a student explains a code violation.

Audiovisual media are an excellent means for creating demonstrations that enhance the learning experience in the classroom or lecture setting. Chalkboards, transparencies, storyboards, display posters and materials, slides, computer projections, and videotapes are tools that help a presenter demonstrate a message or lesson. Audiovisuals help the instructor maintain the learners' attention and help the learners better retain what is said and understand the message more clearly.

Although demonstrations of all types are usually considered to be an important part of the learning process, they are often overlooked. Such oversight can result in a gap in the student's learning experience—a gap that can take considerably more effort to fill in at a later date than would have been expended immediately after the acquisition phase.

Instructional Techniques for Personal Application of New Knowledge or Skills

The instructional techniques used in a classroom-type setting (as distinct from field practice) during the personal application phase of the learning experience are based, in one way or another, on case studies. All case studies are basically descriptions of real or imaginary situations that vary from single paragraphs or brief verbal statements to greatly detailed books.

Case Studies or Case Method.   The applications of the case method to learning situations vary as widely as their types. Sometimes they serve as foundations for simulations and role-playing, but most often they are used more directly.

   • *Simple cases.* Learners are expected to review a case (the description of a situation), decide what issues it raises, and then draw conclusions about them. For instance, fire fighters may be given a prefire plan of a particular building with instructions to review and comment on it. Or questions written by the instructors may be presented for learners to answer.

   • *Staged cases.* Learners are provided with only part of the information necessary to thoroughly analyze a case. They must first decide what additional information is needed so that they can work on the case. After they have gathered the additional information, the learners can work on the case. For instance, in a staged fire situation, the instructor might delete such information as distances from the hydrants, wind direction, and other conditions, even though they are important to the case. The students must then figure out the missing elements to reach an effective conclusion.

Cases can be used with one learner, with small groups, or with large classes. Cases help stimulate creative thinking because learners are required to determine what is essential about the situation and then decide what actions, if any, they should consider. Cases are easy to use because much of the material available to instructors (such as records of previous fires, prefire plans, histories of inspections, articles from fire magazines, and newspaper stories) can form the basis for case studies. Also, instructors may use cases spontaneously when it becomes clear that learners need to explore specific issues, especially after a lecture in which many questions were asked or in situations where a change of pace is needed.

Many of the study questions and activities in these chapters and, to some extent, the scenarios are simple and staged cases.

Simulations.   When learners are asked to imagine that they are one of the people described in a specific scenario, they can participate in a simulation and solve specific problems. Simulations are an ideal means for learning in the fire service. They can dramatically enhance interest in training sessions and contribute greatly to awareness of hazards in the district. In addition, simulations can help to make valuable use of existing prefire plans and, coincidentally, serve as a basis for continual improvement in the prefire plans and in inspections.

A company officer or instructor who builds most instruction around exploring strategies and tactics at local properties can expand them into simulations. These simulations can concentrate on rescue, fire streams, forcible entry, ventilation, or ladder evolutions as well as on specific fire-fighting tactics and strategies based on fire location, weather conditions, or time of day. The inevitable outcome of simulations is greater interest in learning, greater knowledge of local conditions, enhanced interest in inspections, and improvement of prefire plans.

Simulations can be supported with various media or can be worked out on paper. Some can be programmed for use with computers, some can be expanded into field work, and still others can become actual fire drills at real locations (with the owner's or official's permission).

Simulations, like case studies, can be elaborate descriptions of a situation, with detailed data, pictures, charts, graphs, or floor plans, or they can be simple verbal descriptions created spontaneously by the instructor. During simulations, learners can either assume that they are all the same person or take on different roles in the situation. Fire drills on live fires are an extreme example of simulation—the most realistic possible.

Role-Playing.   When the instructor sharpens the "What should be done in this situation?" simulation to encompass "Exactly what words and actions should be used?" the simulation is likely to involve role-playing. In role-playing, learners act out specific roles that are assigned to them. Role-playing is applicable primarily to instructional situations in which the learning goals or objectives involve communicating between individuals or groups.

Even though the most common form of role-playing is a live demonstration staged with one person per role and all others as observers, that form is rarely the most effective. Much better for skill development is the type of role-playing that involves all the members of the group simultaneously. Following are some of the ways role-playing sessions can be made more effective:

• *Use of recorders.* Instructors may record role-playing on an audio- or videotape for later analysis. Then members of the group do not feel that they are criticizing actual people, and they are more willing to discuss correct actions or errors they observe. Topics for role-playing tapes include consoling a victim's grieving relative, calming an irate owner, explaining a violation, critiquing a performance discussion between an officer and a fire fighter, and responding to a citizen's question.

• *Role-playing in small groups.* The presence of a supervisor or other observers is often intimidating for the participants. The intimacy of a small group of friends or peers does not create the same anxiety. As a result, role-playing is a much more relaxed activity in which each participant behaves naturally and realistically.

• *Number of participants per role.* Another important feature of successful role-playing is the number of participants who play a given role. In most cases, only one person assumes a given role. Another effective way to analyze specific phrasing or approaches is to have two people consult with each other as they play each role. Role-playing can be used effectively in the fire service to sharpen skills in interacting with the public during inspections and with the people affected by an emergency.

In addition to the major variants of role-playing, other techniques help make it an effective teaching device. For example, an instructor may interrupt a situation after a few moments and tell participants about a change in the scenario—for example, an emergency or information received by telephone. Still other variations can result by having the person playing the learner's role uncover a specific hidden point.

Coaching and Mentoring: A Technique for Continuing Learning

Group or classroom-type training and drills provide most of the knowledge and skills needed by fire fighters or officers for their jobs. However, because of individual differences, officers must provide individual coaching to eliminate shortcomings and prepare for advancement. Such coaching usually involves suggesting specific self-study goals or objectives or demonstrating and providing supervised opportunities to practice the application of particular knowledge and skills. Self-study recommendations (such as guided reading programs) and individualized help (such as advice) for achieving learning goals and objectives are excellent methods for continuing development. A coach or mentor should gear such methods to the particular needs, aspirations, and the knowledge and skill level of the individual.

Many officers do not appreciate the importance of coaching to meet their training responsibilities. Yet, without effective coaching, fire fighters and officers will not necessarily move aggressively toward enhancing their competencies. At lower levels in a department, coaching is usually similar to tutoring, one-on-one training used primarily for repetitious tasks. Coaching is a form of on-the-job training in which the management of learning principles are applied very specifically to one person.

Officers should keep one important consideration in mind when coaching. Officers represent authority and hence are a potential source of negative consequences for a staff member who has knowledge or skill deficiencies. To be most effective in coaching, an officer must avoid words or actions that imply risk or threat and make it clear to the learner that the officer is acting in a different role, a helping role.

The role of the coach is to help the learner acquire selected knowledge and skills through on-the-job training. The coach is a guide who suggests direction and gives advice for reaching learning goals or objectives that are defined primarily by the learner. In providing this guidance, the coach performs all the management of learning functions. Exhibit 11.6 lists the steps in the learning process as practiced by the U.S. Army. The steps are grouped in a way that parallels the four phases of the learning experience illustrated in Figures 11.4 and 11.5.

GUIDELINES FOR A SATISFYING AND EFFECTIVE INSTRUCTIONAL PROGRAM

The guidelines in this section focus on ways to enhance the learning process. Following these guidelines requires an instructor to be willing to experiment with different ways of managing learning experiences and to be thorough in keeping the learning process in mind.

Exhibit 11. 6 The Four Steps of On-the-Job Training

	Step	Purpose	How accomplished
Acquisition	1. *Prepare the learner.*	A. To relieve tension B. To establish training base C. To arouse interest D. To give confidence	A. Put learner at ease. B. Find out what learner already knows about task. C. Tell relation of task to mission. D. Tie task to learner's experience. E. Ensure that learner is in a comfortable position to see you perform the task clearly.
Demonstration	2. *Present the task.*	A. To make sure learner understands what to do and why B. To ensure retention C. To avoid giving more than learner can grasp	A. Tell, show, illustrate, question carefully and patiently and use task analysis. B. Stress key points. C. Instruct clearly and completely, one step at a time. D. Keep your words to a minimum; stress action words.
Personal application	3. *Try out the learner's performance.*	A. To be sure learner has right method B. To prevent wrong habit forming C. To be sure learner knows what to do and why D. To test learner's knowledge E. To avoid putting learner on the job prematurely	A. Have learner perform the task and do not require an explanation of what is being done the first time through. If learner makes a major error, assume the blame yourself and repeat as much of Step 2 as necessary. B. Once learner has performed the task correctly, have learner repeat it and explain the steps and key points as the task is done. C. Ask questions to ensure that key points are understood. D. Continue until you know that learner knows the material.
Feedback	4. *Follow up.*	A. To give learner confidence B. To be sure learner takes no chances and is not left alone C. To be sure learner remains alert D. To show your confidence in learner	A. Make learner responsible; praise as fitting. B. Encourage questions; tell learner where help is available. C. Check frequently at first. D. Gradually reduce amount of checking.

Source: United States Army, civilian personnel brochure.

1. Explanations and lectures should not be too long. Many educators think that uninterrupted lectures that last more than 20 minutes are likely to result in seriously reduced attention.

2. The instructor should carefully follow the sequence of acquisition, demonstration, and personal application (see Figure 11.5) and correction of deficiencies. One way to use this sequence follows:

- Give the explanation or lecture to explain concepts, supported with visuals if possible, that offers learners as much opportunity as possible to ask questions.
- Assign an appropriate activity to be done individually, in small groups, or both.
- Review the activity through reports and individual or group discussions.
- Give learners an opportunity to ask questions.

3. As much as possible, the atmosphere should be informal and relaxed so that learners can ask questions freely and explore points in terms that are familiar to them. Questions are very important, especially in a group that contains several highly motivated learners. The instructor may think that the entire group is exploring a subject in detail when, in reality, only a few are doing so, while others who may be seriously confused by the subject are reluctant to speak. To help create an informal atmosphere, explanations and lectures must almost continually allow for questions, and instructors must answer questions without being condescending or belittling the learners. At the same time, instructors must be careful not to allow questions to draw the lecture away from the logical sequence of thought necessary for an orderly presentation. Furthermore, the instructor must be honest and open with learners. An instructor who cannot answer a question should admit it and either offer to find an answer or assign one of the learners to research it.

4. Slower learners (or those whose attention has strayed) should have opportunities to catch up with the class, and fast learners (or those who are more familiar with the subject) should be challenged by the learning process. Individual and small-group activities can provide these opportunities because the slower learners or those inclined to attention lapses feel more free to ask questions in smaller environments. At the same time, learners who have a better understanding of the topic can verify their command of it by helping to explain it to others. Providing explanations further strengthens understanding and helps to achieve full mastery. Few topics are as well understood as those that are explained to people who are free to ask questions.

5. Discipline must be effective but polite. Discipline prevents distractions or reduces them to a minimum so learners can pay attention to the subject. Instructors who have difficulty maintaining order can find helpful techniques in many books on conference leadership or classroom discipline.

6. Officers should feel comfortable allowing learners to play an active role in the group. Much can be gained from allowing fire fighters to take turns in leading the discussion during training sessions. Officers should make assignments for leading the discussion at an earlier session so that fire fighters can prepare. Rotating the leadership of discussion with team or class activities can be especially valuable if instructors use prefire plans or records of properties in the district as foundations for the training sessions.

7. Officers should plan training topics for an entire year. They should then evaluate the list of topics so that each topic will be presented effectively to meet the needs of the particular learners. Officers may supplement the list with carefully written plans that specify the learning experiences that will be used for each topic. The use of written plans, such as the one shown in Exhibit 11.7, is strongly recommended as a way to help ensure that these guidelines are met:

- Topic segments are arranged in a logical sequence.
- Appropriate techniques are used for each topic segment.
- Acquisition, demonstration, and personal application phases are considered for each topic segment.
- Techniques are presented in sufficient variety to create a stimulating environment for the learners.

Exhibit 11.7 Sample Learning Plan for Communications Training

Topic segment	Acquisition	Demonstration	Personal application	Correction of deficiencies
Receiving and recording fire calls	Lecture (communications officer)	Simulated call in dispatcher room and discussion	Receiving and recording four simulated calls	Comprehensive drill and supervised work assignment as dispatcher
Establishing location and nature of emergency	Discussion	Demonstration role play by instructor and experienced dispatcher	Role-playing by students (as dispatcher)	Supervised work assignment as dispatcher
Dispatching equipment and apparatus	Assigned reading	Discussion and equipment demonstrations	Individual using equipment for specific tasks	Written and verbal tests

8. Instructors and officers should constantly strive to become acquainted with a wide range of instructional techniques and should experiment with the use of these techniques as tools for helping students learn. The more varied the instructional techniques an instructor uses effectively, the more interesting and stimulating the learning experience will be for learners.

9. Company officer should do their best to combine training with day-to-day activities and thereby enhance the appeal of the activity and of the training at the same time. Some examples follow:

- Training can be done during informal get-togethers during the day, if the officer can guide the discussion to a topic with a smooth transition. Examples of this approach are using the report of a fire to introduce a discussion of various specific aspects of fire suppression activities, or using an article in the newspaper on fire prevention to turn a discussion toward inspection procedures, and using a report on activities at a local school to introduce aspects of fire and life safety education.
- Training can be done concurrently with inspections, when both aspects of codes and identification of hazards can be used to sharpen the skills of one or two fire fighters at a time.
- Prefire planning is one of the most fertile grounds for training because it touches on so many aspects of fire prevention and suppression.
- Training of new recruits can bring many opportunities for including experienced fire fighters who can be asked to explain, demonstrate, observe, and offer suggestions to the new fire fighters.

The design and scheduling of learning experiences can contribute significantly to a learner's motivation. In a similar way, maintaining and strengthening a satisfying climate for learning are important in stimulating positive attitudes toward competence development.

CHAPTER 11
STUDY QUESTIONS AND ACTIVITIES

If you are working alone, prepare your own written responses to these questions. If you are studying with a team or working in class, discuss the questions with the group and write a consensus answer.

1. What is different about the way a manager of learning approaches the instructional goals and objectives from the approach of a teacher with

traditional views? Also discuss the distinction between the process and content of learning.

2. Think about the various techniques your officer (or you) apply to motivate fire fighters to learn and to maintain or sharpen knowledge and skill levels. Then discuss the strengths of the techniques and ways to increase the motivational impact.

3. What could you do to convert negative attitudes toward routine drills to positive cooperation?

4. What can be done to maintain pump operator competence at peak level? What about ladder evolutions? Describe at least three common objections fire fighters have to pumper operation drills and how you would overcome them.

5. Prepare a knowledge/skill profile for your job and for that of a fire fighter with EMT responsibilities, and then review either your or an individual fire fighter's learning priorities based on the profile.

6. For one of the priorities identified in question 5, prepare a complete learning experience, including acquisition, demonstration, personal application, and feedback.

CHAPTER 11
REFERENCES

Bass, Bernard M., and Vaughan, James A. 1966. *Training in Industry: The Management of Learning.* Belmont, CA: Wadsworth.

Bloom, Benjamin S., ed. 1956. *Taxonomy of Educational Objectives, The Classification of Educational Goals, Handbook I: Cognitive Domain.* New York: David McKay.

Bruner, J. S. 1966. *Toward a Theory of Instruction.* Cambridge, MA: Harvard University Press.

Gagne, Robert M. 1968. "Context, Isolation and Interference Effects on the Retention of Fact." *Journal of Educational Psychology* 60:408–414.

Gagne, Robert M. 1977. *The Conditions of Learning,* 3rd ed. New York: Holt, Rinehart and Winston.

International Fire Service Training Association (IFSTA). 1998. *Essentials of Fire Fighting,* 4th ed. Oklahoma State University: International Fire Service Training Association (IFSTA).

Knowles, Malcolm. 1968. "Andragogy, Not Pedagogy." *Adult Leadership.* April, in Kirkpatrick, Donald L. 1971. *Supervisory Training and Development.* Reading, MA: Addison-Wesley, p. 45.

Knowles, Malcolm. 1990. *The Adult Learner, A Neglected Species,* 4th ed. Houston, TX: Gulf.

NFPA 1001, *Standard for Fire Fighter Professional Qualifications,* National Fire Protection Association, Quincy, MA.

NFPA 1002, *Standard for Fire Apparatus Driver/Operator Professional Qualifications,* National Fire Protection Association, Quincy, MA.

NFPA 1003, *Standard for Airport Fire Fighter Professional Qualifications,* National Fire Protection Association, Quincy, MA.

NFPA 1021, *Standard for Fire Officer Professional Qualifications,* National Fire Protection Association, Quincy, MA.

NFPA 1031, *Standard for Professional Qualifications for Fire Inspector and Plan Examiner,* National Fire Protection Association, Quincy, MA.

NFPA 1033, *Standard for Professional Qualifications for Fire Investigator,* National Fire Protection Association, Quincy, MA.

NFPA 1035, *Standard for Professional Qualifications for Public Fire and Life Safety Educator,* National Fire Protection Association, Quincy, MA.

NFPA 1041, *Standard for Fire Service Instructor Professional Qualifications,* National Fire Protection Association, Quincy, MA.

NFPA 1051, *Standard for Wildland Fire Fighter Professional Qualifications,* National Fire Protection Association, Quincy, MA.

NFPA 1061, *Standard for Professional Qualifications for Public Safety Telecommunicator,* National Fire Protection Association, Quincy, MA.

NFPA 1401, *Recommended Practice for Fire Service Training Reports and Records,* National Fire Protection Association, Quincy, MA.

NFPA 1402, *Guide to Building Fire Service Training Centers,* National Fire Protection Association, Quincy, MA.

NFPA 1403, *Standard on Live Fire Training Evolutions,* National Fire Protection Association, Quincy, MA.

NFPA 1404, *Standard for a Fire Department Self-Contained Breathing Apparatus Program,* National Fire Protection Association, Quincy, MA.

NFPA 1410, *Standard on Training for Initial Fire Attack,* National Fire Protection Association, Quincy, MA.

NFPA 1451, *Standard for a Fire Service Vehicle Operators Training Program,* National Fire Protection Association, Quincy, MA.

NFPA 1500, *Standard on Fire Department Occupational Safety and Health Program,* National Fire Protection Association, Quincy, MA.

OSHA 1910.134 (g) (4), Occupational Safety and Health Administration, Washington, DC.

Rausch, Erwin. 1985. *Balancing Needs of People and Organizations—The Linking Elements Concept.* Washington, DC: Bureau of National Affairs, 1978. Reprint, Cranford, NJ: Didactic Systems.

Senge, Peter M. 1990. *The Fifth Discipline—The Art and Practice of the Learning Organization.* New York: Currency and Doubleday.

Skinner, B. F. 1968. *The Technology of Teaching.* New York: Appleton-Century-Crofts.

U.S. Army. 1969. *The Four Steps of On-the-Job Training.* Washington, DC: U.S. Government Printing Office.

Wohlking, Wallace. Ithaca, NY: Cornell University, New York State School of Industrial and Labor Relations.

CHAPTER 11

ADDITIONAL READINGS

Lefrancois, Guy R. 1995. *Theories of Human Learning,* 3rd ed. Pacific Grove, CA: Brooks/Cole Publishing.

Mager, Robert F. 1968. *Developing Attitude Toward Learning.* Palo Alto, CA: Fearon.

McLagan, Patricia A. 1978. *Helping Others Learn.* Reading, MA: Addison-Wesley.

ADDITIONAL INSIGHT 11.1

•••|••

LEARNING CONCEPTS IN SUPPORT OF MANAGEMENT OF LEARNING AND COACHING

Studies themselves do give forth directions too much at large, except they be bounded in by experience.

—*Francis Bacon*, Of Studies

Why include a section on these topics in a book on management in the fire service? The answer is both obvious and rarely considered: officers want to ensure that their organizations will always have the necessary competencies. To assess an organization's competence and to develop the plans for overcoming any shortcomings, officers should have a basic awareness of relevant theories and an understanding of management of learning. This section presents such relevant theories and concepts to enrich understanding of the foundations for managment of learning and for coaching.

RELEVANT THEORIES OF LEARNING

Theories based on research on human learning have understandably concentrated on determining the validity of various hypotheses that attempted to identify what constitutes learning. To a large extent, the assumption has been that, once knowledge is available, instructional strategies can be developed to bring about the most effective learning.

Most of the early work on learning theories was conducted with animals and children. Implications of the theories for the design of learning programs, especially for adults, were addressed only tangentially. Research concentrated mostly on relatively simple concepts appropriate for the grade-school level. Research into the learning of highly complex skills, such as those needed for professionals, was left to the professional societies. Members of the Management Education and Development Division of the Academy of Management have done the most extensive work in this area, much of it concentrating on environmental factors and on transfer of learning.

The critical issue in management of learning, and one that does not seem to receive adequate priority in staff development programs, is arranging information so that it will have most meaning to learners. The clear arrangement of information is especially important in the critical nontechnical competencies, including management and leadership development, interpersonal skills enhancement, conflict resolution, and client and customer relations programs. Training in technical competence, in contrast, usually is based on reasonably sound arrangement of information.

Several theories support the position that clear topic arrangement is of great importance to learning and that guidelines can provide structure for the organization of information. At an early date, Bloom expressed this idea most clearly (1956, p. 35):

> Our general understanding of learning theory would seem to indicate that knowledge which is organized and related is better learned and retained than knowledge which is specific and isolated. By this we mean that learning a large number of isolated specifics is quite difficult simply because of the multiplicity of items to be remembered. Further, they cannot be retained very well if they are kept in such isolation.

The stimulus–response theory suggests that a solid foundation can encourage additional learning and discovery of new concepts (Skinner 1968). For example, the need to prepare a plan or to make an important decision can act as a stimulus to the recall of guidelines (the response), especially the basic, easy-to-remember ones. Gradually, the guidelines in turn act as stimuli for the more detailed, more complex responses (the recall of relevant, more detailed guidelines and the concepts that support them).

Among the cognitive theorists, Bruner (1966) recommends a spiral curriculum, which presents the simplest concepts at earlier stages and revisits them at progressively more advanced conceptual levels. This concept is used for expanding knowledge of the issues behind each of the basic 3Cs guidelines.

Early research by the integrative theorist Gagne (1968) confirmed the benefits of presenting material with simple concepts first. Later he reported (1977) on the need for prerequisites organized in learning hierarchies.

Bass and Vaughan (1966, p. 31) expressed the idea as follows:

> To sum up, the ease of learning will depend upon how well the material lends itself to perceptual organization so that it can be connected with previously learned responses.

CONCLUSIONS FROM THEORIES

With respect to adult competence development, Malcolm Knowles (1968) has suggested five principles in the form of the characteristics of adult learners. All, in one way or another, point to the need to create a motivational climate, which becomes an umbrella principle for the other five.

Adult learning is a matter of motivating the learners to accept responsibility for the outcome of the learning—the broader knowledge or the enhanced skill. Unlike with students in grade school, high school, and college, where the educator often has considerable authority, directing adult learners to learn is nearly impossible. Motivation is not a button that can be pushed or a handle that can be turned. The manager

ADDITIONAL INSIGHT 11.1

of learning must create a climate in which the learner can find the desire to achieve greater knowledge or higher levels of skill. Creating this climate requires that trainers understand the nature of adult learners—the characteristics that distinguish them from other learners. Knowledge of these characteristics can help managers of learning select and design training and learning experiences that are meaningful to learners and thereby stimulate them to seek more learning.

Knowles's five major characteristics reflect the nature of the adult learner:

1. They are motivated by previous experience and knowledge. Adults are motivated to learn by their previously experienced needs and by interests that learning will satisfy. They want the manager of learning to be aware of previous knowledge they possess and to adjust the program accordingly.

2. Their orientation is life centered; learning must be relevant. The focus for organizing adult learning should be on life situations, not on the subject. When working with adults, instructors should make sure that learners see a principle or theory as relevant to their needs and work. As much as possible, officers should provide opportunities for learners to apply new knowledge and new skills to the job as quickly as possible.

3. They want learning to be self-directed. Adults want to know why they should learn a specific subject, and they want to be consulted about how much they should learn about it. Therefore, competence development should be a process of mutual decision making.

4. They are likely to be more skeptical and apprehensive than younger students. Adult learners need to know that the new learning is correct in a conceptual sense and that the new information is adequate to cope with the situations to which it applies. Adult learners are also apprehensive about learning and about their ability to learn because they are no longer in the routine of learning and because they often have more at stake if they fail to learn adequately and quickly.

5. Differences between individuals increase with age. Competence development as well as education must make optimal provision for differences in thinking patterns, speed, and ability to grasp complex concepts and the amount of material a learner can handle at one time. When a learner feels that the material is too simple, he or she may lose interest. If not enough material is provided, the learner may procrastinate or not consider it important enough to devote the necessary effort. If the material is too much or too difficult, then the learner is likely to give up trying to master it.

Other principles apply to adult learners as well as to other learners:

• Receiving encouragement and support is important for overcoming apprehension and for stimulating motivation.

ADDITIONAL INSIGHT 11.1

• Reinforcement aids retention. Very little that is learned only once is retained and used; all important points have to be reinforced, sometimes repeatedly. The relationship between reinforcement and encouragement is close. Instructors reinforce competence sharpening and even acquisition of knowledge when they provide encouragement immediately after correct or appropriate application.

• Recognition, such as praise, approval, other encouragement, and attention, is usually effective in stimulating learners to continue. Large rewards are not more effective than small ones. Regular, periodic reinforcement rather than continuous reinforcement is considered to be the most efficient strategy. Instructors should reward only appropriate behavior and avoid confusing the learner by giving recognition for inappropriate action.

• Careful organization of material helps to bring understanding; a conceptual framework can be helpful. Organizing the material so that it flows logically from simple thoughts to more complex ones aids understanding and retention. The instructor should present simple or foundation concepts first and use reminder devices like a conceptual framework whenever possible. A conceptual framework is basically a picture or format of related things or ideas that helps a learner remember information, a concept, or a set of related concepts. A conceptual framework could be an acronym, a refresher booklet, a picture, pictures, diagrams, a simple table, or a list of steps. The 3Cs model is a conceptual framework.

• Prompt feedback is essential to learning and retention. Feedback is a two-way street. The manager of learning needs feedback from learners, to know how they are progressing, but learners need feedback even more. Feedback must refer to specific behaviors so the learners can make decisions about additional study or practice that may be needed.

IDENTIFYING AND OVERCOMING BLOCKAGES TO LEARNING

ADDITIONAL INSIGHT 11.1

The model of the continuing process for achieving learning objectives, shown in Figure 11.5, can be useful in helping learners overcome knowledge/skill deficiencies only if learning is not prevented by some type of blockage. Generally, fire fighters have a positive attitude toward learning because they instinctively realize that learning is essential to effective performance. However, the learning process can be obstructed if someone has physical or emotional obstacles or negative attitudes toward specific topics or instructions. These obstacles are referred to as blockages. Some learning blockages are unique to individual learners. For example, learners are sometimes psychologically opposed to training, or a subject can be so complex that comprehension is difficult. Figure 11.6 depicts most of the common blockages as well as some of the basic strategies for overcoming them.

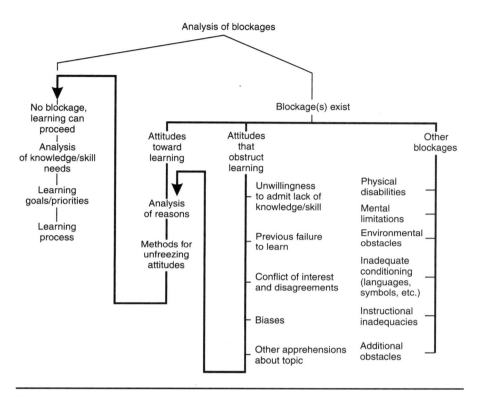

Figure 11.6 The flowchart shows the usual steps in acquiring new knowledge and skills and some of the blockages that can interfere with the learning process. (*Source:* Rausch 1978/1985)

Identifying Blockages

• *Unwillingness to admit lack of knowledge/skill.* People are sometimes reluctant to admit that they have a specific knowledge or skill deficiency, especially if they still do not understand an idea, principle, or procedure even after training. That reluctance is strongest if others seem to understand. The person with the deficiency may try to avoid contact with the topic.

• *Previous failure to learn.* Failure to learn in the past can make learners think that the subject is too difficult; thus, learners often fail to devote the effort necessary to master a particular subject. Also, if learners believe that some knowledge is simply common sense, they will sometimes ignore its value. If the subject is perceived to be too theoretical or irrelevant, learners might not pay adequate attention. Any of these beliefs can lead to failure to learn a topic.

ADDITIONAL INSIGHT 11.1

• *Conflict of interest and disagreements.* Learning can be blocked by attitudes. For example, a fire fighter who believes that a good fire fighter is a "smoke eater" thinks that a person who can last the longest in the dense smoke of a fire is, in some way, the best fire fighter. This fire fighter might view instruction in the use of protective breathing apparatus as negative or unnecessary because it conflicts with the fire fighter's beliefs about what a good fire fighter should do.

• *Biases.* A person might have biases against an instructor based on race, religion, area of origin, personal appearance, or other characteristics. A fire fighter who holds a bias against an instructor is not likely to learn much from that instructor.

• *Other apprehensions about a topic.* Other apprehensions can come from concerns generated when other learners describe a topic as difficult or when the topic requires contact with materials that are unpleasant to the person—for example, odors, slimy materials, wounds, and so on. All these can lead to insufficient learning.

The extreme right side of Figure 11.6 lists blockages to learning that result from limitations that are not based on attitudes or emotions.

• *Physical disabilities.* The instructor should be sensitive to any learners' physical disabilities and be aware that some learners might be embarrassed about them. For example, a fire fighter might find it increasingly difficult to climb a ladder, because of an old knee injury and might have successfully hidden this minor disability during the physical examination and the probationary period. This fire fighter might resist training sessions on using ladders. If an instructor is not aware of a fire fighter's disability, considerable effort can be wasted in misdirected attempts to achieve learning. A more common example is people with hearing and sight impairments, who often are reluctant to obtain and wear hearing aids or glasses. The obvious blockages to learning that result cannot be resolved unless an instructor is aware of them.

• *Mental limitations.* Mental limitations can include perceptual problems and minor emotional problems, such as fear of heights, unreasonable resentment of authority, or superstitious beliefs. As with physical disorders, learners might be reluctant to face their mental limitations objectively. Only when an instructor is aware of such problems can the learning program be adjusted to take them into account. Often a slower pace and some extra help (counseling or coaching) are all that is needed to help a learner achieve the same learning goals and objectives as other members of the group.

• *Environmental obstacles.* These obstacles include noise, movement, and other distractions that interrupt the thought process of instructors and learners alike.

• *Inadequate conditioning.* Some actions must be practiced repeatedly so that learners can perform them automatically when signaled. For example, fire fighters have a specified routine to get ready to answer an alarm. If they practice this routine often enough in response to the alarm, they will react automatically when a real alarm sounds. In this way, certain tasks will be performed almost instinctively in the shortest possible time. Successful conditioning requires that all fire fighters understand and use the same jargon, symbols, and signals.

• *Instructional inadequacies.* These obstacles include illogical presentation of topics and other instructor-related deficiencies, such as talking above the learning level of students, talking condescendingly to students, and ineffectively using the chalkboard and other visuals. All of these blockages can detract from a learning experience. Instructor-related deficiencies primarily apply to lectures and demonstrations. Similarly, participative types of learning experiences that provide personal application might be poorly designed so that they are not perceived as relevant or stimulating to the learning of a specific topic.

• *Additional obstacles.* Other blockages to learning involve time conflicts or illness, which can prevent full comprehension of the material, especially if a learner is absent when critical topics are covered.

Overcoming Blockages

In attempting to overcome learning blockages, a manager of learning must first determine the actual reasons behind the learning difficulty. If the reasons involve the perception or attitude of the learner, the instructor can use counseling to "unfreeze" the belief. The following specific devices can be used as an extension of the counseling process for unfreezing attitudes:

• *Peer pressure.* When learners become aware that their attitudes are different from those of their peers, or if they find that acting on their own views and attitudes actually hurts the team effort, then they are likely to question and modify their views. An instructor can therefore make individuals aware that their attitudes either are inappropriately different from those of their peers or might hurt team achievement.

• *Words of respected people.* When learners become aware that they are in conflict with respected and successful people in the field, they are likely to question their attitudes. Presenting irrefutable evidence of the attitude of successful professionals in the field (through guest appearances in classes or testimonials) can help a manager of learning cultivate better student attitudes toward a subject.

• *A direct order.* Giving a direct order is sometimes the only way to oppose learners' strongly held attitudes. Many behavioral scientists claim that training cannot be effective against strong attitudes without the organizational requirement that learners must follow. Training can then help people satisfy the requirement more easily, and that in turn frequently leads to revised attitudes toward the subject.

ADDITIONAL INSIGHT 11.1

The effects of any of these obstacles and attitudes can be destructive to the learning process. Instructors who are effective managers of learning must be alert to problems or inadequacies in the learning abilities of individuals and take prompt corrective steps as soon as they notice the problems or inadequacies.

Fire Department Services Beyond Fire Fighting

INTRODUCTION

Today's fire service faces an array of challenges not envisioned at the beginning of the 20th century. Whereas the prevention and suppression of fire and possibly some limited medical/paramedic services were once the only missions of the fire service, a host of new services have emerged during the past twenty years. These new roles have expanded to consume a large share of fire department time and resources. Fire officers need to be aware of those areas that are relevant to their departments now and in the foreseeable future:

- Emergency medical/paramedic services
- Confined space and specialized rescue operations, including vehicle and railroad crashes and structure collapses
- Hazardous materials
- Airport rescue/fire-fighting services
- Community disaster planning, preparedness, and response, including such diverse emergencies as floods, earthquakes, tidal waves, riots, tornadoes and hurricanes, terrorist attacks, and other natural or human-caused incidents
- Customer service concept, including the use of facilities and human resources to assist the community in ways such as providing shelters for abused children or neighborhood medical centers

In this chapter we will discuss these different activities, but it is beyond the scope of this book to offer specific guidelines for each function. Using the guidelines suggested in the preceding chapters as models should make it fairly easy for officers to develop guidelines for any functions that are relevant to the department. The 3Cs guidelines, of course, apply in each of the functions.

The best way for an officer to write the appropriate fire department function guidelines is to base them on the goals for the function, as was the case with all the fire service function guidelines in earlier chapters. Therefore, to prepare a meaningful guideline, the officer in charge of the function could follow these steps:

1. Formulate statements of goals with appropriate participation of those stakeholders that should be involved.
2. Use the goals to prepare a general (basic) guideline statement (and objectives, where appropriate).
3. Obtain agreement on the guideline.
4. Prepare specific supporting guidelines based on the goals and objectives for the function.

5. Review and finalize the specific guideline statements with the team of stakeholders that were involved in writing the goals and objectives.

EMERGENCY MEDICAL/PARAMEDIC SERVICES: PREHOSPITAL CARE FOR THE INJURED AND ACUTELY ILL

Overview

The purpose of emergency medical services (EMS) is to save lives and reduce human suffering. EMS provides the victims of life-threatening injuries and illnesses, or lesser trauma, with critical medical intervention.

EMS varies from state to state and from province to province. Depending on state and local regulations, there may be several levels of training, certification, and licensing for fire personnel who serve as emergency medical technicians. Generally, the first responder's role can be divided into three levels of service:

1. Basic medical care and rescue and extrication
2. Basic life support (BLS) first response
3. Advanced life support (ALS) first response

To provide the levels of services, fire fighters must be trained and certified by state and sometimes also by local authorities either as emergency medical technicians (EMT) or as paramedics. Certification as a paramedic requires a higher level of competence in both medical knowledge and medical techniques and skills. Every state that has an EMS program requires that technicians be recertified every few years.

Basic Medical Care and Rescue and Extrication. This level of EMS is usually provided by the first response apparatus. The role of fire fighters is generally limited to extricating victims from a vehicle, other accidental confinement (such as a building collapse or a fall into a confined space), or fireground search and rescue. Basic first aid, including cardiopulmonary resuscitation (CPR), is administered by fire fighters who have first response training and certification. Hospital transport is usually provided by an ambulance that may be supplied by a different fire service company, the police, or a paid or volunteer ambulance company.

Basic Life Support (BLS). Basic life support (BLS) is provided by EMT personnel who may arrive at the scene on a fire department vehicle, in the ambulance of another organization, or in a private car. BLS offers more than basic first aid; it uses EMS life support equipment such as oxygen, defibrillators, and other first aid materials. Some municipalities provide life support through the use of BLS engine or ladder companies, which are regular engine or ladder companies equipped with BLS equipment. Operating from neighborhood fire stations, these units can be on location quickly. They can then initiate and sustain EMS until a transport vehicle arrives. Some communities use dedicated fire service transport ambulance units. Except when a dedicated unit is available, transport to the hospital is provided by other sources, as in the case of the rescue and extrication level of service.

Advanced Life Support (ALS).   Advanced life support (ALS) services are delivered by personnel trained to the paramedic level, for which the educational and training requirements are more stringent. The level of service provided is technically complex and includes the administration of medications that are carried on the apparatus and administered on the authority of a physician who is contacted by telephone or radio. The paramedic "usually is the most thoroughly trained and highly qualified category of prehospital EMS personnel. Advanced life support (paramedic) services are obliged to operate within the policies and procedures established by their medical director, who may or may not be employed by the fire department" (Cote 1997, p. 10-166).

In a fire department ALS program, each aspect of the service is performed by department personnel. When they are not in their own ambulances, fire department ALS personnel may ride in the ambulance of another public- or private-sector operation, with the fire department providing the actual paramedic-level service.

History of EMS

It is commonly thought that EMS and paramedic services are a product of the late 20th century, but this belief is far from true. Ambulance service has existed in North America for a long time. Funeral homes, hospitals, and police and fire departments have all provided EMS services for many years. Not too long ago, someone from the local funeral home might drive up in a one-person ambulance (which looked much like a hearse) with no medical equipment to transport a seriously ill or injured patient to the hospital as fast as possible (Lidwig 1995, pp. 62–63). This service by funeral homes continued into the 1970s.

Local hospitals generally provided ambulance service in larger cities. In other communities, it was provided by police departments, and in still other places, ambulance service was the responsibility of the local fire department.

In the United States, some fire departments began providing first aid and medical rescue services to the public as early as the 1920s.

Somewhere along the way, volunteer groups began to offer first aid and even medical emergency services to their communities. Today volunteer ambulance units provide BLS services in many areas. Concurrently, hospitals have begun to offer ALS, BLS, and ambulance transportation services.

Gradually, fire departments assumed a larger role. The greatest increases of EMS in the fire service came during the 1960s. Federal grant money became available to the fire service to develop EMS programs. A number of fire departments took advantage of funding sources to either bring or expand EMS protection in their communities. Federal involvement in enhancing EMS operations in the fire service slowed during the 1980s as grant money became more difficult to obtain.

Although the fire service is a tradition-bound "industry," its role in EMS is expanding steadily and sometimes quite rapidly (Brahme 1992). The reason for this greater role is the decreasing need for fire suppression services. Fewer fire calls, however, do not lessen the importance of having a fire-fighting force that is prepared for the fires that do occur. With more time available, the fire service is in a unique position to offer a variety of other emergency services. Whereas EMS was originally considered to be a supplement to fighting fires, the emphasis has shifted to viewing EMS as a major component of fire service activities. Today the vast majority of fire departments provide EMS on a routine basis. In many departments, calls for EMS account for three-fourths or more of all alarms received (Cote 1997, p. 10-165).

The Future of EMS in the Fire Service

Conflicting forces have an impact on the future of fire service EMS:

- The continuing changes in health care that require frequent training to maintain competency and new equipment
- The emergence of new profit and nonprofit local organizations that provide complete or partial EMS
- The cost of government-supported EMS
- Conflicts about appropriate compensation for the higher training and skill required of EMS personnel
- Conflicts with the police over which agency (fire service or police) should provide the various components of EMS, especially in light of the potential for higher compensation
- Quality assurance with medical control

Possibly the most important of these influences are in some way related to the budget. As a result, fire departments use various innovative financial approaches. Some departments are charging for EMS or at least the transportation component; some may even offer nonemergency transportation services. Other departments research costs and develop economic and moral justification for EMS. Armed with that information, a public information program can generate the support so that the local government can appropriate the necessary funding.

The development of data on cost effectiveness is influenced by the specific manner in which a fire department operates its EMS unit, the ways in which it budgets for expense and capital items, and the outside influences of private EMS operators.

According to Brahme (1992), it is possible to evaluate the *financial* benefits of saved lives and reduced disabilities with the following steps:

1. Make direct financial comparisons between public and private service providers;
2. Evaluate lives saved and cost benefits to the community from medical services saved through more rapid intervention;
3. Set fair, reasonable, and competitive values on services provided;
4. Evaluate the performance of fire agency EMS programs based on costs and benefits to the community;
5. Perform comparisons with other agencies of comparable size and circumstance that apply the same model;
6. Communicate more effectively with public officials and customers (Brahme 1992).

To retain and even enhance their future role in EMS delivery, fire departments need to be creative within the constraints of local, state, provincial, and federal law. They may be able to increase their operating revenues by providing nonemergency transport. Departments might also be able to deliver preventive medical services, much as a fire department provides code enforcement or public education. Such programs have been developed and are currently in service. Springfield, Oregon, for instance, currently provides a wide range of BLS, ALS, and nonemergency services (Lipowitz 1995). The programs include basic life support ambulance service staffed by personnel trained to the EMT-B level. Advanced life support units are staffed by paramedics capable of performing more advanced aid under the control of a hospital-based physician. Some fire agencies deliver nonemergency transport service to their communities and supplement their budget by billing patient insurance companies.

Support for Fire Service EMS

Fire departments have allies in their efforts to establish sound, effective EMS. The International Association of Fire Fighters (IAFF) and the International Association of Fire Chiefs (IAFC)* are two organizations that have developed strong policies in support of maintaining an effective EMS presence within the fire service. They have developed working member interest sections and committees to strengthen the ability of fire service EMS to maintain a strong and fully justifiable presence in the emergency medical field. The position of the IAFF is that fire departments should be the primary providers of EMS in North America.

Many of the national fire service organizations have created staff positions to assist their members in the EMS field. EMS committees have been created by the International Association of Fire Fighters (IAFF) and the International Association of Fire Chiefs (IAFC). The IAFC also has a membership section for EMS personnel. EMS issues have been addressed in technical standards from the National Fire Protection Association, such as NFPA 1500, *Standard on Fire Department Occupational Safety and Health Program*, and NFPA 1581, *Standard on Fire Department Infection Control Program*. "In addition, at least one network has been created by graduates of the National Fire Academy's EMS administration course. It publishes a periodic newsletter to keep members up to date on each other and current EMS issues and challenges" (Cote 1997, p. 10-167).

Other forces are at work at the state and national levels to assist officers who are assigned to operate and supervise fire department-based EMS operations. Primary among these is the EMS management training program at the National Fire Academy in Emmitsburg, Maryland. The academy provides training on an ongoing basis. "This program is revised and updated periodically and offers to fire service administrators an opportunity to develop a working knowledge of current and future issues that affect the delivery of EMS through a fire department" (Cote 1997, p. 10-166).

Fire Officer Issues in EMS

An important challenge for the officer in charge of a company with EMS responsibilities is to maintain full competence in both fire suppression and

---

*Contact these professional associations at the following addresses: International Association of Fire Chiefs, 4035 Fir Ridge Drive, Fairfax, VA 22038; International Association of Fire Fighters, 1750 New York Avenue, Washington, DC 20505.

EMT (or possibly even the higher paramedic) knowledge and skills. These challenges include the following:

- Budget issues
- Private- and public-sector service delivery controversies
- Mandatory continuing training requirements
- Possible morale issues related to an increased work load
- Labor relations issues, including salary differentials, work schedules, and outside training
- Infection control issues

CONFINED SPACE AND SPECIAL RESCUE OPERATIONS

Confined space and special rescue operations present management challenges to the officer similar to those presented by EMS. These challenges include all the items listed above as well as the need for expensive specialized tools, training for these emergencies, and specialized drills.

Rescue services have always been part of the work of fire departments. In the classic sense, rescue involves searching for and removing endangered persons from fire situations. Because fire departments own extrication equipment, however, they are the natural agencies to come to the rescue of people trapped in motor vehicles. Along with the increase in automobile traffic, this function has grown in importance in the past two decades.

As society has become more technologically complex, the need for more specialized forms of rescue has also grown (see Figure 12.1). Today fire officers are expected to become proficient in many diverse situations:

- High- and low-angle rescue
- Industrial extrication and entrapment rescue
- Vehicle rescues
- Confined space entry and rescue operations
- Trench and evacuation rescue
- Below-grade rescue
- Structural collapse rescue
- Urban search and rescue
- Swift water rescue
- Ice rescue

To assess a community's need for special services, fire officers should review the records for their community and determine what has occurred in the past. They should evaluate the current and future needs for these specialized services. Only after a community has prepared a thorough needs as-

Figure 12.1 Fire fighters may be expected to perform many specialized forms of rescue.

sessment can it decide what to accomplish next. "The development of specialized rescue teams requires...a thorough understanding of the inherent hazard potential present at specific technical rescue responses; the degree of commitment required in the way of financial support, training, and skill enhancement" (Cote 1997, p. 10-131).

Special rescue operations frequently require their own equipment and unique skills. Money must be budgeted to pay for these services. Like all government activities, fire departments have to search continually for greater efficiency to reduce costs. One promising approach is the pooling of people and equipment by contiguous communities into regional service arrangements. Challenges exist, of course. In addition to jurisdictional disputes over control and cost sharing, pooling presents the need for joint planning and for cooperation with a long-term, sincere commitment to the concept.

HAZARDOUS MATERIALS

Hazardous materials present two challenges to the fire service: preventing emergency incidents involving hazardous materials and responding to emergencies. To clarify these roles, a fire officer must understand the diverse nature of hazardous materials and the laws and regulations governing their use.

Understanding Hazardous Materials

Fire service personnel usually lump all materials that have unique potential dangers beyond those of structural fires under the generic label of hazardous materials. Still, legislation and regulations often separate the materials into smaller, more specific groups of substances. Fire officers should be aware of the definitions used for the groups of substances relevant in their area. The definitions have considerable overlap because legislation was passed and regulations were issued at different times and by different levels of government. For instance, radiological, biohazardous, other toxins, explosives, and incendiaries are often included in several groups. The following grouping categories are widely used for substances.

Hazardous Materials.   The U.S. Department of Transportation defines a hazardous material in Title 49 of the Code of Federal Regulations as "any substance or material in any form or quantity that poses an unreasonable risk to safety and health and property when transported in commerce" (49 CFR 171.8).

Hazardous Substances.   The basis for this definition is found in the Clean Water Act and the Comprehensive Environmental Response, Compensation and Liability Act (CERCLA). A hazardous substance is any substance that poses a threat to waterways and the environment when it is released into the atmosphere. The list of such materials is quite long, and reference is also made to these materials under the Occupational Safety and Health Act (OSHA) regulations (Cote 1997, p. 10-91).

Extremely Hazardous Substances.   These substances are chemicals that have been listed by the Environmental Protection Agency (EPA) as possessing properties that are extremely hazardous to a community during an emergency spill or a release. The hazardous nature occurs as a result of the chemical and physical properties of the substance (40 CFR 355).

Hazardous Chemicals.   These substances are defined by OSHA regulations as "any chemical that would be a risk to employees who are exposed to it in the workplace" (29 CFR 1910). The materials pose a risk because of the health, flammability, and reactivity hazards.

Hazardous Wastes.   These wastes are the discarded materials regulated by the EPA for health and safety reasons. Authority to regulate these materials comes from the Resource Conservation and Recovery Act (RCRA) (40 CFR 260).

Marine Pollutants.   A number of materials designated as marine pollutants can have a negative impact on the waterways. Fire officers can gain valuable assistance with problems involving these materials by contacting the U.S. Coast Guard.

Dangerous Goods.   The term *dangerous goods* is used in Canada to describe substances called *hazardous materials* in the United States. Information on the classification of dangerous goods can be found in the International Maritime Dangerous Goods Code.

Federal Legislation and Regulations Involving Hazardous Materials

During the 1980s, the fire department's role in responding to emergency incidents involving hazardous materials expanded dramatically, largely as a result of federal legislation. Fire departments have been forced to assume a responsibility for which many were not prepared. "Public interest in the environment and major catastrophes involving hazardous materials have resulted in the promulgation of a wide range of laws, regulations, and standards that mandate fire department involvement" (Cote 1997, p. 10-90).

Most laws involving hazardous materials have been enacted in response to specific catastrophes. Events such as the contamination of Love Canal in New York State and the explosion in Bhopal, India, have served as the stimulus for laws to protect the environment in the United States and Canada (Cote 1997, p. 10-95). The modern fire officer must be familiar with the laws as they affect a fire department's ability to protect the community. The following significant laws and regulations involve hazardous materials.

Federal Water Pollution Control Act (FWPCA).   The first of many laws to protect the environment was enacted in 1970. The FWPCA gives to the EPA and the U.S. Coast Guard the authority to control the release of oil and other hazardous substances that could pollute the marine environment. One primary provision of this law mandates facility operators to provide a spill control plan that will indicate how to control toxic releases that could cause harm to the marine environment. A spill control plan with toxic release management requires the following provisions:

• Hazard assessment
• Emergency notifications

- Personnel requirements
- Equipment inventory
- Protective clothing
- Plan implementation
- Plan termination

Toxic Substances Control Act (TSCA).   In 1976 the U.S. Congress passed the TSCA, which gives the EPA responsibility for these functions:

- Develop a uniform listing of all chemical substances
- Establish a testing procedure for chemicals already in use and any one of approximately 1,000 new chemicals developed each year
- Determine if these chemicals present an unreasonable risk to the public's health or the environment
- Prohibit or limit the manufacture, processing, use, application, and concentration of such chemicals
- Recall or seize by civil action hazardous substances that are determined to be imminently harmful to the public's health or the environment (Cote 1997, p. 10-90)

Resource Conservation and Recovery Act (RCRA).   The year 1976 also produced the RCRA legislation, which was an omnibus approach to handling the problems inherent in the disposal of solid and hazardous waste. Four major programs were established by this legislation:

- Solid waste
- Underground storage tanks
- Medical waste
- Hazardous waste

Comprehensive guidelines were developed that gave the U.S. Department of Transportation and the EPA the authority to regulate the transportation of hazardous waste materials. The original law was modified in 1984 to broaden the guidelines and bring a wider range of the literally thousands of hazardous materials under the law's umbrella.

The Comprehensive Environmental Response, Compensation and Liability Act (CERCLA).   In 1980 this law, which came to be known as the Superfund, was passed by Congress to address the problems caused by the release of hazardous materials into the environment. CERCLA identified procedures for reporting and responding to hazardous materials incidents. It also provided money to fund the cleanup of major hazardous waste sites. Superfund established procedures for finding those responsible for the releases so that fiscal liability could be established. This legislation set up a procedure for the EPA to inventory uncontrolled hazardous wastes in the United States. The law serves as the basis for a large number of cleanup operations that have occurred since its passage.

Superfund Amendments and Reauthorization Act of 1986 (SARA). Among the laws and regulations enacted for environmental protection, "SARA has had the greatest impact upon hazardous materials emergency planning and response operations" (Cote 1997, p. 10-93). In this legislation, the original Superfund law was revised to provide for a stronger response to emergencies involving the release of hazardous materials. National policies and procedures were developed to assist in hazardous materials planning, training, and response. Title III of this legislation, or the Emergency Planning and Community Right-to-Know Act, defined the planning approach to be used by industry, government, and emergency response personnel. It also defined release notification, inventory reporting, and toxic release inventory procedures. A large part of this law involved the development of training programs for emergency response personnel.

Occupational Safety and Health Administration (OSHA) Hazardous Materials Regulations. OSHA regulations ensuring personnel protection have an important impact on the way fire departments respond to incidents involving hazardous materials and hazardous waste. The Hazardous Waste Operations and Emergency Response Operations regulation (29 CFR 1910.120), which has become known as HAZWOPER, spells out fire department operations during emergencies involving hazardous materials and hazardous wastes. The regulation addresses the following items (Cote 1997, p. 10-96):

- A hazardous materials response plan
- The use of an incident management system
- The need for buddy system operations
- Backup personnel requirements
- The need for a safety officer
- Specific training requirements
- Refresher training requirements
- Medical surveillance programs
- Post-emergency termination procedures

The HAZWOPER regulation specifies the training requirements for the various levels of emergency responder. Fire departments are required to train their personnel to a given level and then conduct periodic refresher training to maintain skills. For further information, see the NFPA's *Fire Protection Handbook* (Cote 1997, p. 10-96).

Prevention of Emergency Incidents Involving Hazardous Materials

Just as with fire prevention, code enforcement efforts have long been a part of the fire department's role in preventing both exposure of people to toxic substances and fire incidents involving hazardous materials. Enforcement efforts are based on the laws, ordinances, and regulations in effect, including those of federal, state, and local authorities. For instance, in many states, the manufacture, storage, and transport of hazardous materials are covered in either state or provincial fire codes or federal regulations.

Two activities permit fire officers to assist in reducing the risk of emergency incidents involving hazardous materials.

- Ensuring that fire fighters (or inspectors) are fully aware of the codes pertaining to these materials or at least sufficiently aware that they know when to obtain clarification on whether a specific situation might present a risk
- Ensuring that codes are strictly enforced when finding violations during inspections, when updating prefire plans, and when investigating citizen complaints about odors that were referred to the fire department

Emergency Responses Involving Hazardous Material Incidents

Fire departments have long played an important role in responding to emergencies involving flammable and combustible liquids and gases. They have had far less involvement with other hazardous and toxic materials.

Responses to hazardous materials are on the rise. Regardless of a community's size, fire officers must always consider the possibility of encountering hazardous materials during emergency operations. Ensuring compliance with the applicable state, provincial, and federal regulations for planning, training, and emergency response and adherence to NFPA codes is the best way to prepare for this eventuality. Additional guidance on hazardous materials response can be found in NFPA documents:

- NFPA 471, *Recommended Practice for Responding to Hazardous Materials Incidents*
- NFPA 472, *Standard for Professional Competence of Responders to Hazardous Materials Incidents*

- NFPA 473, *Standard for Competencies for EMS Personnel Responding to Hazardous Materials Incidents*

Any time fire fighters are exposed to hazardous materials, they must undergo a complete medical evaluation to ensure that they have not been adversely affected and to take remedial steps if necessary.

AIRPORT RESCUE/FIRE-FIGHTING SERVICES

"With the advent of the 'flying machine,' a new challenge was created for those in the fire protection field. As the airplane became larger, and passengers more numerous, the challenge became more complicated" (IFSTA 1978, p. 1). In answer to this need, specialized fire-fighting techniques and equipment have been developed to cope with fires and emergencies involving aircraft operations.

Aircraft emergencies occur infrequently; however, fire departments must consider the potential for such events. Personnel have to be trained to know and understand the various types of aircraft that might fly over their area. Departments can start to prepare for emergencies by contacting the nearest aviation facility and exploring what type of aircraft is in use and which might present the possibility of an incident. The hazards and challenges of such incidents should form the basis for preparing to deal with aircraft emergencies at off-airport locations.

Saving lives is the primary reason for providing airport rescue and fire-fighting services. Incidents involving aircraft have the potential for catastrophic outcomes. The opportunities for saving lives are usually greatest when the emergency occurs at or near an airport. For this reason, specialized crash/fire/rescue equipment is concentrated at those facilities that have ongoing aircraft operations.

Fire protection and suppression present problems at each location where aircraft are operated or stored. Fire officers must be aware of the following hazards:

- Flammable liquid use and storage on the aircraft and at the airport
- Storage of flammable and combustible materials on the aircraft
- Hazardous materials use on aircraft, such as liquid oxygen

Fire departments should also consider training and preparing for aircraft emergencies that can occur at any location that may be overflown by commercial or general aviation traffic. Planning and preparation involve the following considerations:

- Identification of potential hazard levels for aircraft emergencies
- Identification of available fire-fighting foam stocks for deployment to such incidents
- Provision of initial and periodic fire fighter training to prepare for aircraft emergencies

NFPA 403, *Standard for Aircraft Rescue and Fire-Fighting Services at Airports*, provides the requirements against which airport fire protection is evaluated.

COMMUNITY DISASTER PLANNING, PREPAREDNESS, AND RESPONSE

Planning and Preparedness

Many of the wide range of emergencies that occur each year have nothing to do with fire. The fire department is usually the first agency that people call when there is a serious problem; it receives calls for nonfire emergencies for several reasons:

- Fire department response is rapid.
- Fire department personnel are trained and equipped to handle a wide range of emergencies.
- Fire stations are geographically distributed.
- The ability of a department to respond is not limited by geographic or municipal boundaries.

The fire department is also the logical focal point for emergency preparedness.

The public looks to the fire department to help with even relatively minor emergencies, such as flooded basements or electrical and heating system malfunctions that create potential hazards to property or health. Storm-related damage leads people to call the fire department about incidents that range from the occasional downed utility pole to massive destruction wrought by tornadoes, hurricanes, floods, or major snow and ice from winter storms.

The fire service's role in these emergencies naturally draws it into the complex process of emergency and disaster planning that involves federal, state, and other local agencies. At the federal level, disaster planning and response are the responsibility of the Federal Emergency Management Agency (FEMA). Training in disaster response planning and operations is available from its Emergency Management Institute (EMI) located in Emmitsburg, Maryland. States have emergency management organizations that assist local communities in developing plans and programs to meet their specific needs. Guidance is also available from NFPA 1600, *Recommended Practice for Disaster Management*. This document "establishes minimum criteria for disaster management for the private and public sectors in the development of a program for effective disaster mitigation, preparedness, response, and recovery" (NFPA 1600, p. 1600-4).

In recent years, as the need to prepare for acts of terrorism has become more urgent, fire departments have had to consider how they can mitigate the potential destructive effect of incendiary, radiological, chemical, and explosives incidents. Here, too, FEMA provides information, guidance, and training.

Specific response planning is practically impossible because of the variety of potential types of disasters and the vast range in the severity of disaster outcomes. Government agencies therefore concentrate primarily on general issues of preparedness. A fire department can start preparing for disasters by networking with those agencies and private groups that cooperate when meeting an emergency, including the following:

- Local, county, regional, and state emergency management agencies
- Engineering resources at local, county, and state governments that can be drawn on to assess the structural stability of damaged buildings, shorelines, or infrastructure elements
- Government and private organizations that provide heavy construction equipment for earth-moving, temporary dams, building collapse, demolition, and other possible pre- and postdisaster contingencies
- Telephone, gas, electric, and water utilities
- Disaster relief agencies like the Red Cross and Salvation Army as well as religion-affiliated agencies that provide food and shelter
- Schools, churches, and synagogues and National Guard and Army Reserve armories that can be opened as emergency shelters
- Police to provide traffic and crowd control (The speed with which an area can be evacuated depends to a great extent on the ability of the police to control roads and direct traffic. A police presence may also be needed to prevent violence and looting.)
- Volunteer organizations for services such as sandbagging, food service, messenger service, and transportation
- Communications facilities in the event that the fire department's system is inadequate to meet all the demands of the emergency

If a fire department establishes and maintains contact every few months with all relevant groups, it will be as prepared for the unexpected as it can be.

Some natural disasters are not totally unpredictable or unexpected, such as hurricanes, tornadoes, earthquakes, volcano eruptions, and flooding.

When a department is in an area where such disasters occur, more specific planning and preparedness for them can and should be done.

With the increasing likelihood of terrorism, potential target areas can prepare for even these disasters. FEMA publishes emergency response materials, such as the *Emergency Response to Terrorism Self-Study Program,* which was developed in cooperation with the Department of Justice, the U.S. Fire Administration, and the National Fire Academy (NFA 1997). It is available from U.S. Fire Administration Publication Office at 1-800-238-3358, extension 1189; website: *www.usfa.fema.gov.*

Specific planning and preparation for disasters are best done by a committee composed of all the agencies involved, including the fire department. A disaster plan should spell out each agency's roles and responsibilities. When disasters are a serious, expectable threat, periodic drills are part of thorough preparedness. Drills identify problem areas and tighten procedures.

Disaster education programs can be a valuable tool in community disaster preparations. They can motivate people to prepare for disasters (Cote 1997, p. 2-21). Fire departments can perform a distinct service by arming their community with the skills to prepare for and survive a disaster. Educating the public involves developing literature and classes that impart the appropriate knowledge and behaviors.

Community Disaster Response

Proper planning improves a community's ability to respond to and recover from disaster situations. By anticipating what might occur and preparing an appropriate response plan, fire departments will be better prepared to help citizens deal with the interruptions occasioned by natural disasters or those that are caused by humans.

A fire department's disaster response plan should involve most or all of the following elements:

- Providing early warning, when possible
- Responding to specific emergency conditions as notifications are made
- Staffing emergency operation and communications centers
- Assisting in opening shelters organized by cooperating agencies, institutions, and groups
- Providing logistical support as required
- Coordinating the response of mutual aid and outside agency assistance according to the roles developed by the multiagency committee
- Assisting in the recovery phase with personnel, tools, and equipment

Community Emergency Consultation

Assisting in the development of communitywide emergency preparations is an important fire service function. Interaction among citizens, community officials, and fire department officers can improve response during times of crisis. Comprehensive response plans can define roles and relationships, provide for coordination during times of crisis, and identify problem areas. All of these nonemergency tasks can improve a community's reaction to future emergency situations.

THE CUSTOMER SERVICE CONCEPT

Classically, fire departments have seen themselves as reactive organizations. When a problem involved fire, the fire department responded. For more than 200 years, this was the accepted view. Fire officers did not pay much attention to community needs that were not related to fire suppression. During the latter part of the 20th century, however, the number of fire-related responses has dropped significantly. To justify existing staffing levels, fire departments have had to search for additional services to deliver.

The trend toward a broader view of the role that fire departments could, and probably should, assume became an issue of critical importance during the 1990s. In this chapter we have discussed how far away from fire-related services the fire department has reached. Fire departments use facilities and human resources to assist the community by providing shelter for abused children. They may staff neighborhood medical centers and even offer comfort-type assistance to accident victims who are not hurt but need other support.

Effective fire protection, EMS, and related other emergency services are still the primary purposes of fire departments. More and more, however, like other government agencies, fire departments are coming to see the citizens of the community, their guests, and other visitors as the "customers" of all the services that the departments deliver. The needs of the people who are being served has thus become a major focus of progressive fire officers throughout North America. This new approach uses the name *customer service* to dramatize its perspective. "There is nothing mysterious or revolutionary about it. Mostly, it's common sense and a conscious effort to 'be nice' and 'put yourself in the other guy's shoes'" (Hall 1998, p. 10).

Customer-based thinking and policies bring fire departments into a closer relationship with their communities. When properly implemented, the customer service approach can lead to greater public support for fire departments, which is helpful during times of budget restrictions when that support may be needed to avoid cuts in department resources. Public support is equally crucial whether a department is career, volunteer, or a combination.

At the same time, the broadening of services brings new challenges for fire officers. Departments have to provide additional training, maintain more equipment, and deal with sudden work overloads that place new stresses on staff members. All these tasks make it more important for officers to do all they can to satisfy the 3Cs guidelines for superior organizational performance.

CONCLUDING REMARKS

The growing complexity of services rendered by fire departments makes management competence even more important than it was in the past. Officers must make a wider variety of decisions, many of them more complex, concerning more difficult choices in an environment of greater ambiguity. Guidelines and the thought habits that make effective use of them can therefore become key elements in ensuring the most effective operations possible.

CHAPTER 12
STUDY QUESTIONS AND ACTIVITIES

If you are working alone, prepare your own written responses to these questions. If you are studying with a team or working in class, discuss the questions with the group and write a consensus answer.

1. If you are not already familiar with the training and certification requirements for EMT and paramedic work, obtain that information and become acquainted with it. Also determine where and how the training and certification can be obtained.
2. Review information available in your department on which hazardous materials you might encounter in your district and in districts and departments where you might have to provide mutual aid services. Prepare a list.
3. Review the hazardous materials on the list you prepared in question 2, and then prepare a page for each one that gives the following information:
   a. Special protective equipment that may be required when encountering that material

    b. Extinguishing or neutralizing materials that would be needed for that hazard

    c. Aid that would have to be rendered to citizens exposed to the material

    d. Other precautions that need to be observed

4. Prepare a list of all agencies and private groups, with telephone numbers, with which your department is, or could be, in contact to plan for disaster emergencies. If your department has no formal plan with such contacts, then prepare a list, with telephone numbers, of all agencies and private groups that should, or could be, contacted for joint planning or that could be called on in an emergency.

5. Review the meaning of the term *customer service* as it is used in your department. From the list, at the beginning of the chapter, of activities not related to fire, determine what services your department offers to the community. How else might the customer service concept affect what your department does for the community and how it performs its core services (fire-related services such as those discussed in Chapters 3, 4, and 5)?

6. Prepare goals and objectives and guidelines for three of the services listed at the beginning of this chapter, with emphasis on those provided by your department.

CHAPTER 12
REFERENCES

Brahme, Kenneth. 1992. "A Fire Officer Comments." *Fire Chief,* May, p. 50.

Bruno, Hall. 1998. "Fire Politics." *Firehouse,* January, p. 10.

Cote, A. E., ed. 1997. *Fire Protection Handbook,* 18th ed. Quincy, MA: National Fire Protection Association.

Didactic Systems. 1977. *Managing and Allocating Time: A Didactic Simulation Exercise.* Cranford, NJ: Didactic Systems.

International Fire Service Training Association (IFSTA). 1978. *Aircraft Rescue and Fire-fighting Services at Airports.* Stillwater, OK: International Fire Service Training Association.

Lidwig, Gary. 1995. "EMS into the 21st Century." *Fire Chief,* January, pp. 62–63.

Lipowitz, Sara. 1995. "Taking the E out of EMS." *Fire Chief,* May, p. 37.

National Fire Academy. 1997. *Emergency Response to Terrorism.* Emmitsburg MD; National Fire Academy.

NFPA 403, *Standard for Aircraft Rescue and Fire-Fighting Services at Airports,* National Fire Protection Association, Quincy, MA.

NFPA 471, *Recommended Practice for Responding to Hazardous Materials Incidents,* National Fire Protection Association, Quincy, MA.

NFPA 472, *Standard for Professional Competence of Responders to Hazardous Materials Incidents,* National Fire Protection Association, Quincy, MA.

NFPA 473, *Standard for Competencies for EMS Personnel Responding to Hazardous Materials Incidents,* National Fire Protection Association, Quincy, MA.

NFPA 1500, *Standard on Fire Department Occupational Safety and Health Program,* National Fire Protection Association, Quincy, MA.

NFPA 1581, *Standard on Fire Department Infection Control Program,* National Fire Protection Association, Quincy, MA.

NFPA 1600, *Recommended Practices for Disaster Management,* National Fire Protection Association, Quincy, MA.

U.S. Code of Federal Regulations, 29 CFR 1910.

U.S. Code of Federal Regulations, 40 CFR 260.

U.S. Code of Federal Regulations, 40 CFR 355.

U.S. Code of Federal Regulations, 49 CFR 171.8.

A

Goals and Decision Guidelines

For your convenience, this appendix contains all the decision guidelines presented throughout the book. The guidelines are intended to be used as starting points in the decision process. Individual officers should adapt them to their personal style, to the needs of the department, or to any specific situation in which they must make a decision. In some cases, the versions given in this appendix are more comprehensive than those presented earlier in the text to show how officers can expand the guidelines to include more specific issues and topics in making decisions.

INCIDENT COMMAND

Goals

The primary goals of incident command are to ensure that all is done that can be done to protect people—the civilians at risk and the fire fighters—and that everything is done to preserve and protect property by confining the fire and extinguishing it as quickly as possible.

Guidelines

To protect people—both the civilians at risk and the fire fighters—and to preserve and protect property, what do I have to consider in the sizeup and during the attack to ensure that I will use the most appropriate strategy and tactics? Specifically, did I regularly consider the following three basic questions and their critical elements?

1. What's there?
    • Threats to life and safety of both the civilians at risk and the fire fighters

- The involved structures, including location, contents, and water sources
- The fireground
- The fire itself
- Special hazards
- The exposures
- The weather and time of day
- The terrain

2. What does the situation need?
   - Rescue
   - EMS support
   - Exposure protection
   - Confinement
   - Extinguishment

3. What have I got?
   - Apparatus
   - Personnel
   - Equipment of all types (protective, lighting for access and evacuation and for applying extinguishing materials, communications, etc.)
   - Water and other extinguishing agents
   - Hose

FIRE PREVENTION AND CODE ENFORCEMENT

Goals

The primary goals of fire prevention and code enforcement are to create a community safe from fire, through adherence to codes, construction plan reviews, and field inspections.

Guidelines

What else needs to be done to ensure thorough adherence to codes? Specifically:

1. What should be changed with respect to relations with architects to enhance code compliance?
2. What should be changed with respect to enforcement to enhance code compliance?
3. What should be changed with respect to competence development of inspectors?

4. What should be changed with respect to communications with interested parties (architects, engineers, property owners, fire fighters, and contractors) to enhance code compliance?

PREFIRE PLANNING AND RELATED LOSS REDUCTION FUNCTIONS

Goals

The primary goals of prefire planning and related loss reduction functions are to ensure that department members have thorough plans for attacking fires most effectively, with the available water supply and other extinguishing agents, and that preparations can be made so that the members are knowledgeable, skilled, and equipped to implement the plans.

Guidelines

What else needs to be done to ensure that prefire planning for fire suppression is as thorough and useful as possible? Specifically:

1. What else needs to be done to provide adequate information, appropriately analyzed and formulated into plans, to responding companies?
2. What else needs to be done so that prefire plans are used most effectively in staff development (i.e., training and drills), in fire investigations, and in water supply review and testing?
3. What else needs to be done to ensure that relevant information from inspections, fire investigations, water supply changes, and community changes affecting apparatus routes are used to update prefire plans?
4. What else needs to be done so that the related information management systems are as effective as possible?

FIRE AND LIFE SAFETY EDUCATION

Goals

The primary goals of fire and life safety education are to provide comprehensive fire and injury prevention programs designed to eliminate or mitigate situations that endanger lives, health, property, and the environment, and to motivate changes in behavior by members of the public, through various programs and delivery methods, using all mobilizable community and department resources.

Guidelines

What else needs to be done to increase public awareness of fire and injury prevention issues and to motivate more widespread and effective fire and life safety behavior? Specifically:

1. What else can be done to enlist more community resources to reach more people in the community effectively with fire and life safety messages?
2. What arrangements can be made with nonfire resources (including the print, radio, and TV media) to get more fire and life safety messages to more people?
3. What else can be done to motivate those who have been exposed to fire and life safety messages to act on them?
4. What else can be done to evaluate to what extent fire and life safety is being practiced in residences of all types, businesses, and institutional occupancies?

MANAGEMENT OF PHYSICAL RESOURCES

Goals

The primary goal for management of physical resources is to ensure the avoidance of wasteful practices and the acquisition of the best possible equipment and facilities, obtainable with available budget funds and maintained and deployed effectively, to support the department's functions and mission statement so the department can respond effectively to all fire service requirements.

Guidelines

What else needs to be done to avoid wasteful practices, acquire the best possible equipment and facilities obtainable with available budget funds, and maintain and deploy them effectively to support the department's functions and mission statement so it can respond effectively to all fire service requirements? Specifically:

1. What else should be done to identify and eliminate wasteful practices, so that the department's resources are used most effectively?
2. What else should be done to ensure that the department has the optimum number of fire stations considering available standards (such as NFPA and ISO), that these fire stations are located to meet changes in the community's needs, and that they are in condition to meet apparatus and personnel needs?

3. What else should be done to ensure that the department has the optimum number and types of apparatus for fire and other emergencies considering available standards (such as NFPA and ISO), that the equipment meets the needs of the situations that arise and are likely to arise in the community, and that each piece of apparatus is maintained in top operating condition and appearance?

4. What else should be done to ensure that the department has optimum administrative and logistic facilities, repair facilities, alarm receipt and dispatch facilities and other communications equipment, personal protective equipment, and supplies for fire and other emergencies, considering available standards [such as NFPA, ISO, and FEMA (Fire Equipment Manufacturers Association)]; that the equipment meets the needs of the situations that arise and are likely to arise in the community; and that all are maintained in top operating condition?

MANAGEMENT OF FINANCIAL RESOURCES

Goals

The primary goals of management of financial resources are to ensure that the department will have the necessary funds and will administer them effectively to pay fair and equitable compensation to department staff members, maintain facilities and apparatus in full operating condition, obtain the needed equipment and supplies, procure additional facilities when needed, purchase additional apparatus, and refurbish existing apparatus to stay abreast of the needs of the community to be protected.

Guidelines

What else needs to be done to obtain funds for needed facilities, apparatus, equipment, and supplies and to ensure that the available funds are appropriately allocated to the various needs of the department? Specifically:

1. What else can be done to obtain needed funds for the operating budget?
2. What else can be done to obtain approval and funds for the purchase of needed facilities, apparatus, and equipment?
3. What else can be done to ensure that all purchases are made at the best possible prices?
4. What else can be done to correct wasteful practices in all operations and activities—those that use more resources than necessary, including materials, staff time, and utilities?

FIRE SERVICE PERSONNEL MANAGEMENT

Goals

The goals of the fire service personnel management function are to ensure that an adequate staff exists for the emergency prevention and response needs of the community, that the work environment provides all members of the department fair and equitable tangible rewards for their efforts, that vacancies are filled with highly qualified candidates, and that human resource policies and practices ensure a satisfying work climate for all members of the department.

Guidelines

What else needs to be done to ensure that the staff is adequate for the needs of the community, that the most highly qualified candidates are selected for vacancies, that compensation and fringe benefits are fair and equitable, and that the department offers a climate that provides a high-quality work life for all members? Specifically:

1. What else can be done to ensure that recruiting, selecting, and hiring policies are in line with all applicable laws and regulations and that the procedures used, whether administered by members of the department or by a government agency, will bring the most qualified candidates?
2. What else can be done to ensure that the compensation and benefits offered to members of the department are equitable in comparison to each other and to those in other communities?
3. What else can be done to ensure that policies and procedures relating to the quality of work life are appropriate to the department and that all officers have the will and the competencies to ensure a satisfying work climate.

TRAINING

Goals

The primary goals for training are to ensure that all members of the department have high-level competence for all their functions and that officers are competent in the management of learning and training and in coaching fire fighters.

Guidelines

What else needs to be done to ensure that all members of the department have high-level competence for all their functions and that officers are competent in the management of learning and training and in coaching fire fighters?

3 Cs GUIDELINES

The Control Guidelines

Are things going right? What else needs to be done to ensure effective control and coordination, so that the decision will lead us toward the outcome we seek, and so we'll know when we have to modify our implementation or plan because we are not getting the results we want? In other words, how can we gain better control and coordination over this process of "getting there"? Specifically:

1. Are the goals and objectives of high quality?
   1a. Are the unit's goals and objectives in line with the larger organization's goals?
   1b. Do they address matters that are important, rather than those that are urgent?
   1c. Are they both challenging and realistic (achievable)?
   1d. Is it possible to determine whether or not they were achieved?
   1e. Are they true goals and objectives, or are they action steps?
   1f. Are they for a meaningful time span?

2. Are the goals and objectives being communicated effectively to all stakeholders?
3. Is the number of goals or objectives appropriate for the organizational unit and for each of its members, considering their abilities and work load?
4. Is there appropriate participation by stakeholders in setting the goals and objectives?
5. Have you, the manager/leader, accepted your share of responsibility for achieving the goals and objectives?
6. Do the goals and objectives address not only the functional achievements but also the management/leadership aspects of control, competence, and climate?
7. Is the award/reward system of the organization coordinated with performance on achievement of goals and objectives?

The Competence Guidelines

Does everyone know what to do, and can they do it? What else needs to be done so that all those who will be involved in implementing the decision and those who will otherwise be affected (all the stakeholders) have the necessary competencies to ensure effective progress toward excellence in fire department operations and service to the community? Specifically:

1. Are changes needed in recruiting and selecting for vacancies?
2. Are management of learning concepts applied effectively?
3. Are coaching and counseling on self-development being used to their best advantage?

The Climate Guidelines

How will the stakeholders react? What else needs to be done so that the various groups and individuals who have to implement the decision or plan and those who will be affected by it (all the stakeholders) will be in favor of it or at least have as positive a view as possible, so there will be a favorable climate? Specifically:

1. Are appropriate psychological and tangible rewards offered and provided effectively and efficiently to bring the highest possible level of satisfaction from the creation or use of the product/service?
2. Are policies in place to help reduce work-related stress?

Detailed List of Items to Consider in the Definition of the Problem

The three questions from the section in Chapter 3 entitled "Defining the Problem and Obtaining the Information" are discussed in greater detail in this appendix.

1. WHAT'S THERE?

1a. Threats to Life and Safety of Both the Civilians at Risk and the Fire Fighters

Although all fires, if not confined, are a potential threat to life and safety, some obviously represent a more immediate danger than others. Responding units may be somewhat aware of the seriousness of a fire situation from prior data and from on-site reports relayed by courier or computer. Still, the officer who arrives first must obtain a complete picture by sizing up the situation and by questioning knowledgeable individuals regarding any people who might be inside a structure.

In addition to the direct danger to people in a burning structure or vehicle, a fire can indirectly threaten the lives of other people in adjacent buildings. If the fire is spreading rapidly, occupants in adjoining structures must be made aware of the emergency and helped to evacuate if necessary. Bystanders who gather to watch a fire can also be endangered by explosions, debris, or falling parts of structures.

The officer in command must also consider the potential risk of injury to fire fighters when deciding on the method of rescue or attack.

1b. The Involved Structures

During the sizeup process, the officer must rapidly assess the problems presented by the involved structure, vehicle, or materials. While doing this, the officer should consider several important factors:

473

- Construction features and materials that might contribute to fire spread or intensity
- Height of structures
- Layout (separation and compartments, access routes, ventilation options, etc.)
- Protective devices (fire-fighting resources that are part of the location, such as fire curtains, fire doors, sprinklers, and other automatic fire-extinguishing systems)
- Routes by which a fire could travel and spread, even in sprinklered premises (raised floors, suspended ceilings, cable ducts, air ducts, open stairwells, or elevator shafts)
- Contents and the special problems they might present (highly combustible materials, flammable liquids, other hazardous materials such as toxic materials, poisonous gases, and smoke)
- Location (terrain features that might influence decisions)
- Special problems of outdoor fires (dry brush or timber, unknown contents in trash, storage areas, trucks, etc.)
- Water sources, including flow from hydrants

1c. The Fireground

- Residential
- Commercial
- Industrial
- Warehouses
- Barns
- City neighborhood
- Suburban neighborhood
- Rural
- Livestock

1d. The Fire

The fire itself must be considered in relation to the following factors:

- Location (Is the fire in the only staircase of a building or in the corner of a junkyard?)
- Spread potential
- Type of combustible
- Intensity of the fire on arrival

1e. Special Hazards

- Flammable liquids
- Explosives
- Toxic materials
- Radioactive materials
- Collapse of a structure

1f. The Exposures

Exposures that need immediate attention to prevent the spread of fire must be seen in light of all the problems presented by the involved structures themselves.

1g. Weather and Time of Day

- Wind direction and intensity
- Temperature
- Precipitation (rain, snow, or the effect of dry conditions)
- Time of day (as it gives information about traffic volume, number of people to be evacuated, and other related problems)

1h. Terrain

- Topography (sloped or flat)
- Rock formations
- Vegetation, such as dry bush, dense forest, etc.
- Natural obstructions, such as rivers, gullies, or rock formations

2. WHAT DOES THE SITUATION NEED?

2a. Rescue

Life-saving operations require:

- Rapid containment of the fire
- Evacuation equipment (aerial, elevated, and extension ladders, ropes, and rigging)
- Equipment to gain access to trapped people
- EMS
- Adequate number of skilled personnel

2b. Exposure Protection, Confinement, and Extinguishment

Attack on a fire requires the following:

- Adequate amounts of extinguishing agents
- Equipment for applying extinguishing agents in adequate quantities
- An adequate number of skilled personnel
- Proper ventilation techniques
- Special equipment, such as lights, cutting tools, etc.
- Protective equipment for fire fighters
- Communications equipment
- Support services for maintenance, medical aid, utilities, information, traffic, and spectator control

3. WHAT HAVE I GOT?

3a. Apparatus

What resources are available at the fireground? The officer in command must consider both currently available and expected resources. Apparatus includes pumpers, ladder and rescue trucks, and specialized vehicles.

3b. Personnel

- Fire-fighting personnel
- EMS personnel
- Personnel with special skills

3c. Equipment

- Standard equipment on apparatus
  - Protective
  - Lighting
  - For access and evacuation
  - For applying extinguishing materials
  - Communications
- Additional equipment for rescue, ventilation, suppression, salvage, and overhaul

3d. Water and Other Extinguishing Agents

- Water on pumpers
- Hydrant water flow
- Sprinkler and standpipe flow
- Other water sources
- Other extinguishing agents

3e. Materials for Toxic and Hazardous Conditions

3f. Hose, Nozzles, and Supplementary

- Size and number of lines
- Number and types of nozzles
- Supplementary equipment (suction basins, etc.)

Officers in command must compare the resources at the fireground with what is needed to determine what additional resources should be obtained and, if not enough can be reached, what strategy is needed to bring the best results with the resources that are and will be available.

Examples of Goals and Objectives (and Action Steps) for Fire Departments and Officers

The goals, objectives, and action steps in the following list are not clearly matched because, in many cases, action steps support more than one goal or objective. For instance, supplying the sprinklers and ventilating are useful for combating the fire but may also be essential to rescue operations.

1. Examples of goals, objectives, and action steps at the emergency scene
    a. Goals and objectives
        - Rescue the residents on the second floor.
        - Prevent the spread of the fire to adjacent structures.
        - Revive the victim.
        - Don't let the flammable liquid spread.
    b. Action steps
        - Supply the sprinklers.
        - Go to the roof and ventilate.
        - Administer CPR.
        - Cut the car door to get at the victims.

2. Examples of very long range goals for the department
    - Locate sufficient stations to provide the best possible service to the community.
    - Provide the necessary apparatus and equipment for fire fighters to carry out their responsibilities.
    - Minimize cost to taxpayers.

3. Examples of chiefs' long-range goals
    - Establish and achieve apparatus and equipment performance goals and objectives.
    - Establish and achieve goals and objectives related to new apparatus and to the location and structure of approved new fire stations.

- Establish and achieve public information goals and objectives on the need for apparatus or equipment.
- Establish and achieve stringent but realistic budgets.

4. Examples of chiefs' short-range goals and objectives
   - Develop improved techniques to reduce by 10% the total response time needed to start actual fireground operations, by (date).
   - Establish coordination procedures to increase the effectiveness of mutual aid companies through better communications during response and through clearer delineation of responsibilities, by (date).
   - Develop standards for overhaul procedures, by (date).
   - Develop standard times for all major hose evolutions, by (date).
   - Ensure testing to evaluate alternative hose loadings and breathing apparatus mountings by (date).
   - Arrange to publish three articles on fire safety in local newspapers before Fire Prevention Week, by (date).
   - Achieve agreement with the municipal manager on department budgetary needs by the end of next month, by (date).
   - Submit a report on the impact of the proposed new hospital on department fire protection capability, by (date).
   - Reduce tieup time by 10%, by (date).
   - Establish revised regulations for use of airpacks and ensure adherence, by (date).
   - Reduce excess water usage in private home fires to cut down on water damage, by (date).
   - Expand coverage of postfire critiques to include setting goals and objectives to meet training needs uncovered at fire fighter and officer levels, by (date).
   - Include paper and pencil fireground simulations in training sessions, by (date).

5. Examples of station long-range goals (for officers in charge of stations)
   - Establish and achieve goals and objectives related to adapting apparatus and equipment to the district's needs.
   - Establish and achieve goals and objectives related to reducing turnout time.
   - Establish and achieve building maintenance goals and objectives.
   - Establish and achieve goals and objectives related to greater efficiency in attack preparation.

6. Examples of station short-range objectives (for officers in charge of stations)
   - Analyze available breathing equipment and recommend specifications, by (date).
   - Obtain commitments to reduce turnout time, by (date).
   - Without reducing cleanliness or building operation efficiency, reduce building housekeeping and maintenance time by 10%, by (date).
   - Obtain data and prepare specific recommendations on changes in the layout of the crew's quarters, by (date).

7. Examples of company officer short-range goals
   - Finish three time study tests of experimental hose layout changes, by (date).
   - Reduce turnout time by 20 seconds, by (date).
   - Ensure that fire fighter Smith thoroughly understands revised prefire planning symbols, by (date).
   - Take a course in advanced fire tactics and obtain a grade of B or better, by (date).

8. Examples of company officer action steps
   - Write and submit a report on the completed time study tests, by (date).
   - Meet with each company member individually for suggestions on changes in quarters, by (date).
   - Have fire fighter Smith study prefire planning symbols on Wednesday afternoon. Prepare and administer an exam within the next week, and set up another study session if Smith does not score at least 80% on the test.
   - Register for an advanced fire tactics class at an accredited college, by (date).

Officers rarely write out a complete set of goals, objectives, and action steps, except possibly when using them for the first time. Usually, they write out only two or three, and then note only important action steps on a calendar or in a notebook. If working with goals and objectives is to be a way of life, an officer must make the mechanical aspects informal. Paperwork, though important, should be kept to a minimum.

D

Defensive Strategies

Defensive strategies have many possible variations. Following is a list of some of the alternatives a command officer may consider. They are presented in increasing degrees of fire fighter involvement.

1. Allowing the fire to burn itself out. In some cases the question is not whether resources are sufficient for an attack, but whether an attack is worthwhile. If the involved structure or vehicle is no longer salvageable, and if attempts to put the fire out would be potentially dangerous, then it might be better to let the fire burn itself out. In this type of situation, a small detachment of fire fighters should stand by to make sure the fire does not spread.

2. Taking a stand at a fire wall and allowing everything on the other side to burn. This option may be chosen when forces on the scene are inadequate for direct attack (see the far left side of Figure D-1) and adequate forces cannot reach the scene.

3. Holding the perimeter of the fire while making encroachments gradually. Encroachments could include cooling a specific area to prevent an explosion or partial collapse of a structural element, preventing extension of the fire in a predictable direction, or allowing rescue operations to take place at the expense of protecting other extensions. This strategic option may be chosen when resources are somewhere left of center in Figure D-1.

4. Protecting exposures and attacking the fire in the most crucial areas or gradually from all sides. This strategic option may be chosen if the availability of resources is close to the center of Figure D-1.

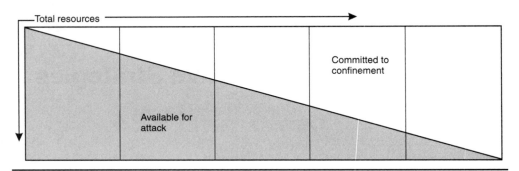

Figure D.1 Options available to an officer during the sizeup process.

The officer in charge must choose the strategic combination that is most likely to meet the overall fire incident goals of safety for civilians and fire fighters and preservation of property. Identifying what can be saved without undue risk to personnel thus becomes an important aspect in choosing an option (Cote 1997, pp. 10–12).

Selecting a strategy should focus not only on the attack alternatives that will hold fire damage to a minimum but also on those that will cause the least damage from the fire-fighting operations themselves. For instance, a poorly trained company might apply such an excessive amount of water to a weak structure that the weight of the water added to the existing load collapses the building. Less extreme are actual cases in which excessive damage to contents (for example, furnishings, carpets) by water is greater than would have been caused by smoke and fire if a less aggressive attack plan had been followed.

E

Seven Reasons for Using Multiple Alarms

Years of research and practical experience have shown that officers in charge of fires commonly order multiple alarms for seven main purposes:

1. To obtain extra personnel and equipment to aid in rescue operations
2. To obtain the personnel and equipment to run and staff additional large-caliber hand lines (for example, to cover additional positions)
3. To obtain additional personnel and equipment needed to place heavy streams in service
4. To set up and help staff a command post or field headquarters at major fires
5. To obtain additional help principally for truck duty, such as forcible entry, ventilation, and salvage
6. To provide relief personnel in situations where the fire fighting is unusually exhausting
7. To cover exposures downwind from the fire when the main body of fire presents a possible flying brand hazard

One or more of these reasons may be involved in any decision to sound a multiple alarm. The officer in charge must anticipate the need for extra help and not allow the situation to get ahead of the resources at hand.

Index